PERGAMON INTERNATIONAL LIBRARY
of Science, Technology, Engineering and Social Studies
The 1000-volume original paperback library in aid of education, industrial training and the enjoyment of leisure
Publisher: Robert Maxwell, M.C.

Introduction to Nuclear Techniques in Agronomy and Plant Biology

Other Pergamon titles of interest

DILLON
The Analysis of Response in Crop and Livestock Production,
2nd Edition

FAEGRI and VAN DER PIJL
Principles of Pollination Ecology,
3rd Edition

GOLLNICK
Experimental Radiological Health Physics

GOODWIN and MERCER
An Introduction to Plant Biochemistry

HALL
Radiation and Life

KASE and NELSON
Concepts of Radiation Dosimetry

KENT
Technology of Cerals,
2nd Edition

LOCKHART and WISEMAN
Introduction to Crop Husbandry,
4th Edition

NASH
Crop Conservation and Storage

WAREING and PHILLIPS
The Control of Growth and Differentiation in Plants,
2nd Edition

Introduction to Nuclear Techniques in Agronomy and Plant Biology

by

PETER B. VOSE
International Atomic Energy Agency, Vienna at
Centro de Energia Nuclear na Agricultura Brazil

PERGAMON PRESS
OXFORD ● NEW YORK ● TORONTO ● SYDNEY ● PARIS ● FRANKFURT

U.K.	Pergamon Press Ltd., Headington Hill Hall, Oxford OX3 0BW, England
U.S.A.	Pergamon Press Inc., Maxwell House, Fairview Park, Elmsford, New York 10523, U.S.A.
CANADA	Pergamon of Canada, Suite 104, 150 Consumers Road, Willowdale, Ontario M2J1P9, Canada
AUSTRALIA	Pergamon Press (Aust.) Pty. Ltd., P.O. Box 544, Potts Point, N.S.W.2011, Australia
FRANCE	Pergamon Press SARL, 24 rue des Ecoles, 75240 Paris, Cedex 05, France
FEDERAL REPUBLIC OF GERMANY	Pergamon Press GmbH, 6242 Kronberg-Taunus, Hammerweg 6, Federal Republic of Germany

First edition 1980

British Library Cataloguing in Publication Data

Vose, Peter B
Introduction to nuclear techniques in agronomy
and plant biology.—(Pergamon international
library).
1. Agricultural research 2. Nuclear physics
3. Botany
I. Title
630'.7'2 S540.N/ 79-40778

ISBN 0-08-024924-8 (Hard cover)
ISBN 0-08-024923-X (Flexicover)

Printed in the United States of America

CONTENTS

Preface

NUCLEAR techniques have undergone great development during the past twenty years, and methods which at best were experimental have now become routine. Nowhere is this more true than in the field of agriculture and biology, where Cyril Comar published his classic as long ago as 1955.

Probably the author of any textbook primarily dealing with well-established facts requires to justify himself! However, I have had the feeling that existing texts did not do justice to the subject, and/or were unnecessarily theoretical. It is very easy to make a text concerned with nuclear techniques difficult and complex, while hiding the essential fact that the principles and methods are usually quite straightforward. It is not so easy to estimate what to leave out as being unnecessary for a basic understanding, and I hope that in attempting to keep the text straightforward I have not missed out anything essential.

The text material has arisen in various ways: some had been previously written for other purposes e.g. lectures on various occasions, about half has been especially written. As the book is not a "committee book" obviously the author knows some parts less well than others, but I hope that specialists will not be too unhappy at my treatment of their subjects. It need hardly be said that the references are not meant to be exhaustive. They have been chosen to illustrate a point, to amplify and to show the scope of techniques. In many cases, others could have equally well been chosen.

Hardly any scientific textbook stands alone: it is complementary to other texts. For those instructors organizing a laboratory class in soil-plant relations and seeking ideas for class experiments then the IAEA Tech. Rept. Series No. 171 (IAEA, 1976) can be recommended, while plant breeders will find the FAO/IAEA *Manual on Mutation Breeding*, 2nd Ed. (IAEA, 1977) an invaluable sourcebook.

I am happy to acknowledge my debt to past and present colleagues from whom I have learned much. However, I have had a little difficulty with specific references, due to the material having been put together over quite a long period, so if any people feel that their work has not been properly acknowledged I trust that they will accept my apologies. I should particularly like to thank Dona Diva Athié for invaluable help with the manuscript.

Acknowledgments

THE following individuals and companies are sincerely thanked for supplying illustrations and permission to publish:

Panax Nucleonics, Redhill, U.K. (Figs. 4.1, 4.7, 4.10, 10.3), Beckman Instruments Inc., Irvine, Calif., U.S.A. (Fig. 4.10), Mullard Ltd., London, U.K. (Fig. 4.5), Nuclear Enterprises Ltd., Reading, U.K. (Fig. 4.9), Rank Hilger Ltd., Margate, U.K. (Fig. 9.2), EG & G Ortec, Oak Ridge, Tenn. U.S.A. (Fig. 9.3), Takashi Muraoka, CENA, Piracicaba, Brasil for autoradiographs (Figs. 10.3, 10.4 and 10.5), Dr. D. M. Silva, CENA, Piracicaba, Brasil (Fig. 10.6), Prof. Dr. W. Kühn, Institut für Strahlenbotanik, Munich, FDR (Fig. 13.5), Dr. Y. Watanabe, Institute of Radiation Breeding, Ohmiya-machi, Ibaragi, Japan (Fig. 15.1), Dr. B. Sigurbjörnsson, Agricultural Research Institute, Reykjavik, Iceland (Fig. 15.3), Drs. M. L. Petersen and J. N. Rutger, University of California, Davis, U.S.A. (Fig. 15.5), Dr. A. Tulmann Neto, CENA, Piracicaba, Brasil (Fig. 15.6), Dr. C. Broertjes, ITAL, Wageningen, Netherlands (Fig. 15.7). Other photographs are by the author, but I should like to thank Dr. V. F. Nascimento Filho (Figs. 3.1, 3.2, 5.10), and Dr. K. Reichardt (Fig. 14.13) of CENA, for their assistance.

Introduction to Nuclear Techniques in Agronomy and Plant Biology

CHAPTER 1

The Nature of Isotopes and Radiation

THE ATOM

All matter is composed of atoms. An atom has a structure resembling the solar system, consisting of a positively charged *nucleus*, occupying little space but containing nearly the whole mass of the atom, while around the nucleus revolve the negatively charged planetary *electrons*. The diameter of the atom is about 10^{-8} cm or 1 Ångstrom Unit (*Å*), while the diameter of its nucleus is about 10^{-12} cm. The nucleus consists of *protons* (symbol *Z*), particles having a positive charge, and *neutrons* (symbol *N*) without any charge but with a mass nearly that of a proton. The proton is identical to the nucleus of the hydrogen atom.

The electrons, which are 1/1840 the mass of a proton, are arranged in a series of orbits and balance the positive nuclear charge due to the protons, thus giving a neutral atom. We speak of the orbits of the electrons being arranged in *shells*, and we identify them as *K, L, M, N, O* and *P* shells, from the innermost orbit outwards. There is only one electron in each orbit but there is more than one orbit in each shell, *K* containing 2 orbits, *L* containing 8 orbits, *M* with 18 and *N* with 32 orbits. If an electron is within its own orbit it is not radiating energy, but if an external force acts on it the electron jumps into another orbit with the liberation of a quantum of energy. The lowest energy orbits are the inner ones and these are the most stable. An electron may pass into successive orbits each nearer the nucleus, losing energy at each jump until it achieves the smallest possible orbit, when the atom is in the normal state. It will be seen that an atom consists largely of empty space, its overall size being determined by the outermost orbit.

In the neutral atom the charge on the nucleus, the *atomic number* (symbol Z = number of protons), always equals the extranuclear electrons. The extranuclear electrons determine the chemical properties of an element, and therefore an element may be defined as a substance composed of atoms with the same net positive charge on the nucleus, i.e. having the same atomic number. The number of protons in the nucleus is characteristic of a particular element, though the atoms of an element need not necessarily have the same number of neutrons in the nucleus. The sum of the protons and neutrons is known as the *mass number* (symbol *M*) and corresponds to the atomic weight of the element. The term *nuclide* is a general expression describing a species of atom as characterized by the number of protons and neutrons in its nucleus. Atoms of an element which have a different number of neutrons, *N*, but the same number of protons, *Z*, that is they are nuclides having the same atomic number but with a different mass number, are called *isotopes*.

The relationship of neutrons and protons in the constitution of isotopes is well illustrated by the simplest case of the isotopes of hydrogen. There is common *hydrogen* with one proton, but no neutrons; *deuterium*, or heavy hydrogen with one proton and one neutron; and *tritium*, a radioactive form of hydrogen with one proton and two neutrons. Thus these nuclides have the same number of protons but a different number of neutrons. Having the same number of protons they naturally have the same number of extranuclear electrons, and are therefore isotopes having the same chemical properties. The nucleus of the deuterium atom is known as a deuteron and is an important particle in certain reactions (Chapter 2). Figure 1.1 illustrates the classical example of the isotopes of hydrogen and their comparison with helium, while Fig. 1.2 contrasts the structure of some of the isotopes of carbon and nitrogen.

ISOTOPES

Isotopes are therefore slightly different forms of the same element, having the same chemical properties and characteristics, but each isotope having a slightly different atomic weight or mass number. It is this vital difference which enables us to make such good use of them, as a naturally occurring element either exists in the one isotope form, or if it exists in more than one form we know the characteristic properties. If therefore we take a minute amount of a rare isotope of an element we can use it as a *tracer* to follow the behaviour of much larger amounts of the common isotope of the same element.

We distinguish between the isotopes of an element by writing the mass number as a superscript alongside the symbol of the element, e.g. ^{12}C and ^{14}C for carbon, ^{31}P and ^{32}P for phosphorus, and ^{14}N and ^{15}N for nitrogen. Formerly, and still in quite common practice, the mass number was written as right superscript e.g. C^{14}, P^{32}, and N^{15}. Occasionally the atomic number, *Z*, and the number of neutrons, *N*, are also given with the symbol of the element, the former as a left subscript and the latter as a right subscript according to the general formula $^{M}_{Z}X_{N}$ where X is the chemical symbol, e.g. $^{11}_{6}C_{5}$ for carbon-11 and $^{60}_{27}Co_{33}$ for cobalt-60. Isotopes may be of two kinds, *radioactive* and *stable*.

A number of radioisotopes occur naturally in very small amounts, such as potassium-40, but the major contribution to biological research has come from radioisotopes artificially produced in nuclear reactors.

Radioactive Isotopes

Radioisotopes are unstable, that is, they undergo spontaneous disintegration and give off atomic particles, as a stream of *radiation*, which can be of different types. The emission of atomic particles can be visualized as flashes of invisible "light". The atomic particles given off can be recorded by means of X-ray film or usually

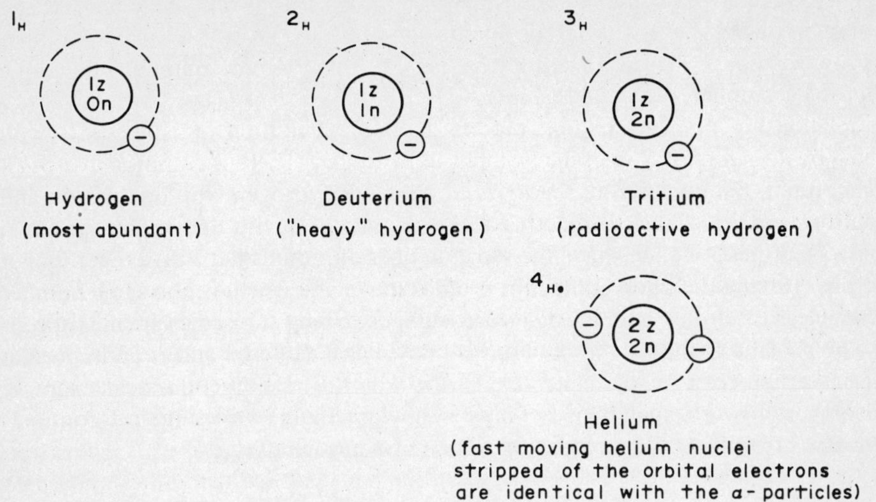

FIG. 1.1 The three hydrogen isotopes have the same number of extranuclear electrons balancing the positively charged protons of the nucleus, consequently they have the same chemical properties. Helium differs from ^{3}H in having 2 protons in the nucleus and hence is chemically different.

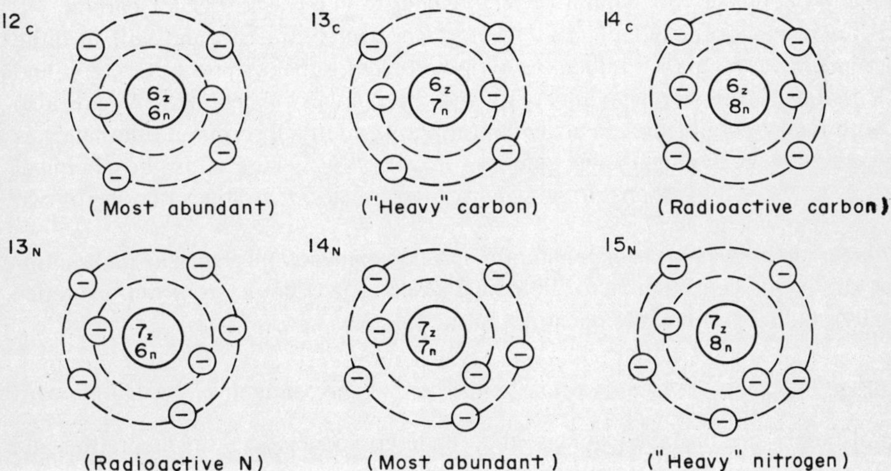

FIG. 1.2 Carbon and nitrogen differ by one proton in the nucleus and they have widely different chemical properties. Contrast the isotopes of both elements, which differ in the number of neutrons in the nucleus.

more conveniently and precisely by special electronic devices. The flash of energy in the form of an atomic particle enters a gas-filled Geiger-Müller tube or other type of detector, where it is converted into electrical energy and is registered by a counter.

Units

The rate of spontaneous disintegration, or *decay*, of an isotope is used as an indication of the amount of radioactivity present, and from this derives the unit of radioactivity, the Curie (Ci). The Curie is defined as: *The amount of any radioactive material in which 3.7 × 10¹⁰ atoms disintegrate per second.* In experimental practice in biology, a Curie is quite a large amount of radioactivity and smaller fractions are usually referred to: e.g. the *millicurie* (mCi) which is 1/1000 of a Curie and the *microcurie* (μCi), equivalent to 10^{-6} Ci. Absolute activity, or disintegration rate, is expressed as disintegration per second (d.p.s.), or per minute (d.p.m.). These relationships are summarized in Table 1.1. It will be apparent that using the relationship lpCi = 2.22 d.p.m. or 1μCi = 37,000 d.p.s., activities expressed as "disintegrations" can readily be converted to Ci units.

TABLE 1.1
The relationship of units of radioactivity to absolute disintegration rate

Decimal	Units of radioactivity		Disintegration rate
1	1 Ci	= 1 Curie	= 3.7×10^{10} d.p.s. or 2.22×10^{12} d.p.m.
1×10^{-3}	1 mCi	= 1 millicurie	= 3.7×10^{7} d.p.s.
1×10^{-6}	1 μCi	= 1 microcurie	= 3.7×10^{4} d.p.s.
1×10^{-9}	1 nCi	= 1 nanocurie	= 3.7×10 d.p.s.
1×10^{-12}	1 pCi	= 1 picocurie	= 3.7×10^{-2} d.p.s. or 2.22 d.p.m.

The S.I. unit of radiation is the becquerel (Bq) based on the reciprocal second, as the physical dimension of activity is time to the power minus one (s^{-1}). There is some resistance to adopting the becquerel because of its inconvenient dimension: thus 1 Ci = 3.7×10^{10} Bq, or 1 μCi = 37 kilo Bq, and 1 mCi = 37 mega Bq. The curie-related units are retained in this book.

Specific Activity

In radioisotope experiments the absolute amount of radioactivity is seldom required and is therefore not measured, but comparative activity is recorded as pulses or counts per minute (c.p.m.) or as counts per second (c.p.s.). The counts from the "unknown" sample are then referred back to a "standard" of known composition and count rate. At this point we should understand the concept of *specific activity*. A radioisotope is most often accompanied by stable isotope, either in the initial preparation that is used for the experiment or when it is subsequently incorporated into biological material. Specific activity is then the amount of radioactivity per unit weight (or volume) of total element present, including both active and stable isotopes. Various expressions may be used, such as Ci/g, μCi/g, Ci/mole, μCi/ml, c.p.m./mg, etc.

Radioactive Decay and Half-life

An important decay characteristic of a radioisotope is its *half-life*. The half-life of a radioisotope is defined as the time required for half of the radioactive atoms to undergo decay, or in other words for the radioisotope to "lose half its radioactivity". After the first half-life only half the original number of radioactive atoms remain; after the second half-life only a quarter remain; after the third only an eighth of the original

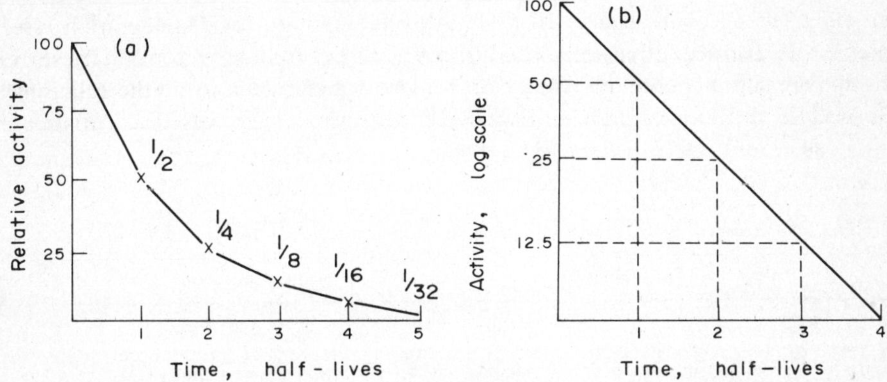

FIG. 1.3 Half-life of a radioisotope: the relationship of radioactivity to time, (a) linear plot (b) semi-log plot.

activity remains, and so on, as shown in Fig. 1.3(a). The half-life of an isotope may vary from seconds to hundreds of years, e.g. ^{11}C ($t_{1/2}$ = 20.4 minutes), ^{42}K ($t_{1/2}$ = 12.44 hours), ^{32}P ($t_{1/2}$ = 14 days), ^{14}C ($t_{1/2}$ = 5568 years). The rate of decay of any isotope is a basic property and cannot be altered by any treatment such as freezing or heating. Given the initial radioactivity of a preparation and the half-life of the isotope it is easy to determine graphically the activity at any subsequent time by plotting the decay curve: activity V time. If plotted on semi-log paper a straight line will be obtained due to the exponential nature of radioactive decay as shown in Fig. 1.3(b).

A more fundamental but often less convenient manner of expressing the decay characteristics of a radioisotope is by means of its *decay constant*, λ. The decay constant is the fraction of the number of atoms of a radioisotope which decay in unit time, and is expressed in terms of reciprocal time. It is established as follows from the fact that the number of disintegrations per unit of time is a constant fraction of the number of radioactive atoms present at that time:

The *activity*, A^* of a substance is in effect its decay intensity, and this is proportional to the number of radioactive atoms which are present. Thus if $\frac{dN}{dt}$ is the disintegration rate, N the number of radioactive atoms present at time t, and λ is the decay constant, then

$$A^* = \frac{-dN}{dt} = \lambda N \qquad (1)$$

This equation is known as Rutherford's equation and the minus sign is used to indicate the decrease in the number of atoms with time.
Rearranging to obtain λ:

$$\lambda = -\frac{1}{N}\frac{dN}{dt} \qquad (2)$$

The decay constant is directly related to the half-life. If the differential equation (1) is integrated between the limits of N_0 and N, and t_0 and t, where N_0 and t_0 respectively represent the number of radioactive atoms present at zero time, then

$$\int_{N_0}^{N} \frac{dN}{N} = -\lambda \int_{t_0}^{t} dt \qquad (3)$$

and $$1n = \frac{N}{N_o} = -\lambda t \qquad (4)$$

giving, $$2.3 \log \frac{N}{N_0} = -\lambda t \qquad (5)$$

or in exponential form $$N = N_0 e^{-\lambda t} \qquad (6)$$

$e^{-\lambda t}$ is known as the *decay factor*, *f*.

The expressions decay constant and half life are readily convertible. From equation (6) it is apparent that the time required for half the original activity to decay is independent of the initial number of atoms. So if the time required for the original activity to decrease by a half is $t_{½}$. then:

$$½\, N_0 = N_0 e^{-t_{½}} \qquad (7)$$

and $$\lambda t_{½} = 1n2 = 0.693 \qquad (8)$$

or $$\lambda = \frac{0.693}{t_{½}}, \text{ and } t_{½} = \frac{0.693}{\lambda} \qquad (9)$$

For convenience in practical tracer work, the half-life is mostly used, rather than the decay constant. It may be determined graphically, or alternatively if it is already known, the graph may be used to determine the proportion of radioactivity remaining after a given period of time. Equation (5) shows that log N plotted against t on a linear scale will give a straight line, Fig. 1.3(b). The slope of this line is $\frac{\log N_0 - \log N}{t}$ or $\frac{0.3}{t_{1/2}}$. λ can be determined as 2.3 times the slope.

TABLE 1.2

Basic numerical data for calculations of specific activity, half-life, attenuation and other functions

Avogadro's number	$N = 6.025 \times 10^{23}$	atoms / g atom, or molecules /g mole
Base of natural logarithms	$e = 2.71828$	
	$ln\ 2 = 0.693$	
	$ln\ x = 2.3026 \log x$	
	$\log e = 0.4343$	

Conversion data

		Hours	Minutes	Seconds
1 year	=	8.760×10^3	5.525×10^5	3.154×10^7
1 day	=	24	1.440×10^3	8.640×10^4
1 hour	=	1	60	3.600×10^3

Relationship of Activity to Specific Activity and Half-life, etc.

Practical work with radioisotopes is continuously requiring the calculation of specific activity, weights of reacting substances, amount of activity remaining at a given time, minimum detectable amounts of radioactivity etc. Table 1.2 gives some basic numerical data often required in such calculations.

Equations (1) $A^* = \lambda N$

and (9) $\lambda = \frac{0.693}{t_{1/2}}$; provide the foundation of many of these

calculations.

The total number of radioactive atoms, N, in a carrier-free radioactive isotope, i.e. one not containing any stable isotope, can be calculated by means of Avogadro's number (6.025×10^{23}) which is defined as the number of atoms in the atomic weight of an element expressed in grams, or in the case of a compound the number of molecules in the gram molecular weight. Thus N for 1 g of a pure radioisotope will be Avogadro's number divided by the mass number, e.g. for ^{32}P, $N = \frac{6.025 \times 10^{23}}{32}$ atoms/g.

Example: Calculate the specific activity of (a) a sample of pure ^{35}S, and (b) a sample with 75% stable *S*, (half-life ^{35}S = 87 days).

$$\therefore \lambda = \frac{0.693}{87} d^{-1} = \frac{0.693}{87 \times 1.44 \times 10^3} \text{min}^{-1}$$

note: conversion to minutes, as it is necessary later to express d.p.m. as Ci

$$\text{and } N = \frac{6.025 \times 10^{23}}{35} \text{ atoms/g}$$

$$\therefore Sp\,A^* = \frac{0.693}{87 \times 1.44 \times 10^3} \times \frac{6.025 \times 10^{23}}{35} \text{ d.p.m./g}$$

$$= \frac{0.693 \times 6.025 \times 10^{23}}{87 \times 1.44 \times 10^3 \times 35} \times \frac{1}{2.22 \times 10^{12}} \text{ Ci/g}$$

$$= 4.9 \times 10^4 \text{ Ci/g } S$$

and a sample with 75% stable *S* would clearly only have 25% of the activity per gram total *S*

$$= 1.225 \times 10^4 \text{ Ci/g } S$$

Example: Calculate the weight of 5 mCi of pure ^{32}P (half-life ^{32}P = 14 days).

$$5 \text{ mCi} = 5 \times 3.7 \times 10^7 \text{ d.p.s.}$$

$$\therefore A^* = 5 \times 3.7 \times 10^7 \text{ d.p.s.}$$

$$\lambda = \frac{0.693}{14 \times 8.640 \times 10^4} \text{sec}^{-1}$$

but

$$N = \frac{A^*}{\lambda} = \frac{5 \times 3.7 \times 10^7 \times 14 \times 8.640 \times 10^4}{0.693} \text{ atoms}$$

$$= \frac{5 \times 3.7 \times 10^7 \times 14 \times 8.640 \times 10^4}{0.693} \times \frac{32}{6.025 \times 10^{23}} \text{ g}$$

$$= 17 \times 10^{-9} \text{ g}$$

$$= 17 \text{ nanograms}$$

Example: Assuming a minimum statistically correct detectable count rate of 10 c.p.m. above background, and a counting efficiency of 25%, calculate the minimal detectable amount of ^3H, (half-life ^3H = 12.26 years).

$A^* = \lambda N$ can be used to calculate the number of ^3H atoms giving this activity.

The minimum detectable disintegration rate will be

$$A^* = \frac{10}{0.25} = 40\text{ d.p.m.} = \frac{40}{60}\text{d.p.s.}$$

and

$$\lambda = \frac{0.693}{12.26 \times 3.154 \times 10^7}\text{sec}^{-1}$$

$$\therefore N = \frac{A}{\lambda} = \frac{40 \times 12.26 \times 3.154 \times 10^7}{60 \times 0.693}\text{ atoms}$$

$$= \frac{40 \times 12.26 \times 3.154 \times 10^7}{60 \times 0.693} \times \frac{3}{6.025 \times 10^{23}}\text{ g}$$

$$= 17.7 \times 10^{-16}\text{ g } ^3\text{H}$$

Stable Isotopes

The stable isotopes which are of value as tracers occur naturally in small amounts. They are the so-called *heavy isotopes*, that is they have atoms that are heavier than normal ones, although they are not radioactive, Fig. 1.2. The principle of the use of stable isotopes as tracers is exactly the same as in the case of radioisotopes, but a special instrument, the mass spectrometer (page 153) or a modified emission spectrograph, has to be used to measure the ratio of normal to heavy isotopes. The chief heavy isotopes used in biological research are nitrogen, ^{15}N, oxygen, ^{18}O, hydrogen, ^{2}H, and carbon, ^{13}C. The concept of specific activity is not applicable to stable isotopes, and the presence of tracer quantities of a stable heavy isotope is stated in terms of *atom per cent excess* of the naturally occurring abundance. This is explained further in Chapter 7.

The great advantage of isotopes is that they are detectable in small, in fact minute, quantities, and they can also be detected when mixed with large quantities of the abundantly occurring isotopic form of the same element. Thus radioactive and stable isotopes behave chemically like normal atoms of the element, but either because of radioactivity or weight differences the tracer atoms can be identified and counted. Radioisotope tracer techniques are much more sensitive than those using stable isotopes. The former can trace at concentrations as low as 10^{-11}, in other words one part in 100 billion.

Tracers have many invaluable uses in soil, plant and animal sciences, but their effective use depends on two things: that the behavior of the tracer is identical to the carrier which is being traced and secondly, in the case of radioactive tracers the radiation level is kept sufficiently low to prevent radiation-induced biological side-effects. In practice there is usually little difficulty in meeting these criteria.

IONIZING RADIATION

In the same way that we may visualize the emission of an atomic particle as a flash of invisible ''light'' we may regard *ionizing radiation* as a continuous flow of atomic particles giving a steady beam of invisible energy. It is characteristic of both X-rays and radioisotopes.

Ionizing radiation is so called because, in its passage through the substance being irradiated it *ionizes* (converts to ions*) some of the atoms in its path, causing a permanent alteration of some of the larger molecules. It is this characteristic whereby electrons are removed from certain atoms and attached to other atoms forming new pairs of positive and negative ions, which make all ionizing radiation damaging to living things. However, under carefully controlled conditions this feature can be put to good use, as for example in mutation breeding and in food preservation by irradiation.

It was noted that the Curie is the unit of radioactivity whereby it is possible to state the amount of radioactivity that is present in a given situation. However, the ionizing radiation given off from a radioactive source is not completely absorbed by a biological tissue, air, or other material which it may encounter. It is therefore necessary to have a unit which expresses the radiation dose, that is the amount of energy absorbed by material from the radiation passing through it. The unit of absorbed dose now generally used is the *rad,* defined as *the quantity of ionizing radiation which results in the absorption of 100 ergs of energy by one gram of irradiated material*. Multiples of the unit are the *kilorad* (Krad) equal to 1000 rad, and the *megarad* (Mrad) equivalent to one million rads. The *millirad* (mrad) is 1/1000 of a rad.

In Health Physics and radiation protection work the unit of exposure dose which has customarily been used in the past is the *roentgen* (r) and the *milliroentgen* (mr) equal to 1/1000 of a roentgen. The roentgen is defined as *the quantity of X-rays or gamma rays which will produce as a consequence of ionization, one electrostatic unit of electricity in 1 cm*^{3} *of dry air*. For gamma radiation the rad and the roentgen are almost comparable, as 1 r of gamma radiation results in the absorption of 86 ergs of energy by one gram of air or 97 ergs of energy by one gram of body tissue. As the rad and the roentgen are units of quantity of dose they are usually expressed in relation to time, e.g. as mrad/hr. The significance of these units is considered further in Chapter 3. The *rem* (roentgen equivalent, man) is defined and discussed on page 39. The essential difference between the Curie and the rad must be clearly understood: the Curie is a measure of the total radiation emitted by a source and which may potentially reach absorbing material; while the rad is a measure of the amount (dose) of energy absorbed by the material subject to irradiation.

Ionizing radiation is of different types, such as *X-ray, gamma-ray, alpha-ray, beta-ray* or *neutrons*, and each type has different properties. These properties affect not only potential utility and application, but also procedures for handling and radioassay, and consideration of radiation hazard.

An important property is the *Energy of Radiation*, which is expressed in terms of *millions of electron volts* (MeV). One MeV is equivalent to the kinetic energy acquired

*An ion is defined as any charged particle of nuclear, atomic, or molecular size.

by an electron on being accelerated through a potential difference of one million volts. The energy of radiation of different radioisotopes is now known and is readily available in charts and tables, such as Table 6.1. It will be seen that there is a wide range in energy of radiation between different isotopes. A knowledge of this decay characteristic of an isotope is necessary both for the choice of a radioassay method, and also for determining the type of radiation shielding required. This is because the degree to which radiation is absorbed by matter through which it passes is inversely related to its energy, or in other words, the greater the energy of radiation the higher its penetrating power.

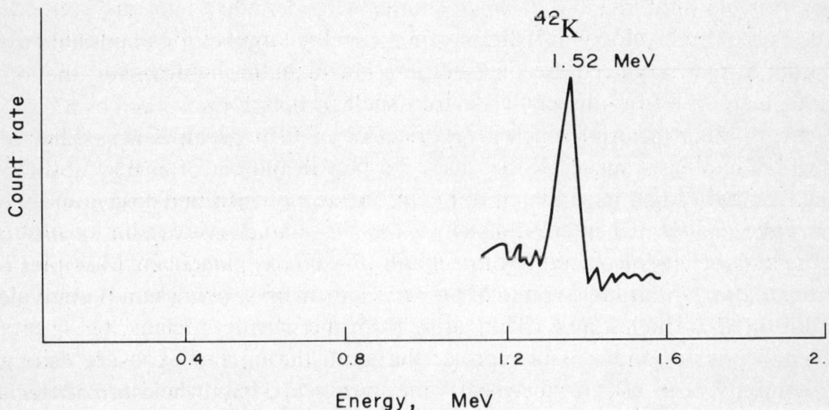

FIG. 1.4 Gamma radiation spectrum of ^{42}K showing a discrete single-energy line. Other gamma-emitters have more than one energy. Thus ^{60}Co has distinct energies of 1.17 and 1.33 MeV, while ^{56}Mn has energies of 0.845, 1.81 and 2.11 MeV (See Figs. 8.2 and 8.3).

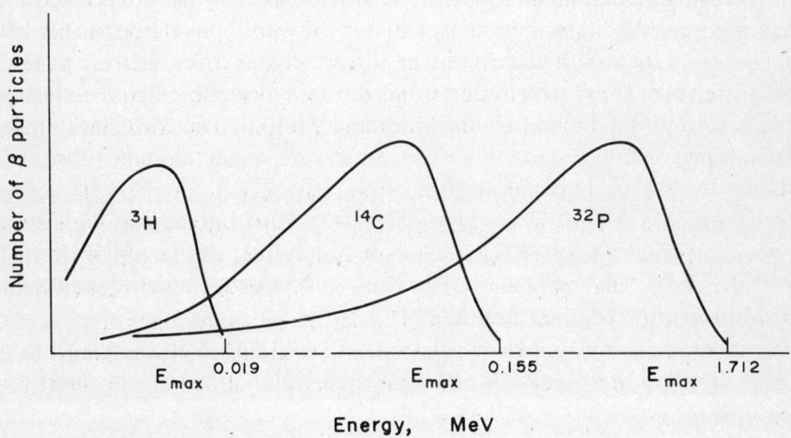

FIG. 1.5 The energy spectra of three frequently used β emitters, ^{3}H, ^{14}C and ^{32}P. Diagrammatic and not to scale.

Alpha, gamma and X-radiation energy spectra have one or two discrete single-energy lines, but the energy of beta particles emitted by a given radioisotope varies from almost zero up to a characteristic maximum energy (E_{max}), as in Figs. 1.4 and 1.5. The beta energies given in reference tables are E_{max} values, the average beta-particle energy being about ⅓ of the E_{max}.

Characteristics of Radiation Particles and Photons

X-rays. These rays are highly penetrating electromagnetic radiations (photons) analogous to visible light rays but of much shorter wave-lengths. They are emitted when cathode rays of high velocity fall directly on a metallic target (anticathode) in a vacuum tube. The X-rays produced are of a definite wavelength characteristic of the target element, and come from the outer electron shell of the atom.

In general any stream of high-energy electrons will produce X-rays when slowed down by a suitable material. X-rays have no part in radioisotope tracer studies, but are used for radiology, as a source of radiation for inducing mutations, and in X-ray fluorescence analysis (Chapter 9), which utilizes the characteristic radiation of different elements as a means of detecting the amount of element present in a sample.

Gamma-rays. γ-rays have identical properties to X-rays, being non-particulate photons, but differ in their origin. They arise from the atomic nucleus, being produced by the collision of β-rays with parts of the atom during their passage through the atom. γ-rays are, in effect, X-rays of very short wave-length and are characterized by high penetrating power, enabling them to pass through several centimetres of lead.

Quite a number of isotopes used in biological studies are γ-emitters, and their energies cover a range of 0.5 to 2.8 MeV. Gamma radiation is also used for inducing mutations, and for food preservation and disinfestation. In these cases the gamma-radiation source usually used is cobalt-60 or cesium-137.

Gamma-rays react with matter in three characteristic ways. Between 0.01 MeV and 0.5 MeV the main mechanism of energy loss is the *photo-electric effect*. A low energy photon collides with a shell electron of an absorber atom, transferring its entire energy to the electron, which is then emitted from the atom as a photoelectron (last electron). Such a photoelectron can cause ionization and excitation in a manner analogous to beta particles.

With higher energies the photo-electric effect has an insignificant role, the *Compton effect* predominating. Thus in the energy range 0.5–10 MeV gamma photons collide elastically with free or loosely bound absorber electrons and lose part of their energy to these electrons. The photons, now reduced in energy, are scattered (*Compton scattering*) from their original direction. The degree of scattering varies, some photons may be deflected as much as 180°, others scarcely at all. Both the photo-electric and the Compton effect are especially important when the absorbing material has a high atomic number.

The third process is known as *pair production*. If the photon energy exceeds 1.02 MeV the photon may interact with the electro-magnetic field surrounding the nucleus of an absorber atom, producing an electron-positron pair. Energy in excess of 1.02 MeV is imparted as kinetic energy to the two new particles.

Gamma (and X-) radiation is absorbed by matter exponentially, and theoretically it is never completely stopped. In practice of course it becomes reduced to a negligible amount. For purposes of radiation protection this has lead to the concept of *half thickness* (or half value layer, see page 36), defined as the thickness of a given absorbing material which will reduce the intensity of a beam of gamma radiation to one half its original value. The absorbing power of a material increases with atomic number and density.

Alpha-rays. α-particles are helium nuclei carrying two positive charges, that is with two protons and two neutrons. They originate in the nucleus and are expelled with a velocity about 1/10 that of light. As they are without the two negative orbital electrons they are sometimes referred to as stripped atoms of helium. When an α-particle loses energy it attracts electrons and becomes a neutral helium atom.

Alpha-rays being doubly charged are characterized by the capacity to ionize intensely and hence are only able to penetrate a short distance. It is therefore easy to shield against them, as even a few centimetres of air, a sheet of paper, or the dead skin on one's fingers will stop the rays. α-emitting radioisotopes such as radium, plutonium and uranium are not used as biological tracers, but we are later concerned with the α-particle in connection with neutron detection in the determination of moisture content (page 330).

Beta-rays. β-rays are high speed electrons emitted from the nucleus of an unstable atom at the moment of disintegration. They may be of two types, either negative (β^-) or positive (β^+), the latter being known as *positrons*. The essential difference between negative and positive beta emitters is that the nuclei of the former contain too many neutrons for stability, while the latter have too many protons. In the first case this leads to a neutron changing into a proton with the emission of an electron, and in the second a proton is converted into a neutron with the emission of a positron. Apart from the difference in sign, β^- and β^+ emitters behave in like manner, and are counted in identical fashion.

Beta particles cause ionization in matter like alpha particles and lose their energy in this way, but as the mass of the beta particle is only 1/7000 of the mass of the alpha particle, and its charge is only half, it has greater penetrating power and a lower specific ionization. Although beta-rays are much more penetrating than are alpha-rays they are still substantially less penetrating than are gamma-rays, requiring only a few millimetres of aluminium or perspex to stop them. Due to the continuous spectrum of energies their absorption in matter is not truly exponential, although it is for the major part of their range. As beta particles have a small mass they are scattered in their passage through matter in a zig-zag manner, with comparatively little loss of energy. This leads to a phenomenon analogous to reflection from a surface, known as *back-scattering*, that is the beta particles may be deflected backwards as much as 180°. Additionally, when beta-rays over 1 MeV pass through matter the rapid deceleration induces the production of "*bremsstrahlung*" or noncharacteristic X-rays. The production of bremsstrahlung is most significant with elements of high atomic weight.

Beta radiation is characteristic of the majority of radioisotopes used in biological tracer work, most being β^- emitters, with energies covering a wide range e.g. 0.018 MeV (^3H) to 3.58 MeV (^{42}K). The relatively poor power of penetration and the readiness with which beta rays are back-scattered can cause problems in practical counting. Unless the sample is very thin then *self-absorption* and scattering within the sample will take place, while *back-scattering* may occur from the sample mount or detector shield.

Neutrons. Neutrons are elementary, unstable particles of mass number 1, with a half life of 12 minutes, without electrical charge and of great penetrating power. The neutron does not produce primary ionization but decays spontaneously into a proton, a negative β-particle and a *neutrino* (a neutral particle with essentially zero mass but possessing energy), which can then excite and ionize atoms of matter. Neutrons are classified according to their velocity as *high energy, fast, slow* and *thermal neutrons*, corresponding to energies of about 10 MeV–20 MeV, 10 keV–20 MeV, 0.03 eV–10 keV and 0.025 eV.

As the neutron does not carry any charge it can only be stopped by collision with other particles. Thus, fast neutrons passing through matter lose energy in a series of elastic collisions, ultimately becoming thermal neutrons, that is neutrons whose average energy is equal to the average kinetic energy of the absorber molecules at room temperature. The absorber nuclei that have been hit are energised, losing one or more of their orbital electrons and giving rise to dense ionization along their paths. It is found that light nuclei like carbon and hydrogen, e.g. substances such as paraffin, are particularly effective for slowing down or ''moderating'' neutrons.

Following elastic and inelastic scattering of fast neutrons, the resulting slow neutrons are rapidly captured by absorber nuclei with an increase in energy of about 8 MeV on average for each capture. The excited nucleus then releases excess energy by emitting a particle or photon. Such radioactive (neutron) capture reactions are of considerable practical importance in radioisotope production and are mentioned in more detail on page 21.

Under appropriate conditions if neutrons collide with the nuclei of certain elements of high atomic number, fission results (page 23). Neutrons are extensively produced in nuclear reactors as a result of the fission process, and may also be produced by nuclear bombardment. The latter process involves the production of neutrons either by alpha-ray reaction on beryllium, usually in the form of a mixed radium/beryllium source (see page 25), or else by a particle accelerator, a machine source known as a ''neutron generator''.

There are no radioisotopes of natural elements in biological systems which emit neutrons, and our interest in neutron radiation is because it may be used as a source of radiation for radiobiological and mutation studies; in the so-called ''neutron moisture meter'' for determining soil moisture (page 329); in radioisotope production; and for inducing radioactivity in stable elements for activation analysis (page 177). The artificially produced element Californium has an isotope ^{252}Cf which is a neutron emitter now often used in small radiation sources.

Attenuation of γ-radiation

γ-radiation is absorbed exponentially, and the absorbing power of a substance increases with atomic number and density. The attenuation of γ-radiation by matter can be described mathematically, and is defined by the *linear absorbtion coefficient* μ', measured in cm^{-1}, and which is the fractional decrease in radiation intensity per unit of distance, its value depending on the nature of the material.

When μ' is the linear absorbtion coefficient and I_o the intensity of an incident γ-beam, then the intensity I of the radiation after passing through absorbing matter of thickness T is given by

$$I = I_o e^{-\mu' T} \tag{10}$$

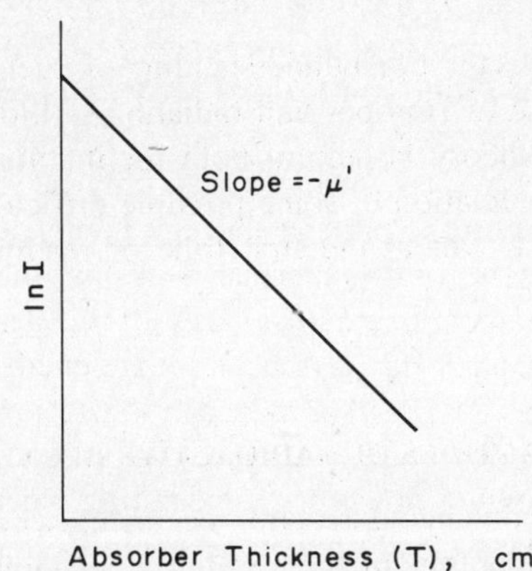

FIG. 1.6 Determination of the linear absorbtion coefficient, μ'. Log radiation beam intensity versus absorbing material thickness.

This is exactly the same form as equation (6) relating to radioactive decay, and is shown graphically in Fig. 1.6.

The *mass absorbtion coefficient*, μ, is the fractional decrease in radiation intensity per unit of surface density (cm^2g^{-1}) and is of most practical importance. It is defined as $\frac{-\mu'}{\rho}$, where ρ is the density of the absorbing matter.

Some practical applications of gamma attenuation theory and the determination of mass absorbtion coefficients are considered further on pages 342–344 of Chapter 14.

REFERENCES FOR FURTHER READING

1. CASARETT, A. P. *Radiation Biology*. Prentice-Hall Inc. (1968).
2. CHASE, G. D. and Rabinowitz, J. L. *Principles of Radioisotope Methodology*. 3rd Ed. Burgess Publishing Co. (1968).
3. GLASSTONE, S. *Sourcebook on Atomic Energy*. 2nd Ed. Van Nostrand, New York, pp. 641 (1960).
4. LAPP, E. R. and Andrews, H. L. *Nuclear Radiation Physics*. 3rd Ed. Prentice-Hall Inc. (1963).
5. WANG, C. H. and Willis, D. L. *Radiotracer Methodology in Biological Science*. Prentice-Hall Inc. (1965).

CHAPTER 2

Nuclear Reactions

ALTHOUGH the purist may disagree, an understanding of nuclear reactions is not really essential to the practical use of isotopes and radiation in biological research. Nevertheless, some background theory is helpful both for intelligent application of some techniques, and for the appreciation of some possible difficulties. It is suggested that this chapter may be passed over at the first time of reading, or read quickly and returned to later.

REACTIONS OF RADIOACTIVE DECAY

Radioactive decay is a spontaneous reaction occurring when there is nuclear instability. Nuclides vary considerably in their stability, and unstable nuclei eject subatomic particles, usually electrons, but also alpha particles, these nuclear rearrangements often being accompanied by the emission of γ-rays.

Stable nuclides are characterized by light elements, such as helium-4, carbon-12 and oxygen-16, having approximately the same number of neutrons as protons ($N = Z$) in their nuclei. With increasing atomic number it is found that the number of neutrons exceeds the number of protons, resulting ultimately in unstable nuclei, due to a too high $N{:}Z$ ratio. Thus the heaviest stable nuclides are lead-208 and bismuth-209, while above these the naturally occurring nuclides such as thorium-232 and uranium-238 are all unstable, with $N{:}Z$ ratios of about 1.5. These elements characteristically emit α-particles on decay.

Certain artificially produced radionuclides may have an excessive number of protons in the nucleus, and are unstable due to a too low $N{:}Z$ ratio. For the stability of each element there is therefore an optimum $N{:}Z$ ratio, and instability results if either neutrons or protons in the nucleus are excessive.

It is also found that the majority of stable nuclei tend to have even numbers of protons and neutrons rather than odd numbers. An additional factor in nuclear stability is the *binding energy* of the nucleus. This, put simply, is the amount of energy required to reduce the nucleus to its constituent particles. It is therefore apparent that nuclear decay processes will be a reflection of the various interrelated factors which may affect the stability of a radionuclide.

Negatron Emission

If the nucleus has an excess of neutrons (high *N:Z* ratio), the number of protons in the nucleus is increased through the emission of a negative β-particle (*negatron*) from the nucleus, and the conversion of a neutron into a proton within the nucleus, generalized as:

$$n \longrightarrow p^{+} + e^{-} + \nu\,(\text{neutrino})$$

This is the typical β^- decay characteristic of many of the isotopes used in biological research. A practical example is the decay of phosphorus-32 to sulphur-32.

$$^{32}_{15}P_{17} \longrightarrow {}^{32}_{16}S_{16} + e^{-} + \nu\,(+1.71\text{MeV})$$

Nuclear fission products frequently contain several more neutrons than is possible for stability, and there may be a chain of disintegration stages. Emission of β^- particles will then occur at each stage, for example:

$$^{90}_{36}Kr \xrightarrow{e^-} {}^{90}_{37}Rb \xrightarrow{e^-} {}^{90}_{38}Sr \xrightarrow{e^-} {}^{90}_{39}Y \xrightarrow{e^-} {}^{90}_{40}Zr$$

Positron emission. With an excess of protons in the nucleus the optimum *N:Z* ratio is achieved by the emission of a positron (β^+ particle), with the conversion of a proton into a neutron, as:

$$p^{+} + (\text{energy}) \longrightarrow n + e^{+} + \bar{\nu}\ (\text{anti-neutrino})$$

Examples of this type of decay are zinc-65 to copper-65,

$$^{65}_{30}Zn_{35} \longrightarrow {}^{65}_{29}Cu_{36} + e^{+} + \bar{\nu}\,(+0.325\text{MeV})$$

and phosphorus-30 to silicon-30,

$$^{30}_{15}P_{14} \longrightarrow {}^{30}_{14}Si + e^{+} + \bar{\nu}\,(+3.3\text{MeV})$$

Associated Positron and Negatron Emission

A few special instances have both positron and negatron emission. Thus copper-64 may decay either to nickel-64 with positron emission, or to zinc-64 with negatron emission:

$$^{64}_{29}Cu \longrightarrow {}^{64}_{28}Ni + e^{+} + \bar{\nu}$$
$$^{64}_{29}Cu \longrightarrow {}^{64}_{30}Zn + e^{-} + \nu$$

and cesium-130 to barium-130 or xenon-130:

$$^{130}_{55}Cs \longrightarrow {}^{130}_{56}Ba + e^{-} + \nu$$
$$^{130}_{55}Cs \longrightarrow {}^{130}_{54}Xe + e^{+} + \bar{\nu}$$

Electron or "K" Capture

As an alternative to positron emission, the *N:Z* ratio can be increased by the capture of an orbital electron from the K-shell, known as "electron" or "K-capture":

$$p^+ + e^- \rightarrow n + \nu$$

The gap in the K-shell is then occupied by an electron from the L-shell with the emission of a characteristic X-ray as a consequence of the energy difference between the L- and K-shell electron. The decay of iron-55 to manganese-55 can be represented as:

$$^{55}_{26}\mathrm{Fe} + e^- \rightarrow {}^{55}_{25}\mathrm{Mn} + \nu$$

K-capture occurs when there is not sufficient energy available for positron emission, although electron capture and positron emission often occur simultaneously.

Emission of Gamma Radiation

Quite often following the emission of an electron or alpha-particle or the processes of K-capture, radionuclide decay does not proceed straight to the normal (ground or stable) state of the daughter, but passes through a transition or "excited" state. In these cases the final nuclear transformation from the excited high energy state to the ground state is accompanied by the emission of a gamma photon. Thus:

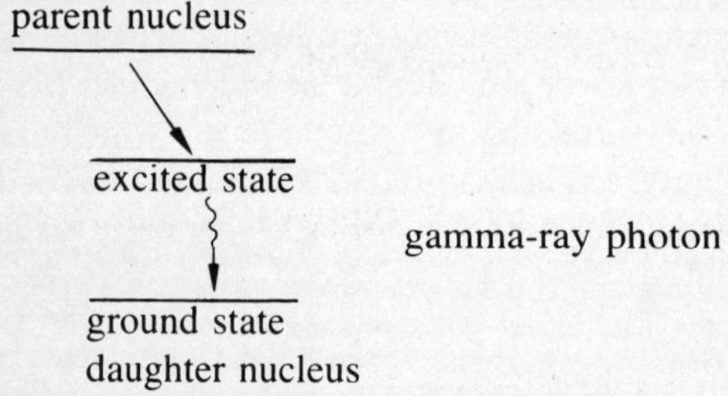

For example, the decay of cobalt-60 to nickel-60:

$$^{60}_{27}\mathrm{Co} \longrightarrow {}^{60}_{28}\mathrm{Ni} \quad + \quad \begin{array}{ll} e^- & (0.5\mathrm{MeV}) \\ \gamma_1 & (1.3\mathrm{MeV}) \\ \gamma_2 & (1.17\mathrm{MeV}) \end{array}$$

Sometimes, e.g. following irradiation, a nucleus exists at a high energy level for a measurable time before finally decaying to the normal ground state with the emission of a gamma photon. Such a nuclear form is known as an *isomer*, the intermediate stage is the *metastable state*, and the final transition to the ground state is known as *isomeric transition*, e.g.

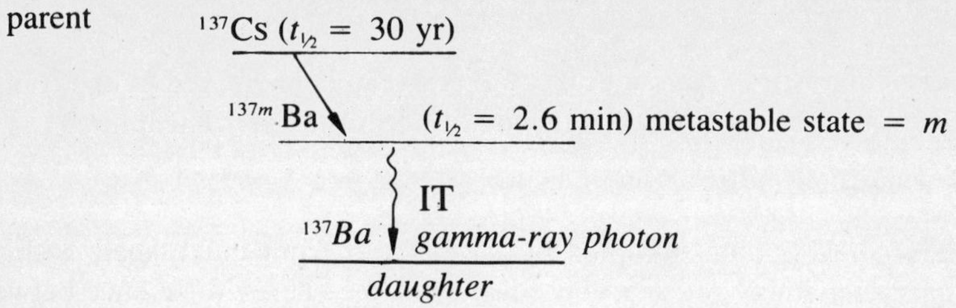

parent ^{137}Cs ($t_{1/2}$ = 30 yr)

^{137m}Ba ($t_{1/2}$ = 2.6 min) metastable state = *m*

IT

^{137}*Ba* *gamma-ray photon*

daughter

Internal Conversion

A proportion of gamma photons emitted from a nucleus may interact with an orbital electron, causing the electron to be ejected from the atom with the photon ceasing to exist. This process is known as *Internal Conversion* because the whole of the energy from the gamma photon has been transferred to the electron. Thus I.C. electrons are those emitted as result of the interaction between a γ-ray and an orbital (valence) electron. These electrons have discrete energy, and such a stream of electrons is responsible for producing line beta spectra as opposed to the more common continuous spectra resulting from β^- and β^+-emission.

Alpha Particle Emission

An alpha particle, being a stripped atom of helium consisting of two protons and two neutrons, is therefore positively charged and of relatively large size. It is one of the most stable nuclear combinations, and as such is always emitted as a single particle. Radionuclides which emit an α-particle are thus changed into another nuclide with mass number four units less and with two fewer protons, i.e. atomic number two units less. Being such a large particle it is not surprising to find that nuclei which eject α-particles are themselves comparatively large. Thus virtually all α-emitters have atomic numbers above 82. As mentioned previously, the majority of naturally occurring radionuclides are alpha emitters:

$$^{238}_{92}U \rightarrow ^{234}_{90}Th + \alpha + 4.19MeV$$

Radionuclide Decay Schemes

It is apparent that radionuclides may decay in one of several main ways, and additional subsidiary processes are also possible. Any individual nuclide may decay simultaneously by more than one process or route. This introduces a certain complexity, but the probability of a nuclide decaying by one process as opposed to another is constant, so that if there is more than one decay process each can be defined in terms of per cent of total decay or disintegration.

The decay process has been established for virtually all radionuclides, and recorded in decay or disintegration schemes. Decay schemes show the characteristic radiations and the energies of the radiations, the half-life of the nuclide, the nature of the decay process and the half-life of the intermediate isomer if there is one. Such a decay scheme is a complete characteristic of the nuclide.

Decay schemes are important for the intelligent measurement of radioisotopes and in health protection. An example of a typical decay scheme is for copper-64, which it will be recalled has both positron and negatron emission:

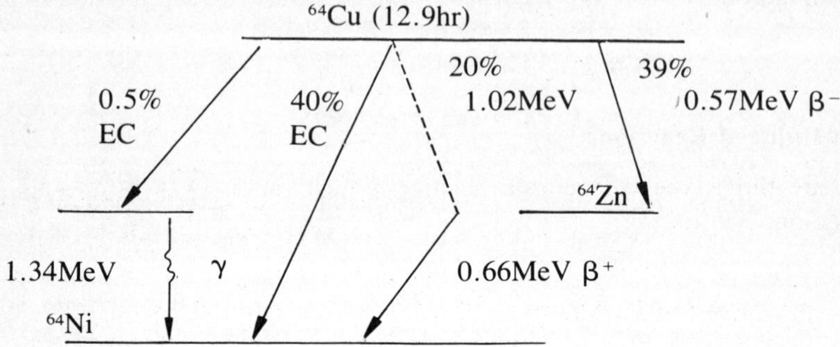

In typical charts of the nuclides very similar information is given, but in more condensed form, e.g.

Cu
12.9hr
EC 1.34
β^-.57β^+.66
E^- .57E^+1.68

Key

hr = half-life, hours
EC = Electron Capture
E = Disintegration energy MeV
% = Probability of decay process

INDUCED NUCLEAR REACTIONS

So far the reactions considered have been those spontaneous ones resulting from nuclear instability giving rise to radioactive decay. Induced nuclear reactions occur when a nucleus is acted upon by another high velocity nucleus or elementary particle. Such reactions may be induced by bombardment with particles originating from a nuclear reactor or a particle accelerator, such as a cyclotron. The significance of these reactions is that all radioisotopes used in tracer work originate in this manner, while activation analysis (Chapter 8) also depends on the measurement of induced radioactivity.

Reactions have been induced by the use of bombarding particles such as neutrons, protons, deuterons, photons, electrons, ^{3}He and ^{3}H. The general nature of induced nuclear reactions is for the incident (bombarding) particle to coalesce with a target nucleus, with the formation of a compound nucleus. Invariably this compound nucleus is unstable due to its excess energy, and consequently it immediately decays to give a product nucleus with the emission of a high energy particle. A type of short-hand notation is usually used to describe such reactions. In this the initial target nucleus is written first and the final product nucleus last. Between them, in parentheses, are written the light bombarding particle and the fragment. This will become clear from the succeeding paragraphs.

For practical purposes induced nuclear reactions may be roughly classified into neutron induced reactions, typical of irradiation by reactors and "neutron generators"; and transmutation by deuterons and other particles following acceleration in a cyclotron.

Neutron Induced Reactions

There are three types of neutron reaction (i) activation by neutron capture; (ii) transmutation; (iii) fission.

(i) *Activation by Neutron Capture*

The majority of elements show *radiative capture* reactions with slow neutrons, where the energy of the compound nucleus is not sufficient to eject a nucleon following the capture of an incident particle, but some of its excess energy is emitted as gamma radiation. This process is the most common method for obtaining artificial radioisotopes, particularly as reactors are such an easily available source of neutrons.

In this type of reaction a neutron is taken up by the target nucleus and the resulting compound nucleus then emits its excess energy as a gamma photon. For example:

$$^{63}_{29}\text{Cu} + ^{1}_{0}n \rightarrow ^{64}_{29}\text{Cu} + \gamma$$

or, in short notation:

$$^{63}_{29}\text{Cu}(n,\gamma)^{64}_{29}\text{Cu}$$

The neutron is an especially suitable particle for disintegrating nuclei because as it carries no charge it is not repelled. The ease with which different elements capture neutrons is defined by the so-called *neutron cross section*, σ. This is effect is the size of the target presented by a particular nucleus to the bombarding neutron, and determines the probability of a particular reaction. Thus if the nuclear cross section is high the probability or efficiency of the reaction is high, while if the cross section is low, then so is the probability. A neutron cross section is defined in terms of the intensity of incident particles, the number of target nuclei per square centimetre, and the number of these target nuclei which undergo interaction in the specified time:

$$\sigma = \frac{A}{NI}\,\text{cm}^2 \text{ per nucleus}$$

where, I is the number of incident particles striking 1 cm^2 of target material in a given time

A is the number of target nuclei undergoing interaction in the given time

N is the number of target nuclei per cm^2, and

σ is the cross section in barns/nucleus (1 barn $= 10^{-24}$ cm^2).

The neutron cross section is a function of energy, and its components directly relate to the reactions of neutrons with matter (Chapter 1, page 14), i.e. elastic and inelastic scattering, and neutron capture. The neutron cross section therefore varies widely for different elements and indeed for the same element with different neutron energies.

The cross section, together with the half-life of the particular isotope and the neutron flux (intensity) of the reactor, determines the length of time the stable target element must be irradiated to achieve a certain degree of saturation with the radionuclide. When these reactions are used for producing radioisotopes the target material must be spectrographically pure, otherwise a part of the radiation emitted by the product may be from a radionuclide other than that desired.

As there is no change in the number of protons (Z), the chemical identity of the target is the same, but the product is a mixture of stable and radioactive isotopes, the longer the irradiation then the higher being the specific activity. Further examples of these reactions are:

$$^{23}Na\ (n,\gamma)\ ^{24}Na$$
$$^{35}Cl\ (n,\gamma)\ ^{36}Cl$$
$$^{41}K\ \ (n,\gamma)\ ^{42}K$$

There are instances where the radioactive product resulting from the (n,γ) process may decay to a daughter radionuclide of another element. This permits the preparation by chemical separation of either a ''carrier-free'' or high specific activity preparation of the daughter radionuclide, e.g.

$$^{130}Te\ (n,\gamma)\ ^{131}T_e \xrightarrow{e^-} {}^{131}I$$
$$^{76}Ge\ (n,\gamma)\ ^{77}G_e \xrightarrow{e^-} {}^{77}As$$

Activation by *high energy neutrons* (>14MeV) makes possible an $(n, 2n)$ reaction, such as,

$$^{14}N\ (n,\ 2n)\ ^{13}N$$

which can be utilized for the activation analysis of nitrogen.

(ii) *Transmutation by Neutrons*

In reactions with fast neutrons another type of reaction, that of (n,p) becomes possible. This is because with the higher energies available the proton is permitted to escape from the compound nucleus. In such reactions the product nucleus has the same mass but possesses an atomic number one unit less than that of the target element, e.g.

$^{32}S\ (n,p)\ ^{32}P$
$^{35}Cl\ (n,p)\ ^{35}S$
$^{14}N\ (n,p)\ ^{14}C$

This last reaction is the means by which cosmic ray neutrons produce ^{14}C in nature, and is also the basis of producing ^{14}C commercially. The chemistry of C and N permits ready separation of carrier-free ^{14}C from the target material.

In certain cases, usually with target elements of atomic number 40 or below, the fragment that is ejected may be an alpha-particle:

$^{40}Ca\ (n,\alpha)\ ^{37}Ar$

This reaction is used in the production from lithium-6 or tritium, the hydrogen isotope of mass number 3:

$^{6}_{3}Li\ (n,\alpha)\ ^{3}_{1}H$
(tritium)

The big advantage of the neutron transmutation reactions in radioisotope production is that they provide an easy means of separating a radionuclide of high specific activity from the stable target element.

Neutron detector tubes containing boron also depend on the $(n,\ \alpha)$ reaction:

$^{10}B\ (n,\alpha)\ ^{7}Li$

(iii) *Fission*

Fission is a high energy process whereby a heavy nucleus captures a neutron, is split into two pieces of approximately equal size and gives off energy. Thus a nucleus of uranium-235 on being struck by and on absorbing a neutron, breaks up completely into two nuclei of medium atomic weight and ejects two or more neutrons, with the release of about 190 MeV of energy. A chain reaction can therefore be set up, whereby the excess neutrons are captured by other isotopes such as the more stable uranium-238 to give more fissionable material. This forms the basis of the operation of nuclear reactors and the production of atomic energy.

The less stable nuclei such as uranium-235 or plutonium-239 are fissionable with either slow or fast neutrons, but more stable nuclei such as uranium-238 and thorium-232 (known as *fertile materials*) require fast neutrons for fission. Nuclides produced by fission may have atomic numbers in the range 30 to 64, but they tend to group near atomic numbers 42 (molybdenum) and 56 (barium).

Although the total number of fission products is very large, about 80 for uranium-235 for example, the number of practical radioisotope tracers produced in this way is comparatively small because of the problems of separation and purification. They are primarily limited to ^{90}Sr, ^{89}Sr, ^{137}Cs and ^{99}Mo. As these isotopes can be separated chemically they can be obtained carrier-free. However, strontium-90 has the short half-life (2.5 day) daughter yttrium-90 present, while strontium-89 preparations are likely to have about 5% strontium-90.

Transmutation by Deuterons and Protons

The reactions which have been considered so far have been those due to neutron irradiation in nuclear reactors. The next series of reactions could equally well be classified as "cyclotron reactions". The initiation of neutron reactions is comparatively easy, because the neutron does not possess a charge and is therefore not repelled by the target nucleus. Deuterons, protons and alpha-particles, all being charged particles, must be accelerated to a sufficiently high energy to overcome the repulsion of the nucleus and permit reaction with the target nuclei.

A number of machines are available for the acceleration of particles, but in the practical production of radioisotopes the cyclotron is normally used. A cyclotron consists basically of a huge magnet which sets up a magnetic field, causing the particles to move at ever increasing speed in a spiral of increasing radius until they are deflected by magnets onto the target material. Two flat semi-circular boxes, known as "dees" because of their shape, are situated between the poles of the magnet and are kept under vacuum. A high-frequency alternating current is applied between the dees and an electric arc at the centre of the dees produces a stream of nucleus, ionizing deuterium for deuterons, helium for alpha particles and hydrogen for protons.

A large number of transmutations can be produced by various cyclotron accelerated particles, but the reactions producing radionuclides of practical interest to the biologist are all initiated by deuterons. These may be of the (d,α), the (d, n) or $(d, 2n)$ type. Thus sodium-22 is produced by the transmutation of magnesium-24 by deuterium bombardment, with the ejection of an alpha-particle:

$$^{24}Mg\ (d,\alpha)\ ^{22}Na$$

also,

$$^{56}Fe\ (d,\alpha)\ ^{54}Mn$$

The short half-life (20.4 min) carbon isotope ^{11}C is prepared by bombardment of boron, in a (d, n) reaction:

$$^{10}B\ (d,n)\ ^{11}C$$

The possibility of using a carbon isotope of such short half-life has made some interesting plant tracer experiments feasible (page 305).

A $(d, 2n)$ type of reaction, in which two neutrons are expelled from the compound nucleus, occurs when the deuterons have energies of 10 MeV or more and the target is of a fairly high mass number. Examples are:

^{65}Cu (*d*, 2*n*) ^{65}Zn
^{55}Mn (*d*, 2*n*) ^{55}Fe
^{52}Cr (*d*, 2*n*) ^{52}Mn
^{56}Fe (*d*, 2*n*) ^{56}Co (together with ^{57}Co and ^{58}Co).

Cyclotron produced radioisotopes are of high specific activity, but the amounts produced are small and expensive compared with reactor irradiated products.

Activation by deuterons and protons is also possible, and the reactions ^{14}N (*p, d*) ^{13}N and ^{14}N (*d, p*) ^{15}N have been suggested as a means of nitrogen determination (Chapter 8).

Transmutation by Alpha-particles

Finally, consider a transmutation by α-particles, a (α, *n*) reaction in which the importance of the reaction lies in the fast neutron emitted rather than the product nucleus:

$$^{9}_{4}\text{Be}\ (\alpha, n)\ ^{12}_{6}\text{C}$$

In this reaction beryllium-9 is bombarded with α-particles and is transmuted to carbon-12 and a fast neutron is emitted. This transmutation need not be cyclotron-mediated, as α-particles can be obtained from the decay of radium-226. The full reactions are:

$$^{226}_{88}\text{Ra} \rightarrow ^{222}_{86}\text{Rn} + \underset{(\alpha\text{-particle})}{^{4}_{2}\text{He}} + \gamma$$

and,

$$^{9}_{4}\text{Be} + ^{4}_{2}\text{He} \rightarrow ^{12}_{6}\text{C} + \underset{(\text{fast neutron})}{^{1}_{0}n} + 5.76\ \text{MeV}$$

This reaction is used in the construction of small sources of fast neutrons of constant flux for laboratory use, as used for example in the measurement of soil moisture content by neutron moderation (page 330). For these sources, radium and finely ground beryllium are combined in a mixture.

Other Induced Reactions

It should be understood that there are very many more types of induced nuclear reaction possible, than those given in this chapter. The intention has been to give examples of the most frequent and significant types of reaction, particularly those of practical importance in radioisotope production.

REFERENCES FOR FURTHER READING

1. ANON. *The Radiochemical Manual: Part I, Physical Data, and Part II, Radioactive Chemicals*. Published by the Radiochemical Centre, Amersham, England (1962 and 1963). (A good summary of radiochemistry from the radioisotope user's point of view).
2. COOK, G. B. and Duncan, J. F. *Modern Radiochemical Practice*. Oxford, Clarendon Press, (1952).
3. CARSWELL, D. J. *Introduction to Nuclear Chemistry*. Elsevier, New York (1967).
4. FRIEDLANDER, G., Kennedy, J. W. and Miller, J. M. *Nuclear and Radiation Chemistry*, 2nd Ed. John Wiley & Sons Inc. (1964).
5. MCKAY, H. A. *Principles of Radiochemistry*. P. 550. Butterworths, London (1971).

CHAPTER 3

Working With Radioisotopes

IT should be understood at the outset that there will be no hazard in working with radioisotopes if the proper precautions and rules are observed. There is no need to be afraid of radiation, but there is a real need to appreciate its potential dangers, to learn safe radioisotope handling practices, to develop careful habits and to work methodically.

The personal rules are quite few:

1. Never eat, drink or smoke when working with radioactive materials.
2. Always wear a laboratory coat. Wearing shoe covers is usually not necessary with the amounts of isotopes handled in the typical procedures in biological and agricultural applications.
3. Always wear rubber gloves when working with radioactive materials. Wash and monitor them before taking them off carefully, without touching the outsides. Disposable polythene gloves are often used.
4. Wear a film badge or a pocket dosimeter.
5. Always use a pipetting device (rubber suction bulb or plunger pipette) when transferring aliquots of radioactive solutions. *Never* pipette by mouth.
6. Always work on a plastic tray lined with absorbent paper, when transferring radioisotope solutions or other active material liable to spill.
7. Use an adequate fume hood if there is likely to be vapour or fumes, e.g. from evaporation or ashing, which might be radioactive.
8. Return any stock radioisotope solution or sealed source to its place of storage immediately after use.
9. See that all containers, volumetric flasks etc. containing radioisotopes are adequately labelled, with details of the isotope and its initial activity at the date of receipt or dilution.
10. It is normally permissible to handle planchets for "counting" without gloves, using forceps. Apply barrier cream such as "Savlon" to the hands before doing so.
11. After removal of rubber gloves always wash your hands with soap and water, and especially carefully before leaving the laboratory for meals at the end of the day.
12. Always work with a beta-gamma monitor beside you, and always monitor your hands, shoes, clothing (coat cuffs etc.) at the end of the day, or after

you have been handling relatively high activity material, as when diluting stock solutions.

13. *Never* take your rubber gloves into the counting room, and it is usually desirable not to wear the same laboratory coat in the counting room that is worn in the preparation laboratories. Failure to take care in this respect will soon raise the general background radiation level and hinder counting of low activity samples.

These personal routine procedures for safe working with radioisotopes have been given first so that the reader may better appreciate the discussions which follow on laboratory design, shielding, avoiding contamination, health and safety requirements, personal monitoring and waste disposal.

DESIGN OF LABORATORIES

Most readers will not be immediately concerned with the initial design of laboratories for radioisotope work, nevertheless laboratory design influences both the safe handling of radioisotopes and also the efficiency of work, so we shall briefly consider the basic requirements.

Radioisotope laboratories are really just normal chemical laboratories that have been given certain special features to permit safe working with radioactive materials. Working space in such laboratories should be more generous than usual with at least 10 m^2 per worker. Floors should be non-porous and sealed so that spilled radioactive solutions can easily be wiped up. Heavy grade linoleum or rubber floor covering provides a good washable surface, but has the disadvantage that if it is not possible to clean off contamination then a piece of flooring will have to be cut out. For this reason rubber or linoleum tiles may be better, as they can be replaced individually. The cracks between the tiles are filled through normal waxing of the floor.

Benches should have a hard non-porous surface and can easily be either of wood or steel, the former being best covered with polished linoleum or P.V.C. (polyvinyl chloride) sheet in case of spills. At least part of the benching should be strong enough to support lead shielding. Ideally, sinks should be of stainless steel, though for low level work porcelain sinks will suffice. In laboratories where high levels of activity are handled it is normal to have taps foot- or elbow-separated, but with the levels of activity we are mostly concerned with in biology and agriculture, normal taps are usually adequate.

The walls of radioisotope laboratories should be covered in high-gloss non-porous paint which can easily be decontaminated by washing. It should also be chemically resistant and easily removed in the very rare event of serious contamination. Ceiling should be similarly painted.

Adequate fume cupboards must be provided. Ideally these should be of stainless steel, but for the levels of activity we are considering wood or sindanyo (transite) sheeting covered with strippable paint is adequate. The exhaust system should be effective, with a minimum air velocity of 30 linear m/min across the opening when in the working position, and the outlet should be well away from windows and air intakes. It is best that all controls e.g. water, vacuum, air, gas are outside the fume cupboard, which is in any case normal practice in modern chemistry laboratories. The base of the cupboard should be strong enough to support lead shielding.

Glove boxes are also useful. These are enclosed cabinets with glass or perspex fronts, provided with arm holes and attached rubber gloves. Glove boxes usually have an exhaust fan and filter, with power and other services fitted. They are used for dusty operations and handling β-emitters.

Where there are a series of laboratories and a great deal of radioisotope work, it is sometimes the practice to designate certain of the laboratories as "hot", "medium level" and "low level" according to the type of work to be carried on in them. Thus a hot laboratory would be used possibly for storage, unpacking and dilution of stock radioisotope solutions, and for any high level radiochemical operations. The bulk of the work, including sample preparation, will be carried out in the medium level laboratory, the low level laboratory being reserved for ^{14}C work of low activity, such as sample preparation, radiochromatography and thin-layer chromatography. Many units are too small to enjoy the advantages of separate laboratories, and in this case the laboratory is in general equivalent to medium level. Any high level activity dilution procedures are carried out in the fume hood.

It is usually desirable to have in addition to the laboratories which are used for experimentation, manipulation and sample preparation using radioactive materials, a separate counting room or rooms where the samples are counted. This is because it is desirable to keep the background count, that is the pulses registered by the counter due to cosmic and other radiation, as low as possible. This is especially so when counting low levels of a soft β-emitter such as ^{14}C. The background activity is usually somewhat higher in any radioisotope laboratory than it is in a separate counting room. Where both ^{14}C and hard β or γ counting are regularly being carried out then it is desirable to have a separate counting room for the ^{14}C work. Where radioisotope work is infrequent and there are only one or two workers involved, it is possible to have counters in the general laboratory, although it is not ideal.

The counting room requires no special provision, except for an adequate supply of electric power points, with simple benches for holding sample changers, counters, calculators etc. that are not free standing. The wiring of the power circuits should be generous to prevent voltage drop. Air conditioning is necessary in tropical countries with high humidity, but is not normally necessary under temperate conditions provided temperatures are constant. For small laboratories with only one or two counters it is not unusual for an ordinary instrument or balance room to serve a dual purpose as a counting room. Where there is air conditioning or ducted ventilation, the design must be such that air from laboratories handling radioisotopes cannot be drawn in, either regularly or through malfunctioning of the system. A constant temperature is desirable to avoid variable performance of some equipment.

Normal practice where there is a suite of laboratories devoted to radioisotope work is to site the counting room nearest to the offices, and have the low level and hot laboratories progressively further away.

The size of a simple counting room obviously varies with the potential work load, but should provide a minimum of 1.5 m of bench space per instrument and a minimum of 12-15 m^2 of floor space. A corner of the benching should be allotted for simple instrument maintenance, such as changing leads and connectors, replacing resistances, soldering etc. In a bigger laboratory a workshop is desirable. Where much counting has to be done it is usually better to have more than one counting room rather than have one very big one, as it makes it easier to separate the different classes of work which may interfere with each other, γ versus soft β counting for instance.

For purposes of determining the maximum amounts of radioactive materials that may either be stored or manipulated, laboratories are classified in three grades, A, B and C, according to their facilities. Table 3.1 gives the maximum permissible quantities of different classes of radioisotopes that may be stored or handled in each grade of laboratory.

For most general purposes in biological and agricultural applications of radioisotopes the Grade B laboratory is most practical, and it is to this grade that most of the previous remarks refer. Impervious floors, covered benching strong enough to take lead shielding, good fume hoods, adequate ventilation and separate high and low level laboratories and counting facilities are expected in this grade of laboratory.

Much good work can be carried out with only Grade C laboratory facilities, and for demonstration-type work, as carried out in many schools for instance, Grade C is quite adequate. Such a laboratory is basically a good chemistry laboratory with impervious linoleum floor covering, and with benches covered with linoleum or disposable polythene sheets.

TABLE 3.1

Radiotoxicity of isotopes	Grade of Laboratory		
	Grade C Good chemical laboratory	Grade B Radioisotope (medium level) laboratory	Grade A High level laboratory
I Very high	10 Ci max.	10 mCi max.	10 mCi
II High	100 Ci max.	100 mCi max.	100 mCi
III Moderate	1 mCi max.	1 Ci max.	1 Ci
IV Slight	10 mCi max.	10 Ci max.	10 Ci

(For classification of isotopes by potential radiotoxicity see Table 3.2, page 31.)

This Table must be interpreted using modifying factors, as a given quantity of radioisotope is obviously more dangerous when being manipulated than when being stored:

PROCEDURE	MODIFYING FACTOR
Storage (stock solutions)	× 100
Very simple wet operations	× 10
Normal chemical operations	× 1
Complex wet operations with risk of spills	× 0.1
Simple dry operations	× 0.1
Dry and dusty operations	× 0.01

Grade A laboratories are not normally met with in the type of work we are considering: they are used for work with fission products, radioisotope preparation and separation work, etc. Stainless steel benching and fume hoods are normal, and in fact most operations will be carried out in glove boxes or lead brick caves with remote handling tools. Provision is made for ventilation with over 50 changes of air per hour, and filtration of the air removed from the laboratory. Changing rooms are provided, special shoes or shoe covers are worn and there is usually a physical ''step-over'' barrier between the low level and hot areas. Special arrangements, including large storage tanks are made for accumulating and monitoring effluent. Continuous monitoring and alarm systems are usually installed.

In addition to normal radioisotope laboratories, specialised work in agriculture may require additional facilities. For instance, work with large animals such as sheep and cows will require the provision of a ''hot animal'' room or barn, in which radiotracer-injected animals can be held in metabolism cages. Arrangements have to be made for the collection and disposal of radioactive excretions, and for easy decontamination in the case of spills.

TABLE 3.2

Classification of some radioisotopes according to their relative radiotoxicity

Class I Very high toxicity	$^{90}Sr + {}^{90}Y$, $^{210}Pb + {}^{210}Bi$ (Radium D)
Class II High toxicity	^{45}Ca, ^{59}Fe, ^{89}Sr, ^{131}I, ^{140}Ba
Class III Moderate toxicity	^{22}Na, ^{24}Na, ^{32}P, ^{35}S, ^{36}Cl, ^{42}K, ^{52}Mn, ^{54}Mn, ^{56}Mn, ^{55}Fe, ^{58}Co, ^{65}Zn, ^{74}As, ^{76}As, ^{82}Br, ^{86}Rb, ^{99}Mo, ^{105}Ag, ^{111}Ag, ^{137}Cs, ^{137}Ba, ^{203}Hg
Class IV Slight toxicity	^{3}H, ^{14}C, ^{51}Cr, ^{197}Hg

LABORATORY PRACTICES

The contamination by radioactivity of glassware such as beakers and pipettes and also planchets and other sample containers cannot be avoided. Contamination of work areas and the laboratory in general *can* be avoided by proper handling and working procedures. *Contamination of work areas is the result of carelessness* and should not occur. Emphasis therefore is on reducing contamination as much as possible.

Very high standards of cleanliness, both of working procedures, glassware and other equipment, and the laboratory must be maintained. Regular monitoring must be carried out of benches and floors. Floors must be regularly wet-mopped and polished to keep down dust. Benches must be regularly cleaned.

Work areas, containers and glassware etc. that actually contain a significant active source should have a self-adhesive radiation symbol on them. Keep the symbol for application to containers and limited areas which do actually have a potential radiation hazard: the symbol should mean what it says. All labels used in the laboratory should be self-adhesive to avoid licking.

In the case of "hot" laboratories, tools, glassware and possibly other equipment are best marked in some permanent manner, such as with red or orange paint, in order to identify them. Remember that equipment from such laboratories should not, if it can be avoided, be taken into or used in a low level laboratory, as it could interfere with results. In any case they must be properly monitored, and indeed it is good practice to be always monitoring equipment and bench areas during the course of daily work.

No equipment, glassware or rubber gloves should ever be taken from an "active" laboratory into a counting room. No radioisotope, other than samples for counting or sealed reference sources, should be taken either. If a substantial number of rather active γ-emitting samples have to be counted it is sometimes advisable to bring only a few at a time into the counting room, especially if the counting room is being shared by another worker carrying out low level counting.

Manipulations involving significant activity must be carried out with the aid of tongs and other devices, making it possible to stand at a distance from the radiation source. Remotely operated can openers, bottle cap removers, pipettes, cutters and holding vices are available.

Normally all manipulations involving radioisotopes in solution should be carried out over a plastic tray in case of spills. Always have a carton of paper tissues alongside and mop up any minor drips as they occur. When pipetting radioactive solutions always use a suction bulb or plunger pipette device; always wipe the tip of the pipette afterwards with a tissue. When a remote pipetting device has to be used, practice with a non-radioactive solution first.

Stock radioisotope preparations are best retained until use in the lead or other container in which they were received from the supplier. Semi-stock solutions should always be held in an unbreakable outer container, i.e. if a dilution of all or part of a radioactive stock has been made into a volumetric flask for further dilution, the flask should be placed in a stainless steel, aluminium or polythene beaker padded with paper tissues.

Heating of solutions or digests containing significant radioactive material is most safely carried out on a sandbath, in case of breakage of glass beakers or flasks. Evaporation of small quantities of radioactive solutions in beakers, or aliquots in planchets, is for preference carried out from above by means of an infra-red lamp, to reduce the risk of splashing. Evaporation must be carried out in a fume hood.

Decontamination

Glassware used in experiments always require decontamination and subsequent monitoring. A sink should be set aside for this especial purpose and should be marked. Normally, liquid waste should not be put down the drain but should be poured into the container(s) provided. Washing should follow normal practice e.g. washing with detergent such as "Teepol" followed by adequate rinsing in tap and distilled water.

Decontamination of glassware and other surfaces is often greatly helped by repeated washing with a strong solution of a stable carrier. In the case of difficult deposits, after a preliminary wash try soaking in 10% nitric acid, ammonium citrate solution or ordinary chromic acid cleaning solution. Chromic cleaning solution made with HCl instead of H_2SO_4 is very effective but extremely noxious and *must* be used in a fume cupboard. Some glassware may occasionally get too active to make decontamination practical. In such cases, if the half-life of the isotope is reasonably short the items should be put on one side (suitably labelled!) for the activity to decay. In the case of isotopes of long half-life the glassware should be regarded as solid waste and discarded.

Where spills occur on *bench, floor or equipment* the liquid should be initially wiped up using paper tissues (wearing rubber gloves!) and the area of contamination marked. Subsequent treatment depends on the material. Linoleum and rubber can have the waxed surface removed with xylene or trichloroethylene, as can the surface of highly polished wooden benches. If wood is not wax polished or the spill has penetrated the wax it is very difficult to decontaminate except by removal by flaming or sanding.

Like wood, concrete is very difficult to decontaminate unless it is sealed with paint or thick waxing. Try treatment with 30% hydrochloric acid. If this fails, sand blasting or complete removal may be necessary.

Lead bricks, containers and castles etc. can be treated with 4 N hydrochloric acid until reaction starts, then neutralized with dilute alkali and washed.

Steel hot-plates, bench tops, apparatus supports etc. can be washed with dilute phosphoric acid or a proprietary inhibited phosphoric acid agent such as Deoxidine 125. Alternatively, sand blasting or steam cleaning is possible.

Paint surfaces may be cleaned with detergent or ammonium citrate solution. If necessary they may be stripped off using normal paint strippers.

Decontamination of clothing is normally adequate by the usual washing procedures. Otherwise it must be discarded.

In general, the activity of the area of contamination should be reduced to 10^{-4}-10^{-3} $\mu Ci/cm^2$, or in practical terms about 500 c.p.m. for radioisotopes of categories I and II (high toxicity) and to 1000 c.p.m. for radioisotopes of categories III and IV (slight to moderate toxicity) when a 25 mm diam. G-M counter is held as close to the area as possible (see Table 3.2 for categories of radioisotopes).

Personal Decontamination

Contamination of skin should not occur, but accidents do happen. Any contamination of the skin should be washed with soap and water, brushing lightly and not to such an extent that the skin is broken. If this is ineffective try washing with EDTA solution or 1 molar solution of the stable isotope. A drastic treatment is to swab the area with strong potassium permanganate solution for a couple of minutes, rinse in water and remove the stain with 5% sodium bisulphite.

In the case of contamination of a wound, wash under a tap with large amounts of water and encourage bleeding. If it is a facial wound do not contaminate the eyes or mouth. Finally wash with soap and water and seek medical attention. In the case of contamination in an eye treat as a normal first aid measure, washing the eye under the tap with copious amounts of fresh water, then wash in an eyebath using 2% sodium bicarbonate of 0.9% saline (NaCl) solution. Seek medical attention.

What may be regarded as ''major spills'' are rare events in biological or agricultural operations as the amount of radioactive material handled is usually quite low. In the event of a major spill the laboratory must be evacuated at once and the windows and doors shut. Contaminated persons should not go far away until they have been monitored. All clothing which has been contaminated should be removed, and skin which has been contaminated should be washed under running water. The laboratory should not be reoccupied until it has been decontaminated and monitored by persons wearing protective clothing, masks and gloves. There is no universal definition of a major spill, but we may regard a spill as ''major'' if it involves quantities of radioactive material in excess of 100 μCi of a category I ratioisotope, 1 mCi of a Class II radioisotope or 10 mCi of a Class III or IV material if wet, and only 1/10 these quantities if dry.

In countries where there is an established Radiation Protection Service or the facilities associated with a reactor site are available, such ''major spills'' are best handled by these personnel. If, however, these facilities are not available, the individual worker must manage on his own wearing proper protective clothing. In general, the more you work on your own the better prepared you must be to handle unforeseen contamination problems.

SHIELDING

The amount of radiation that a worker receives is governed by the amount of radioactivity present, the distance he is from the source, the nature of any absorbing material between him and the source and finally the length of time he is near the source.

Now, the *inverse square law* states that the radiation intensity varies inversely as the square of the distance:

$$D_2 = D_1 \times \frac{(d_1)^2}{(d_2)^2}$$

where D_1 is the dose rate at a distance of d_1 from the source and D_2 is the dose rate at a distance of d_2 from the source.

It follows that distance is an extremely effective means of reducing radiation dose e.g. a source giving a dose of 10 rad/h (i.e. 10,000 mrad/h) will only give 1 mrad/h at 10 cm distance and 0.11 at 30 cm away.

The use of remote-handling tongs and pipettes when opening and dispensing radioactive stock solutions should therefore be regarded as routine. Even sealed sources of relatively low activity should be handled with tweezers.

In addition to distance, at levels of activity about and above 1 mCi shielding by some absorbing material becomes necessary. The shielding required for β-emitters is usually simple, often the thickness of the normal glass bottle or container being adequate. Beware however of radiation from the open top of a bottle or beaker and if it is necessary to look into such a bottle or beaker, wear protective glasses or place a sheet of glass over the opening. Of the common soft β-emitters used in biological work, ^{14}C, ^{45}Ca, and ^{35}S are normally adequately shielded by the glass of the container.

With higher levels of hard β-emitters, such as the frequently used ^{32}P, additional shielding may be necessary. A perspex sheet of 0.5–1.0 cm thick should normally be adequate. Perspex or aluminium should be used in preference to thin lead sheet, because of the production of bremsstrahlung by the latter (high atomic weight!) with β-emitters over 1 MeV. Alternatively β-emitters may be handled in a glove box with a perspex window. A 0.5 cm thick perspex sheet will stop all β particles up to 1 MeV and 2.5 cm thick perspex will absorb β particles up to an energy of 4 MeV.

Lead shielding is routinely required for an operation involving more than 1 mCi of a gamma-emitter and is best used when any stock solution of a gamma-emitter is dispensed, even if less than 1 mCi is involved. As a barrier restricts vision either windows of lead-glass are provided, or with the simple barrier most often used, a mirror is arranged in a suitable position. Standard bricks are normally 5 cm thick, but 10 cm thick bricks are available, or two layers of bricks may be used. There may also be need for shielding beneath the source to prevent stray radiation reaching the feet and legs. This is a point sometimes forgotten.

With elaborate enclosed shielding for very high levels of activity, "through the wall" type remote manipulators are necessary, but such installations are very rarely, if ever, required for biological purposes. For shielding large radiation sources of X-ray machines, massive steel or concrete is often used, but seldom in the laboratory, as it takes approximately twice the thickness of steel and about five times the thickness of concrete to be as effective as lead.

The question arises as to how much shielding is necessary for a given purpose? This is especially so when a consignment of concentrated radioactive stock has been received from a supplier and is being unpacked, diluted and sub-divided. In biological applications this is usually the time of the main external radiation hazard. One should really calculate the intensity of radiation from the radioactive material, calculate the decrease in radiation dose with distance at the point where the operator will stand and then determine the thickness of lead necessary to reduce the radiation dose to an acceptable level. In practice not many workers, using 1–10 mCi say, bother to make this calculation as they know that if they shield the source with a simple barrier of lead bricks, use tongs and a remote pipetting device and complete the operation in as short a time as possible, they are well within an acceptable dose limit. Most biological applications are in fact concerned with less than 10 mCi; nevertheless, a worker should know how to make the necessary calculation, as work with higher levels may become necessary.

The dose of a γ-emitting source can be found by means of the empirical formula:

Dose rate at 30 cm (1 foot), R_f = 5.8 CE mrad/h,

where,

C is the strength of the source in mCi and

E is the total energy of γ radiation in MeV

R_f values are also given in Table 3.4, for some common radioisotopes.

To the dose rate thus calculated is applied the inverse square law which will indicate the potential radiation dose at the point where the operator will stand. From this figure is calculated the thickness of lead necessary to reduce the radiation to an acceptable intensity.

For this we need to know the *half value layer* of lead for a γ-emitter of known energy, that is the thickness of lead required to reduce the radiation intensity by one half. This can be found from tables such as Table 3.3, then the total thickness of lead that is required to reduce the intensity by the desired factor can be calculated by use of the equation: $N = 3.3 \log \chi$ where N = number of half-value layers required to reduce the intensity by a factor of x.

TABLE 3.3
Approximate half-value layer of lead for γ radiation

Energy (MeV)	Thickness (cm)
0.5	0.41
1.0	0.89
1.5	1.22
2.0	1.45
3.0	1.60
4.0	1.60

Example: We wish to determine the shielding required for 20 mCi of ^{24}Na so that the intensity of radiation does not exceed 10 mrad/h at a distance of 45 cm (18 in), the nearest point to the operator. Obtain R_f, the dose rate at 30 cm either from Table 3.4 or by applying the empirical formula given. In the latter case

$$R_f = 5.8 \times 20 \times (1.38 + 2.76)$$
$$= 480 \text{ mrad/h}$$

Applying the inverse square law, the dose at 45 cm, D_2 is:

$$D_2 = 480 \times \frac{30^2}{45^2} = 213 \text{ mrad/h}$$

Therefore, to reduce the radiation intensity to the required level of 10 mrad/h we must lower the intensity by a factor of 213/10 = 21. The number of lead half-value layer is then 3.3 × log 21 = 4.36. Table 3.3 shows that the half-value layer of lead for radiation of 4 MeV is 1.6 cm, so the total thickness of lead required is 4.36 × 1.6 = 7 cm (6.976 cm). As the usual lead brick is 5 cm thick we therefore have the choice of moving the source further away, using two layers of bricks, or alternatively stacking the bricks on their sides, as they are normally at least twice as high as thick.

If a similar calculation is made for a source of 10 mCi of ^{24}Na it will be found that just under 5.5 cm of lead is sufficient for shielding at a distance of 45 cm. As ^{24}Na has one of the highest R_f values of any of the radioisotopes with which we may be concerned, this indicates that the common practice of not calculating the potential radiation intensity of every consignment of isotope under 10 mCi is not without justification. Nevertheless, the actual dose rate where the operator is standing should always be checked by a dose-rate meter both before and during the operation, whether or not a full calculation of the dose has been made.

TABLE 3.4
γ-ray dose rate of some common radioisotopes

Isotope	Dose rate (R_f) in millirads/h per mCi at 30 cm
Arsenic-74	4.97
-76	3.56
Cobalt-58	6.05
-60	14.58
Iodine-131	2.43
Manganese-52	20.84
-54	5.29
Mercury-203	2.40
Molybdenum-99	1.80
Potassium-42	1.62
Rubidium-86	1.30
Sodium-22	14.26
-24	20.43
Zinc-65	3.0

In general, if persons are likely to be exposed to the radiation for a substantial part of the working day, then sufficient shielding will be used such that the radiation 5 cm from the outside does not exceed 2.5 mrad/h. If however a person is only exposed briefly, such as when removing and diluting an aliquot of a stock solution, then up to 25 mrad/h 5 cm from the shielding is permissible occasionally though not desirable. At such a dose rate the weekly permissible maximum of 100 mrem would be accumulated in only 4 hours and such an intensity presupposes that the source is kept in a restricted storage area and is clearly marked.

HEALTH PHYSICS AND PERSONNEL MONITORING

Health physics is concerned with setting safety standards for exposure to radiation, the detection and monitoring of radiation exposure and the development of better methods of protection. We have already touched on certain aspects of health and safety, such as the relative toxicity of different radioisotopes, the basic requirements for laboratories of different categories and the reduction of radiation intensity through shielding.

There are two kinds of radiation hazard: internal from radioisotopes which may have entered the body and external radiation from a source such as a radioisotope or a machine. All radiation is harmful and our exposure to it must be reduced as far as possible. Nevertheless, experience has shown that exposure to very limited amounts of radiation has negligible probability of causing either severe ill-health or genetic injuries, and it has led to the concept of "*maximum permissible levels*" of radiation exposure, whether internal or external. This concept governs the whole approach to radiation protection.

The classification of radioisotopes by radiotoxicity is essentially on the basis of type and energy of radiation; the half-life of the isotope; the absorption pattern of the isotope and whether it is rapidly eliminated from the body or is retained e.g. in bone; and finally the quantity and form in which the isotope is usually handled. Two concepts are of particular concern here, the *biological half-life* of the isotope, that is the time required for one-half the absorbed radioisotope to be excreted from the body or an organ, and the *critical organ* of the body, which is the organ most liable to damage by a particular radioisotope and in which it usually accumulates.

Thus the alpha-emitters such as radium, plutonium and uranium are the most highly toxic radioisotopes, not only because of the intensely ionizing α-radiation, but also because of their long half-lives and the fact that the critical organs are bone, kidney and lung. The maximum permissible body burden of these isotopes is therefore extremely low. Iodine-131 accumulates in the sensitive thyroid gland but it has a short half-life of 8 days. Sodium-24, though of high gamma energy, is of moderate toxicity because of its short half-life and the rapidity of excretion. It thus typifies the moderately toxic radioisotopes which do not accumulate to any great extent in critical organs and have a comparatively short biological half-life. Carbon-14 and tritium are normally only of slight toxicity despite their long half-life, because of their rapid biological turnover.

In biological tracer work we are not directly interested in the extensive data that exists on maximum permissible concentrations of radioisotopes in the human body, critical organs, air or drinking water. These matters are the concern of specialists engaged primarily in monitoring large-scale radioisotope operations at nuclear plants, power stations and the like. Nevertheless, we must be aware of the potential internal radiation hazard that exists, particularly when working with the highly toxic radioisotopes. Thus the working procedures that are adopted must always be such that there can be no possibility of accidental ingestion with food or drink, absorption through skin or wounds or inhalation.

These requirements mainly lie within the power of each individual and there can be no substitute for following the personal rules given at the beginning of this chapter. At the same time the reason for strictly limiting the amounts of the more hazardous radioisotopes that can be stored or handled in the less well-equipped laboratories will now be clear.

We have considerably more experimental data available on the external radiation hazard, and also more effective means of continuously monitoring it, compared with the internal hazard. Such monitoring must ensure that workers keep well within the maximum permissible dose.

The roentgen is the unit of exposure dose of X-ray or gamma radiation (page 10) and was traditionally used in radiation protection work, because this was initially concerned with the medical use of X-rays. Subsequently it became necessary to have a unit of absorbed dose for any kind of ionizing radiation, including alpha and beta particles and neutrons, as well as X and gamma rays, and this led to the establishment of a new unit, the rad (page 10).

Over the years evidence has accumulated that the effects of different types of radiation on biological systems are not all the same, that is they differ in their *relative biological effectiveness* (RBE). The RBE values for different types of radiation are given in Table 3.5, although these values are only generalizations, as the RBE of two radiations cannot be expressed precisely by a single factor because it depends also on subsidiary factors such as the dose, dose rate, fractionation of the dose, temperature and the effect observed.

TABLE 3.5
Relative biological effect values for different radiations

Type of Radiation	RBE
X-rays, gamma rays, electrons and beta-particles	1
Slow neutrons	2.5
Fast neutrons	10
Protons up to 10 MeV	10
Naturally occurring alpha-particles	10
Heavy recoil nuclei	20

The RBE of any radiation is defined as *the ratio of the dose required to produce a certain biological effect by irradiation with 250keV X-rays, to the dose required by irradiation with the radiation under consideration to produce the same biological effect.*

As the biological effect of a particular radiation therefore depends not only on the absorbed dose but also on the RBE of the radiation, a new unit, the *rem* (roentgen equivalent man) was adopted. The rem is the unit of RBE dose and is now the accepted unit for expression of radiation dose for health protection purposes. It is defined as that dose of any ionizing radiation that produces the same biological effect in man as that resulting from 1 rad of X- or gamma-radiation. The relationship is thus:

$$\text{Dose (rem)} = \text{Dose (rads)} \times \text{RBE}$$

The maximum permissible radiation dose that may be received by occupational workers has been reduced over the years, and the present accumulated dose limits have been fixed by the International Commission of Radiological Protection. It is now accepted that some organs of the body are more sensitive than others, so no general single maximum permissible dose can be laid down, as it depends on whether the whole or only part of the body is being irradiated. Thus the hands have a higher long term permissible average dose than the whole body, gonads or eye lenses. Additionally it is recognized that most workers will not receive radiation at a constant dose rate. Therefore, maximum permissible doses are laid down for accumulated dose over fixed intervals of time, but the recommendations also accept the possibility of compensating for excessive dose in one week by reducing the average dose over a longer period.

The basic recommendations are that the average weekly whole body dose should be less than 100 mrem (0.1 rem); the accumulated dose over any consecutive 13 weeks shall be less than 3 rem and for the whole year should not exceed 5 rem. When hands only are involved the maximum permissible accumulated dose is 20 rem for 13 weeks and 75 rem per year. Additionally, a maximum permissible accumulated total body dose in relation to age has been laid down for persons occupationally exposed to radiation, according to the equation $D = 5(N - 18)$, in which D is the accumulated doses of radiation expressed in rem and N is the age of the person in years.

It is apparent therefore that with beta and gamma emitting isotopes, which have a RBE of 1, the maximum permissible average whole body dose should not exceed 100 mrad per week. As pointed out in the section on shielding, this amounts to 2.5 mrad/hour assuming a 40-hour week. These figures must form the basis of planning all radioisotope operations and if it is necessary to exceed them for a short period at any time, then this must be compensated for by reducing the subsequent average dose to the worker.

Personnel monitoring may be done with film badges or pocket ionization chambers, Fig. 3.1. Photographic film is wrapped in opaque paper in the film badges which are normally worn on the coat, but they can additionally be worn on the wrist, if exposure of hands is likely to greatly exceed the whole body dose. Various types of film badge are available, and the type in which part of the film is covered with a cadmium shield to absorb the beta particles permits distinction to be made between the dose received due to each type of radiation. Also, combining two films of different sensitivities makes it possible to determine the weekly dose on the more sensitive one and the quarterly dose (13 weeks) on the less sensitive.

The direct reading pocket dosimeter is necessary when it is desired to know the

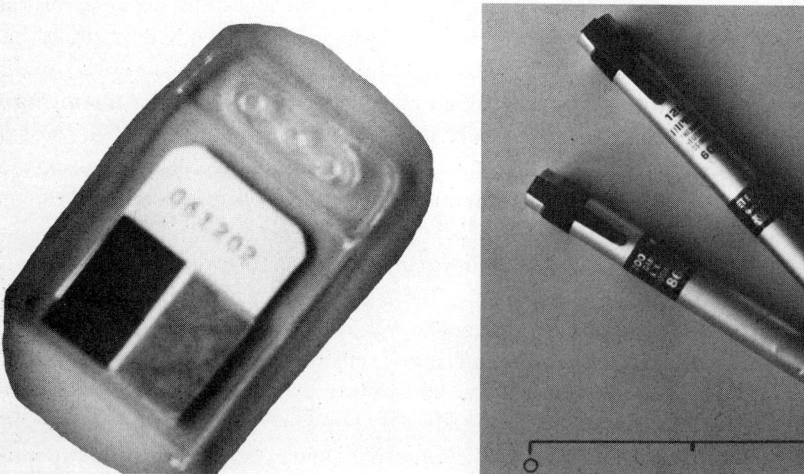

FIG. 3.1 Personnel monitoring is carried out with either or both film badges and pocket dosimeters. The film badge *(left)* is normally worn pinned to the coat but is sometimes worn on the wrist. The film badge provides a long term record of the dose received, while the pocket dosimeter *(right)* can give an immediate reading of the daily dose.

daily dose received. The principle of this instrument is considered in the next chapter, but it consists of a small quartz fibre electroscope with a small magnifier for viewing the fibre and scale. The instrument resembles a fountain pen and is usually worn clipped inside the top coat pocket. Ionization due to radiation leads to a partial discharge causing the fibre to move across the scale. A separate charger is used to charge the instrument before use, the fibre returning to zero on the scale. Dosimeters are available in ranges of 100 or 200 mr, and up to 50 r. When working with high levels of radioactivity it is common practice to wear two dosimeters because of the possibility of accidental discharge.

General monitoring in the laboratory can be carried out with either a portable ionization chamber monitor or by a portable G-M counter (ratemeter). Both are normally battery operated. The mains (line) operated ratemeter is also sometimes used where mobility is not especially important. The instruments are shown in Fig. 3.2. The roentgen being defined in terms of electric charge released in a certain volume of standard air, the ionization chamber (page 54) is the most direct method of measuring dose. This type of instrument is now quite versatile with ranges from fractions of a milliroentgen to several hundreds of roentgen per hour, though the medium ranges are more common (25, 250 and 2500 mr/h). It is less sensitive than the G-M counter and not really suitable for measuring surface contamination, for which specially designed scintillation probes are more effective for gamma emitters.

Earlier ion chamber instruments required different chambers to cover a wide range of radiation intensity, but modern instruments like the Victoreen 470A can cover a range of 3 mR/h–1000 R/h.

The portable G-M counter (page 55) is the most commonly used monitor in the average laboratory because of its all-purpose versatility. It can detect alpha particles or low energy beta particles (e.g. ^{14}C) when fitted with a G-M tube with a thin mica window of 1.5–2.5 mg/cm^2 while more robust thin-walled glass or metal tubes can be used for hard beta and gamma counting. Efficiency is high for beta particles but much less so for gamma counting. Ranges of portable G-M counter-monitors usually go from less than 100 c.p.s. to 3×10^3 or 1×10^4 c.p.s., or with gamma-only counters from 0.1 to 100 mr/h. Ratemeters and monitors usually have a built-in loudspeaker to give audible warning of radiation intensity.

As the G-M counter measures individual particles the counts are most often given in counts per second, but calibration makes it possible to relate the number of counts to dose rate. The relationship between counts per second and dose rate in mrad depends on the instrument, the probe being used and the nature of the radiation. Where no calibration is available about 100 c.p.s. can be regarded as a rough working limit for worker-exposure in the case of experiments, e.g. plant nutrient uptake, laid out on the bench.

In addition to G-M probes, many portable ratemeters will also accept portable scintillation probes (page 64), although the latter have the disadvantage of being more expensive.

Instruments are also available that are designed to monitor the atmosphere, either on a sampling or a continuous basis. For large establishments, wall-mounted or free-standing monitors are available especially designed for monitoring hands and footwear.

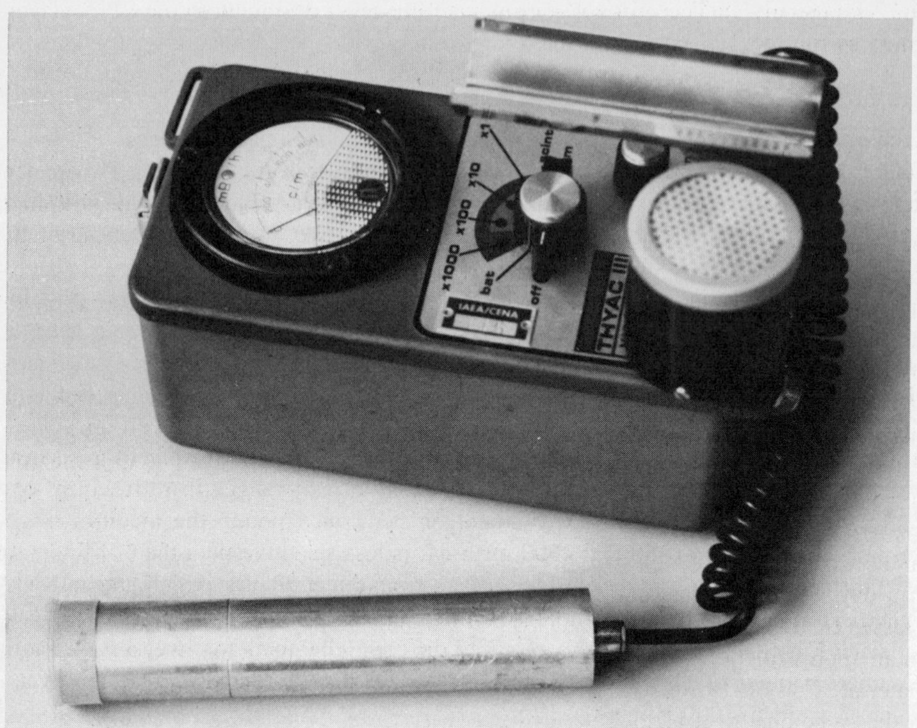

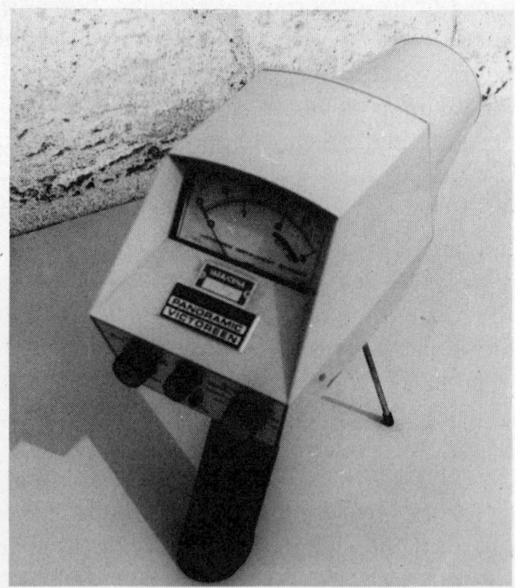

FIG. 3.2 Two of the most commonly used monitoring instruments.

Top: portable battery operated G-M counter with beta-gamma probe. Very versatile and used for monitoring experiments in progress, bench tops, glassware etc.
Bottom: The latest type of battery operated ionization chamber type monitoring instrument, with direct meter read-out. Adoptable to cover a wide range of radiation levels, including neutron radiation.

They are usually sited near the exit from the laboratory and their use becomes established as routine.

DISPOSAL OF WASTE

Most readers will not be greatly concerned with the disposal of waste, except to make sure that liquid waste is poured into the containers provided, and that contaminated solid waste such as paper tissues, disposable gloves, etc. is placed in the special (foot-operated) waste bins. Ideally, different colour-coded containers should be provided for the different classes (long and short half-life) of radioisotopes, or even for individual isotopes, to assist in economical and safe disposal. The ultimate disposal of waste depends on national and local regulations, which have mostly been formulated in line with the recommendations of the International Commission on Radiological Protection.

Short half-life material should be stored until activity has decayed so that it can be disposed of as inactive waste, this being usually after the elapse of about 10 half-lives. Alternatively solid waste of low activity may be disposed of with the ordinary laboratory waste, provided that the total activity put into the waste by any one worker in a day does not exceed 10^{-3} μc for isotopes in Class I, and 1 μc for isotopes in Classes II, III and IV. Long-lived isotopes will have to be stored and ultimately buried on an approved site. Some countries provide a collection and disposal service for such material. Carbon-14 may be disposed of by intimately mixing it with stable carbon in the same chemical form, provided that 10g of stable carbon accompanies every 1 μc of ^{14}C. As the treatment of waste of short and long half-lives is different, it follows that separate containers for each class of waste are necessary in the laboratory.

TABLE 3.6

Maximum permissible concentration of radionuclides in drinking water[1] *(assuming occupational exposure of 40-hour week)*

Radioisotope	μCi/cm³ water	Radioisotope	μCi/cm³ water
^{90}Sr	4×10^{-6}	^{64}Cu	0.01
^{45}Ca	3×10^{-4}	^{55}Fe	0.02
^{59}Fe	2×10^{-3}		
^{89}Sr	3×10^{-8}	^{65}Zn	3×10^{-3}
^{131}I	6×10^{-5}	^{74}As	2×10^{-4}
^{140}Ba	8×10^{-4}	^{76}As	6×10^{-4}
^{22}Na	1×10^{-3}	^{82}Br	8×10^{-3}
^{24}Na	6×10^{-3}	^{86}Rb	2×10^{-3}
^{32}P	5×10^{-4}	^{99}Mo	3×10^{-3}
^{35}S	2×10^{-3}	^{105}Ag	3×10^{-3}
^{36}Cl	2×10^{-3}	^{111}Ag	1×10^{-3}
^{42}K	9×10^{-3}	^{137}Co	4×10^{-4}
^{52}Mn	1×10^{-3}		
^{54}Mn	4×10^{-3}	^{3}H	0.1
^{56}Mn	3×10^{-3}	^{14}C	0.02
^{60}Co	1×10^{-3}	^{51}Cr	0.05

[1]Abstracted from Report of Committee II ICRP and IAEA Safety Series no. 1, 1973 Edition (1973).

Under approved conditions, particularly with isotopes of short half-life, e.g. ^{42}K, ^{24}Na or ^{32}P, some liquid waste can be disposed of to the sewers. Such solutions should be mixed with a large quantity of non-radioactive carrier in the same chemical form. There is no agreed limit as to the amount which may be discharged in this way, but in general for any single isotope in dilute solution it should not exceed, per worker, per day, 10^{-3} μc of a Class I isotope; 1 μc of a Class II material, 10 μc of Class III and 100 μc of Class IV material. Alternatively, a restrictive safe limit is if the concentration of the radioisotope in the liquid discharged to the sewer does not exceed the maximum permissible levels of radioactivity in drinking water, as recommended by the ICRP (Table 3.6) for persons occupationally exposed to radiation.

When new laboratories are being purposely built it is desirable that one or two liquid waste storage tanks should be provided, to ensure that no concentrated radioactivity is released inadvertently to the external drainage system, to permit storage until activity declines through decay, and also to provide for dilution of such waste if necessary. It has been the general experience of laboratories concerned with biological application of radioisotopes that there is rarely, if ever, any need to use methods such as precipitation or ion exchange to concentrate active materials before releasing liquid waste to the sewers.

Laboratories in areas where main drainage and sewage schemes do not exist, as is the case in many developing countries, have especial problems. It seems likely that the practical solution here is to have storage tanks, to permit dilution and monitoring of liquid waste, before leading the effluent into a large capacity gravel soakaway such as is commonly used in septic tank sewage systems, assuming background levels of activity.

The SI units and the proposed dose limitation system for radiation protection

While this book has been in the press new SI units have been approved for radiation dosimetry. The sievert (Sv) will replace the rem as the unit of exposure dose. Thus 1 sievert (Sv) = 100 rem and 1 millisievert (mSv) = 0.1 rem. At the same time the gray (Gy) will become the new unit of absorbed dose, where 1 gray = 100 rad. The new units should not cause too much difficulty as in practical terms they merely represent a difference in scale.

The recommendations in Publication 26 of the International Commission on Radiological Protection (1977) will, if adopted, change the basis of protection. The publication introduces the concepts of Annual Limit on Intake (ALI) and of Derived Air Concentration (DAC) of each radionuclide in one or more chemical forms. The changes in concept will have little practical effect for most biological workers; indeed they cannot be put into practice in some countries, e.g. the U.S.A., without new legislation.

REFERENCES FOR FURTHER READING

1. IAEA. *The Safe Handling of Radioisotopes*. Safety Series no. 1, 1973 Edition, Vienna, (1973).
2. IAEA. *Safe Handling of Radioisotopes*. Health Physics Addendum. Safety Series no. 2, IAEA, Vienna, (1960).
3. Committee of Vice-Chancellors etc. Code of Practice for the Protection of Persons Exposed to Ionizing Radiation in University Laboratories. Assoc. Univ. Brit. Commonwealth, London, (1961).
4. IAEA. *Safe Handling of Radioisotopes*. Medical Addendum. Safety Series no. 3, IAEA, Vienna, (1960).
5. ICRP. Report of Committee II (*Permissible Dose for Internal Radiation*) of the International Commission on Radiological Protection. Pergamon Press, London, (1959).
6. PRICE, B. Y. *et al. Radiation Shielding*. Pergamon Press, London, (1957).
7. FAIRES, R. A. and Parks, B. H. *Radioisotope Laboratory Techniques*. Newnes, London, (1960).
8. BOURSNELL, J. C. *Safety Techniques for Radioactive Tracers*. Cambridge University Press, London, (1958).
9. IAEA. *Radioactive Waste Disposal into the Ground*. Safety Series no. 15, International Atomic Energy Agency, Vienna, (1965).
10. HINE, G. J. and Brownell, G. L. *Radiation Dosimetry*. Academic Press, New York, (1956).
11. REES, D. J. *Health Physics, Principles of Radiation Protection*. Butterworths, London, (1967).
12. IAEA. *The Basic Requirements for Personnel Monitoring*. Safety Series no. 14, IAEA, Vienna, (1965).
13. IAEA. *The Management of Radioactive Wastes Produced by Radioisotope Users*. Safety Series no. 12, IAEA, Vienna, (1965).
14. IAEA. *Basic Standards for Radiation Protection*. 1967 Edition, Safety Series no. 9, IAEA, Vienna, (1967).
15. OVERMAN, R. T. and Clark, H. M. *Radioisotope Techniques*. McGraw-Hill, New York, (1960).
16. MORGAN, K. Z. and Turner, J. E. Eds. *Principles of Radiation Protection*. Wiley, New York and London, (1967).

CHAPTER 4

Detection Systems and Instrumentation

ALL radioisotope tracer techniques require the detection and determination of the amount of radioactivity in samples. The method employed will be determined by the isotope being used, the likely level of activity, and to a certain extent the equipment that may be available. Quite often it is possible to assay a radioisotope by more than one method, although usually one particular method is the most efficient and preferable.

The methods of detection are based on the ionization and other properties of the particles or gamma photons. Thus there are two groups of detectors. Those in which ionization takes place in an enclosed sensitive medium between two oppositely charged electrodes are characteristic of *ionization chambers, Geiger-Müller* and *proportional counters*. Second are systems which do not depend on ion collection but make use of the property that γ-ray photons, and to a lesser extent α- and β-particles, have for exciting fluorescence in certain substances. These systems are known as *scintillation counters*.

Semi-conductor detectors also depend on primary ionization effects, but unlike G-M counters they have the advantage of solid state. Semi-conductor detectors are not at present used for routine counting purposes in biology and lie outside the mainstream of this book. They do however have a number of specialized applications of interest in biology and they are discussed briefly at the end of this chapter.

The distribution, and to a certain extent the comparative amount, of a radioisotope tracer in a biological specimen can also be conveniently found by *autoradiography*. This technique makes use of the fact that ionizing radiations affect the silver halide in photographic emulsions, which on development show a blackening of the areas exposed to radiation. If a specimen has been placed next to an X-ray film or plate for an adequate exposure time, the areas of the specimen which show concentration of the tracer therefore show up on development, in a form of self-portrait. This technique is considered with practical applications in plant science in Chapter 10.

The detector, of whatever type, is mounted in a lead "castle" or shield, normally 2.5–5 cm thick but sometimes more, which reduces the background counts due to both cosmic radiation and the minor contamination of the laboratory, Fig. 4.1. The lead shield also has provision for holding sample containers of various kinds appropriate to the particular type of detector. Every detector, whether G-M tube or scintillation counter, must have an appropriate holder, cable and plug: such an operational assembly is commonly called a "probe".

All detectors require an associated scaling system, basically comprising a power source and an electronic counter to record the pulses detected. As scaling systems are virtually common to the various detection systems they will be considered first.

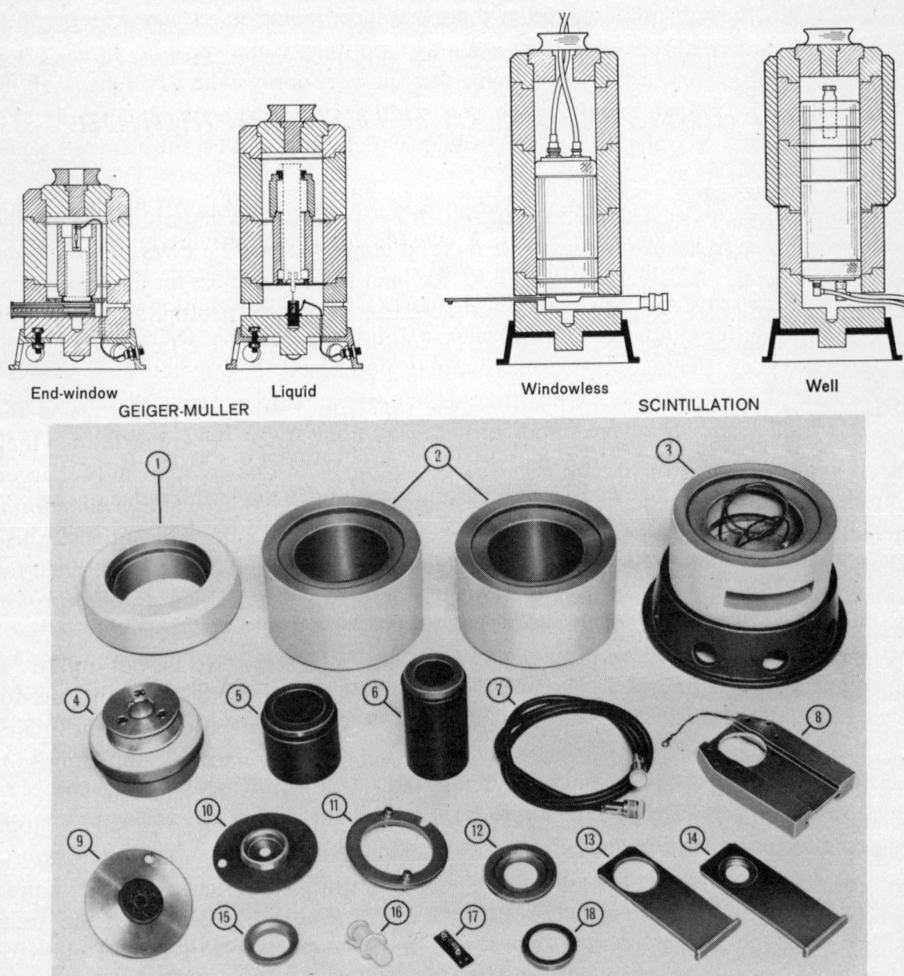

FIG. 4.1 *Top*: The components of an adaptable lead castle; primarily intended for end-window G-M counting, it can be arranged (*Bottom*) for liquid G-M counting, and simple windowless or well scintillation counting for gamma-emitters (Panax).

SCALING SYSTEMS

The *scaler* is the electronic basis of all counting systems and its function is to permit the fast recording of a large number of pulses. In common usage the word "scaler" is used to denote a complete counting assembly, comprising a power supply for the detector, an amplifier, a register to record the pulses and usually a timer. Other refinements are also common and are essential for certain types of counting. At one time separate units assembled in a rack were the rule, but the majority of present day scalers are completely self-contained assemblies. With the coming of transistorized circuits and increasing miniaturization there has been some move towards modular systems, in which the individual units such as amplifier, count register, timer, print-out etc. are plugged into a pre-wired frame with power supply. The advantages of this are compactness, the ease with which any scaling system can be assembled for a particular purpose, and also the readiness with which units can be substituted in case of breakdown.

The *power unit* stabilizes the mains supply and provides variable EHT voltage from a few hundred to a few thousand volts. A standard unit may provide an EHT output of 0–2500 at 250 μA with a fluctuation of $\pm$ 0.2% for mains voltage fluctuations of up to $\pm$ 10%. Such stability is suitable for most G-M and scintillation counting. For very sensitive scintillation counting greater stabilization may be desirable and units giving $\pm$ 0.005% fluctuation can be routinely purchased. Where the mains supply voltage is extremely variable, as in some developing countries, it may be necessary to have a large separate voltage stabilizer in the counting room that can supply several counting systems with pre-stabilized current.

An *amplifier* is necessary in counting systems to amplify the initial pulse or current, which is often too small to otherwise register, into strong low impedance pulses which can be conducted through a cable to the scaler. At one time it was general practice to have a *pre-amplifier* performing this function, connected as close as possible to the detector to avoid pulse losses due to the capacity of the cable. Further amplification was done by a second *linear amplifier*. Linear amplification is especially important for proportional counting where the multiplication factor must be independent of pulses size. Nowadays, general improvements in electronics have made the pre-amplifier, as it was formerly understood, obsolescent.

A *discriminator* may be provided to reject pulses above or below a set level. The object of this is to improve the signal to background "noise" ratio, and is especially important for pulse height analysis.

The *counter* or *register* must be able to record a large number of counts at a very fast rate. It may either be a mechanical register, now virtually obsolete, or an electronic register, of which there are several types. The disadvantage of the mechanical register is its limited capacity to accept pulses. The older-style 4-digit register could not accept more than about 10 pulses/second before it failed to resolve counts, although modern 6-digit registers will take up to about 15 pulses/second before jamming. Often a mechanical register follows, and is activated by, a series of electronic counters, in order to record the higher digits. The advantage of the mechanical register is cheapness and it is also compact.

The original electronic counters worked on a scale of 64, having so-called binary circuits which emit one pulse for every two received. These instruments are now obsolete, though some are still in use. However, the instruments now on the market are decade scalers, that is they count in multiples of ten. In the case of the old scale-of-64 counters there are interpolation lights for 1, 2, 4, 8, 16 and 32 counts. At 63 counts all lights will be on, but with the next pulse 64 counts will be recorded, the lights will then go off and the slave mechanical register advances by one unit. Each unit on the mechanical register therefore represents 64 counts. The disadvantage of this type of counter is the rather tedious totalizing of the counts which must be done.

The modern decade scaler may be completely electronic or like the older scale-of-64 type instruments may have the units and tens of pulses recorded by neon indicators and the hundreds and thousands registered mechanically. There has been a general move towards all-electronic instruments. At one stage the decade circuit scalers were not so reliable as the more simple circuits of the binary type but improvements in electronics have overcome this. An all-electronic instrument with cold cathode dekatron glow transfer tubes will have a count range up to a million counts and a dead time of say, 20 μsec. For very fast counting applications integrated circuits in the first decade can reduce the dead time to about 1 μsec. With these instruments the number of counts can easily be read off from left to right, according to the position of the indicator neons opposite the scales.

The most recent development is that of decade scalers with number tubes or L.E.D. digital display which enable the count to be read directly. They are very fast scalers and undoubtedly the easiest to read, while systems incorporating them can be made very compact.

All modern scalers can be operated to a preset count, that is they switch themselves off when the preset count has been achieved. A typical instrument may offer a choice of six to nine preset counts, covering a range from 100 to 1 million counts. A preset count facility must be operated in conjunction with a *timer*.

Timers may be mechanical as a form of stop watch, or now more usually of the electronic decade type similar to decade scalers. Another form sometimes met with is of mechanical register pattern actuated by a mains frequency derived clock unit. The timer permits recording the time taken to achieve a preset count, as described above, or alternatively can be set to stop the scaler after a pre-set time. A range of pre-set times is provided, usually including 10, 100, 1000 and 10,000 seconds.

On the whole it is usually more convenient to operate a scaler on preset time rather than on preset count. This is because often many of the subsequent calculations will require the activity to be expressed as "counts per minute", and the transformation can be most easily achieved if a preset time of 100 or 1000 seconds is chosen, then appropriately inserting the decimal point in the count and multiplying mentally by six. Additionally, on preset time a timer can usually be set to control several scalers, if so desired.

Scalers intended for the more complicated counting procedures demanded by the assay of samples of very low activity and for pulse analysis are provided with *anti-coincidence* and *coincidence* facilities. An anticoincidence unit has the function of rejecting pulses that arrive simultaneously or within a very short time interval, i.e. about 1 μsec. The anticoincidence facility is necessary for pulse height analysis and is also used for guarding against cosmic radiation in very low background counting systems, Fig. 4.2. In contrast a coincidence unit rejects single pulses but emits one

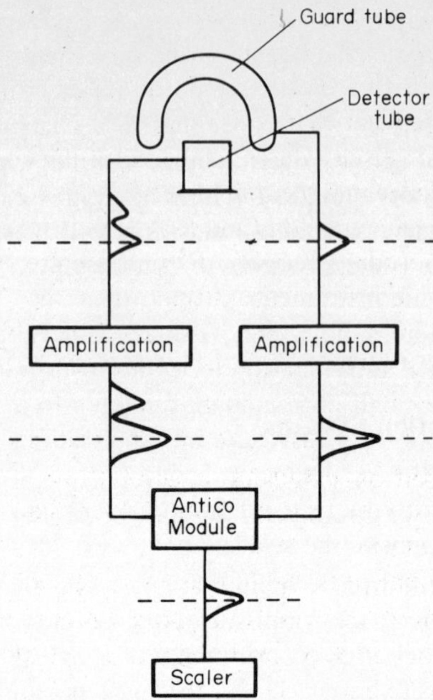

FIG. 4.2 Block diagram of anticoincidence layout for guarding against cosmic radiation. Pulses which arrive from the detector tube and guard tube in coincidence are rejected.

pulse when two pulses arrive in coincidence, i.e. within 1 μsec. Such a unit is used with two scintillation detectors to cut out photomultiplier noise pulses and to reduce the background count.

For *spectrometry* (page 106) a scaler must be equipped with *pulse-height analysis* facilities. These comprise a combination of two variable bias discriminators with an anti-coincidence unit. This assembly permits only pulses within a predetermined pulse-height interval to be recorded. Essentially a spectrometer permits the measurement and analysis of the energy distribution of radiation.

It therefore makes it possible to identify and measure isotopes by reference to their characteristic energy spectra, such as are given in tables. Special instruments called multi-channel analysers are available which can sort according to pulse-height every pulse received, as opposed to the single-channel system just described which registers only those pulses which fall in the single defined range. Multi-channel analysers are considered further in Chapter 5.

Modern scalers usually include optional provision to control automatic sample changers, and the associated punched tape or digital print-out of the results. An additional refinement is a low count rejector unit, which can be used in automatic sample changing systems to terminate the counting of samples which fail to reach a pre-determined count rate.

In addition to the scaler the *count rate-meter* is sometimes used for the recording of sample radioactivity. Equipped either with a G-M or scintillation probe, the rate-meter gives a direct reading of the counting rate on a meter, and is useful for circumstances where truly quantitative results are not required. Thus the rate-meter may be used for monitoring (Fig. 3.2). When coupled to a chart-recorder the rate-meter is used for *in vivo* studies of activity over a period of time, e.g. in thyroid studies with ^{131}I; in recording the activity from radiochromatograph scans; and in making scans during gamma spectrometry. A typical rate-meter will have a choice of six to nine ranges covering less than 100 c.p.s. to over 20,000 c.p.s.

GEIGER-MÜLLER, PROPORTIONAL, AND OTHER ION COLLECTION COUNTERS

Principles of Ion Collection Systems

Detection systems based on ionization and ion collection include electroscopes, ionization chambers with electrometers, the Geiger-Müller counter and proportional counters.

When a gas such as air, nitrogen, helium or argon is contained in a sealed container provided with charged electrodes, ionizing radiation passing through the gas in the tube causes electrons to be removed from the atoms of gas to form ion-pairs, that is pairs of electrons and positive ions. Under the influence of the applied field some of the electrons move towards the anode and some of the positive ions towards the cathode. As a consequence, charges collect on the electrodes and initiate pulses; a continuous stream of these pulses constituting a weak electric current.

With a low potential not all the electrons and positive ions reach the electrodes and therefore recombine to form neutral atoms, as would be the case if no potential was applied. However, if the applied potential is increased the proportion of electrons and positive ions reaching the electrodes without recombination increases proportionately, until saturation occurs. With this applied voltage all ion-pairs are reaching the electrodes and no further recombination of ion-pairs takes place. At this point a moderate increase in potential does not result in any further rise in current.

This can be illustrated if a plot is made of log current (pulse size) as a function of applied voltage, as in Fig. 4.3. Of course there is no single detector which can operate over such a voltage range, and the figure consequently represents a composite situation, derived from different detectors, electrodes and gas filling. Thus region I in Fig. 4.3 represents the situation where the size of pulse produced by an incident particle increases with applied voltage, while region II is where pulse size retains a constant value over quite a range of applied potential. The magnitude of the voltage required to achieve saturation is of the order of 50–150 volts. However, the actual voltage depends on a number of factors, including the nature of the gas being ionized, the pressure of the gas and the shape and spacing of the electrodes. Regions I and II are known as the *regions of simple ionization*, region II being characteristic of the operating conditions for ionization chambers and electroscopes.

Now, if the applied potential is substantially increased a further increase in current will be obtained. This is due to secondary ionization, because with the higher voltage

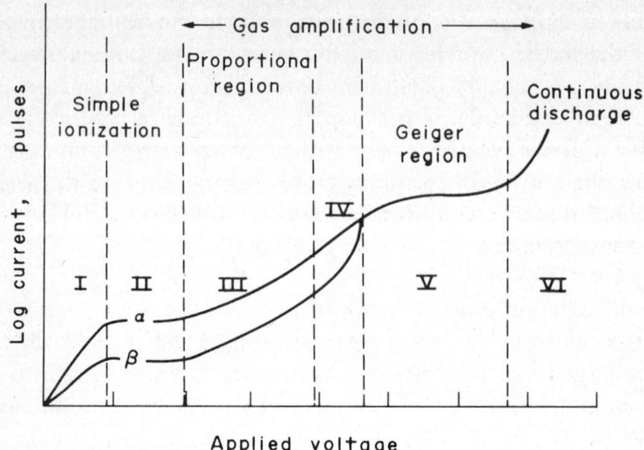

FIG. 4.3 Effect of ionizing particles on current (pulse size) as a function of applied voltage.

some of the primary electrons achieve sufficient energy to ionize additional gas atoms. The electrons so produced may initiate further ionization and so on, the number of ion-pairs available for current production being thereby successively increased. This is known as *gas amplification* and is characteristic of the regions III-VI in Fig. 4.3. Although gas amplification is unity in the regions of simple ionization, it may be as much as 10^5 in the upper part of the proportional region. Thus the process of secondary ionization becomes of increasing significance as the potential difference between the electrodes is increased, when the number of secondary ions produced is proportional to the number of initiating primary ions. Hence this region is called the *proportional region*. Region III is the true proportional region, while region IV is the *limited proportional* region, a transitional stage where the ionization depends not only on the initial number of ions produced but also on the voltage.

As there is a considerable difference in the degree of specific ionization produced by α- and β-particles, the former being about 1000 times as effective as the latter, it follows that for a given applied potential an alpha particle will produce a thousand times as many primary ion-pairs as a beta particle and hence give a greater current, as shown in Fig. 4.3. This provides the basis for differentiation between alpha and beta radiation using a proportional counter.

Although gas amplification increases in proportion to applied voltage a limit is reached in the limited proportional region. The exact point is governed by the size of the counter tube and the number of gas atoms it contains, as these two factors determine the maximum number of ion-pairs that can be produced. However, with further increasing voltage, ionization due to incident particles or photons becomes steady and is independent of the number of primary ion-pairs produced, all radiation now giving the same current flow. This characterizes the *Geiger region*, V in Fig. 4.3.

Gas amplification having reached its maximum in the Geiger region each primary ionization results in an *avalanche* of electrons, and the substantial output pulse is now independent of the initial particle or photon energy. Any further moderate increase in voltage does not greatly alter pulse size. In the Geiger region any primary radiation whether alpha, beta, or gamma, of comparable intensity will produce the same degree of ionization for a given voltage. It is not therefore possible to distinguish between alpha and beta radiation, as in the proportional region. Beyond the Geiger region a voltage is attained where *continuous discharge* takes place. It is of no value for detection or assay purposes.

Detection by ionization and ion collection can therefore be briefly summarized as follows: incident radiation ionizes atoms of gas maintained in an electric field in a detector. Electrons will then move to a positive electrode and positive ions to a negative electrode; the charged particles arriving at the electrodes give rise to an electronic pulse, which can either be registered as such, or merged to form an electric current, which can also be measured. The magnitude of the applied potential governs the degree of primary and secondary ionization in the detector, and hence whether the detection instrument is operated in the so-called simple ionization, proportional or Geiger regions. These regions are in effect a reflection of the influence of ionizing radiation on pulse size as a function of applied voltage.

G-M counters and proportional counters are used in radioassay, and of these the former is by far the most important. Pocket dosimeters (electroscopes) and ionization chambers are used in radiation monitoring and have been referred to in the previous chapter.

Pocket Dosimeters (Electroscopes)

The pocket dosimeter is an elementary electroscope in which the miniature ionization chamber acts as a simple capacitor, the chamber being charged by the application of a potential of 100–200 volts across it for a short time. The wall of the chamber is the negative electrode and a centrally mounted and insulated quartz fibre is the positive electrode.

The quartz fibre serves to indicate the electrical charge remaining, as when the dosimeter is fully charged the deflection of the fibre will be maximal, in the direction of the chamber wall (the cathode). If insulation were perfect and no ions were to reach the element the fibre would stay in this position. However, incident radiation will ionize air in the detector and electrons will move onto the quartz fibre and positive ions will move towards the chamber wall, thus releasing the charge on the fibre which tends to assume its original central position. The fibre passes across a scale graduated in milliroentgen or roentgen, and which in pen-sized pocket dosimeters is read by means of a small magnifier. Provision is made when charging the dosimeter for accurately setting the image of the quartz fibre to zero by means of adjusting the potential.

The dosimeter is of course an integrating type of instrument, that is it records the total amount of radiation absorbed in a given period of time. It does not give an instantaneous reading of radiation intensity. Electroscopes such as the gold leaf electroscope and the Lauritsen electroscope have been used for radioassay in the past but are almost never used nowadays and will not be described here.

Ionization Chambers (Electrometers)

Although the electroscope is in fact a simple ionization chamber functioning as a condenser, in everyday laboratory usage the term "ionization chamber" is usually taken to refer to a more sophisticated type of instrument. In this the ions discharging at the electrodes give rise to pulses which are smoothed to form a weak electric current, which is amplified and registered on a meter. For example, this is basically the type of instrument often used for monitoring in radiation protection, and popularly referred to as the "cutie pie" (Fig. 3.2).

The cutie-pie ionization-rate meter is an instantaneous measuring type of instrument. It consists of a cylindrical ionization chamber of known volume (see definition of roentgen, page 10), the inside surface of which is electrically conducting, fitted with a centrally mounted and insulated electrode rod to which a positive potential is applied. A thin window (2–3 mg/cm) permits the entry of even soft-beta radiation. The production of ion-pairs in the chamber results in a weak flow of current which is led through a resistance. A vacuum tube electrometer connected across the resistance is then used to determine the magnitude of the ion current, which is directly proportional to the rate at which ionizing radiation is entering the ionization chamber. A diagram of an ionization chamber with a simplified circuit is given in Fig. 4.4.

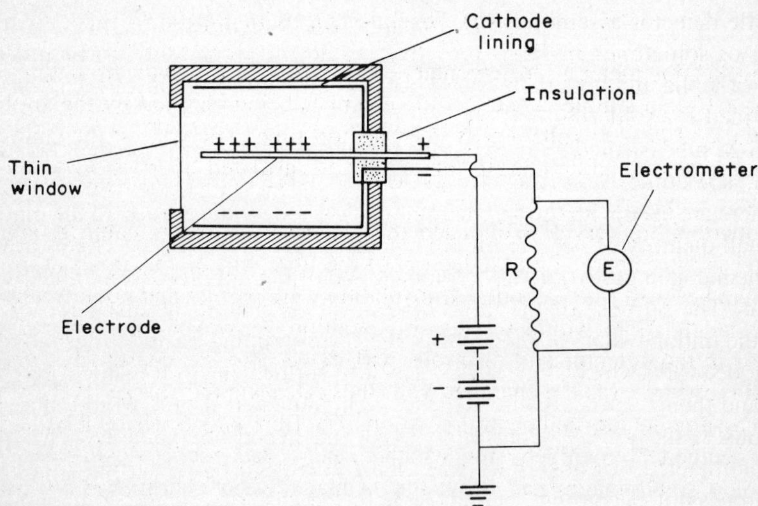

FIG. 4.4 Diagram of ionization chamber with a simplified circuit.

An ionization chamber requires calibrating against a gamma source of known activity. The dose-rate value, R_f, is calculated as given on page 35 or obtained from

Table 3.4. It is applied using the inverse square law to take account of the distance that the instrument is from the source. Cutie-pie instruments may have chambers of different volumes that can be fitted to cover a wide range of radiation levels. Each may require separate calibration.

Ionization chambers can be used for neutron monitoring, either lined with a boron coating or else filled with BF_3 gas (see page 62).

An electrometer that is sometimes used for radioassay is the vibrating reed electrometer. It is particularly used for the specialized measurement of very small quantities of carbon-14 and tritium in gaseous form.

Geiger-Müller Counters

The Geiger-Müller or G-M counter has been the mainstay of many laboratories concerned with the use of radioisotopes in agriculture and biology. It requires the simplest circuit, the least expenditure of money, has the least operating difficulties and is widely adaptable to a variety of counting problems.

A G-M tube consists of a sealed cylindrical tube or envelope made of glass or metal, which is coated internally with conducting material such as silver or graphite. This constitutes the cathode. A thin wire, usually made of tungsten, is sealed in concentrically at one end of the tube and acts as the anode. Alternatively, in some tubes the cathode consists of a wire spiral surrounding the anode. The wall of the tube may have the same thickness of glass or metal over the whole of its working surface, but what are known as "end-window" tubes are most frequently used, especially in lead-castle detector assemblies. In these a much thinner window made of mica (1–3 mg/cm^2) or sometimes mylar is provided at one end of the tube to permit the ready passage of alpha and beta radiation. Tubes usually have a multi-pin plug base or tags are provided for soldering.

The G-M tube is filled with one of the monatomic noble gases such as argon, helium or neon, sometimes at a pressure as low as 1/10 atmosphere, but pressures near atmosphere are used with the fragile thin window tubes. In addition to the main counter gas a small quantity of polyatomic "quenching" gas is also used. This is now usually one of the halogen gases, such as bromine or chlorine, although formerly ethyl alcohol was used. The purpose of the quenching agent is to limit the discharge of the tube, as after the initial avalanche of electrons the corresponding positive ions move towards the cathode wall, which they strike, ejecting further electrons which give rise to a second and thence successive avalanches. An unquenched tube would thus discharge continuously, the pulses being unresolved and hence uncountable. The quenching agent has the effect of absorbing part of the energy of the electrons after ionization, in the process of becoming decomposed. In the case of organic compounds such as alcohol this decomposition is irreversible and reduces the useful life of the tube, but with a halogen quenching agent the atoms usually recombine. The latter type of tube has an indefinite life, though in fact decomposition products deposited on the wall of the tube and on the central electrode wire limit useful life.

Types of G-M Tubes

G-M tubes are particularly effective for detecting beta particles and also alpha particles, if the window is thin enough. However, the very reason for this effectiveness makes them much less effective detectors of gamma photons. Thus radiation can penetrate the tube very readily, too readily in the case of gamma photons, because with their high penetrating power and low specific ionization it means that a major proportion of incident gamma radiation passes straight through the gas in the tube, producing comparatively little ionization.

Specially devised G-M tubes, sometimes called gamma-counters, are available and can improve the overall efficiency of gamma counting more than 6-fold, though they are still less than 10% efficient in comparison to beta-efficiency. These tubes are made of metal or relatively thick glass, sometimes being coated with bismuth, and are of comparatively long active length. The longer tube permits the use of a long central electrode wire which is of some importance in the G-M tube, as the normal working voltage potential results in the whole length of the anode being used in the discharge. The thick walled tube increases the possibility of the gamma photons causing ionization with the ejection of electrons. Gamma-counters are frequently side-window tubes, that is they are used with their side towards the radiation source, in order to make full use of the whole active length of the tube. They are usually held in metal tube counter-holders having a window out in the side.

G-M tubes are available for dipping in radioactive solutions, and also of the liquid-counter type, such as the M6H or MX124, in which the radioactive solution is poured into the tube. Such tubes are not as sensitive as thin end-window tubes, as they must be made more substantial to withstand the frequent counting. They are consequently not suitable for soft-beta counting. Some typical examples of G-M tubes are given in Fig. 4.5.

Operational Characteristics of G-M Tubes

A G-M counter is connected to either a scaler or rate-meter, which provides EHT voltage source, an amplifier, and in the case of scalers both register and timing facilities. The operational voltage of the tube is determined by its "plateau characteristics".

From the earlier discussion it will be recalled that the G-M counter operates in a region where gas amplification has reached its maximum, the output pulse is independent of the initial particle energy and a moderate increase in voltage does not alter the number of pulses per minute. The Geiger plateau where these conditions prevail is shown in more detail in Fig. 4.6. The threshold voltage and the plateau range depend on the design of the tube, the nature of the gas it contains and also its age. The plateau range may extend over 150 to 300 volts, in the region from about 400 to 1500 volts. In general the newer near-atmospheric pressure halogen quenched tubes operate at lower voltages than the older organic quenched tubes.

Although the operating voltage for any G-M tube is usually marked on it, or can easily be obtained from the manufacturer's catalogue, actually determining the plateau of a new G-M tube is a worthwhile exercise. The plateau is virtually never completely

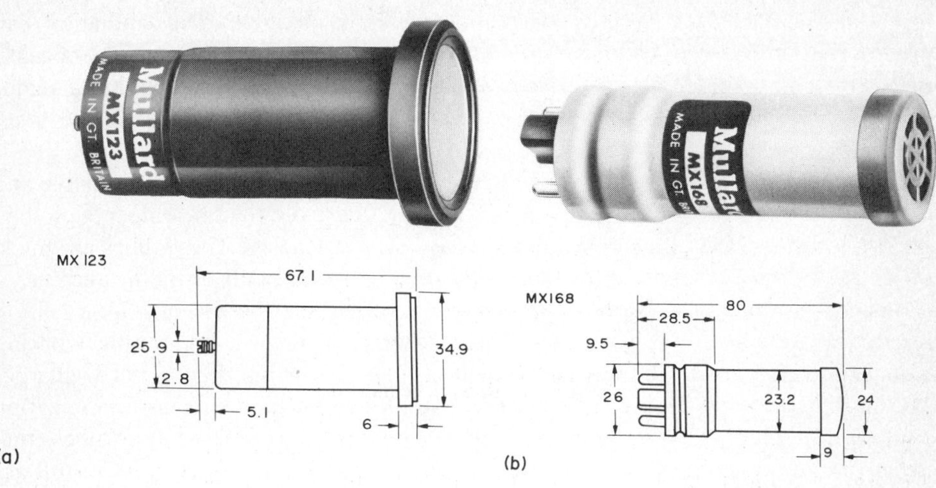

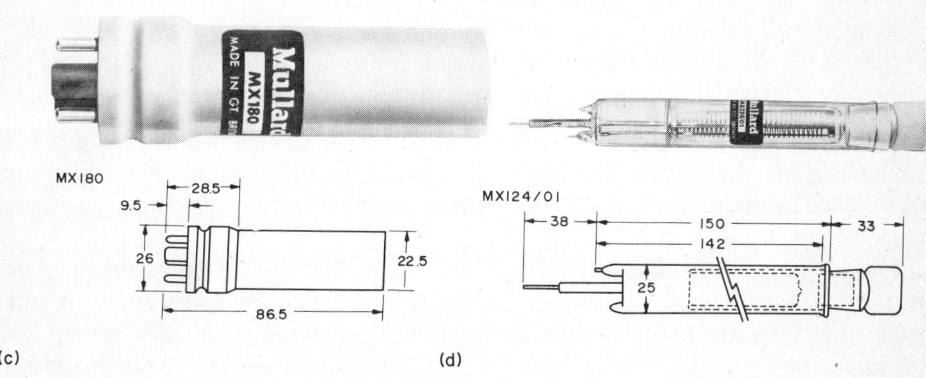

FIG. 4.5 There is a wide range of Geiger-Müller tubes for various specialized applications and a few are illustrated. (a) MX123, a very thin mica end-window tube of wide application but especially for alpha/beta counting. (b) MX180, a metal γ-sensitive tube for gamma counting. (c) MX168, a mica end-window tube for alpha/beta counting but of more robust construction than MX123/01. Used frequently in portable monitors. (d) MX124, a glass counter for liquid samples, of 10 ml capacity. (Mullard Ltd.)

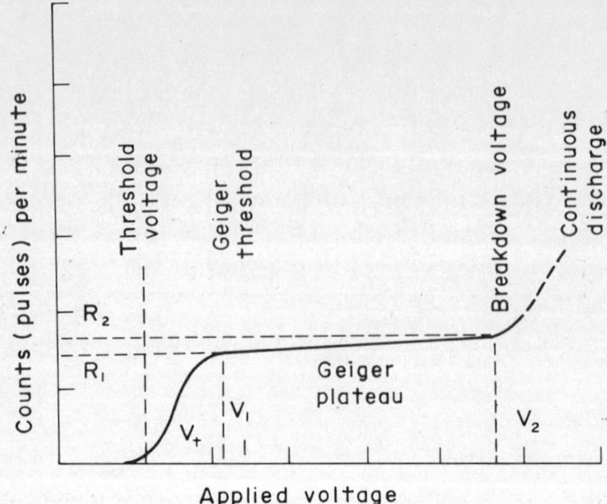

FIG. 4.6 Characteristic curve of the relationship between count rate and voltage of a Geiger-Müller tube.

flat but has a slight slope. This should not exceed 10% per 100 volts, and for a good tube can well be under 5%.

A reference standard giving 5–10,000 counts per minute is placed in the normal sample holder of a detection assembly and the EHT voltage slowly increased until counts are registered. This is the threshold voltage. The count rate is then determined at increasing steps of 25 volts, counting about 10,000 counts at each step, until the continuous discharge region is reached, with a sudden rise in count rate. The figures may then be plotted as a graph to obtain a curve similar to Fig. 4.6. The Geiger threshold, V_1, is the point from which the count rate does not change appreciably with increasing voltage. The slope of the curve expressed as per cent increase in count rate per 100 volts is then given by:

$$\% \text{ increase in count rate/100 volts} = \frac{100(R_2 - R_1)/R_1}{V_2 - V_1} \times 100.$$

The plateau length is obviously V_2–V_1.

It is customary to operate halogen quenched G-M tubes at a voltage corresponding to the mid-point of the plateau. With organic quenched tubes a working voltage is chosen in the first quarter of the plateau, i.e. at the lowest practical working voltage, in order to prolong the life of the tube. With age it is found that the threshold voltage of a G-M tube will rise and the breakdown voltage will become lower, thus effectively shortening the plateau length. Care should be taken not to exceed the breakdown voltage as a tube going into continuous discharge will soon be destroyed, e.g. in plotting a tube plateau do not let the counting rate beyond the breakdown voltage rise by more than 15 per cent of the plateau value.

A G-M tube has a *dead time* or *resolving time* during which it is unable to detect incident particles or photons. This is the time taken from the initiation of each discharge of electrons until the tube is restored to a voltage state in which it can detect further particles. This is a period of from 100 to 350 μsec, though some modern high current tubes have dead times as low as 15 μsec. During this period, if other incident particles should enter the detector they will not be recorded. It follows that the number of pulses registered by a G-M counting assembly is less than the number of particles entering the detector. Thus with high count rates it may be necessary to introduce a correction for dead time, although with count rates below 2000 c.p.m. the dead time can usually be ignored. This is further discussed in the next chapter. The background count of an end-window G-M tube is usually about 15–25 c.p.m. when held in a lead castle detector assembly, and about three times this figure without shielding.

Windowless Gas-flow G-M Counters

This type of counter was developed for the counting of low activity samples of very soft beta-emitters such as carbon-14, calcium-45 and sulphur-35, for which even the very thin mica window of the most sensitive G-M tube may result in significant loss of counting efficiency through the absorption of beta particles. The windowless gas flow counter gives a very high counting efficiency, both through the absence of any radiation absorbing window and also because of the excellent 2π (amost 50%) geometry, with the sample being counted actually inside the counter. (*Geometry* is a function of the distance of the sample from the counter and essentially refers to the per cent radiation emitted from a sample that enters the detector, see page 88). Nevertheless, the great development of liquid scintillation counting for soft beta-emitters has made windowless gas-flow counting obsolescent for solid samples, though it still finds favour for specialized purposes.

In these counters the sample is placed inside the counter which is purged with a gas mixture to expel air. The gas mixture used may be helium and iso-butane, 99%:1%, or argon and propane, 98%:2%, with which the counter is operated in the Geiger region at a potential of 1200–1300 volts. In one well known example of this type of counter (Tracerlab), a 3-position turntable permits pre-flushing of samples before each in succession is turned round into the counting chamber. This pre-flushing with Geiger gas saves time on sample changeover, as when carrying out a series of counts it is not necessary to completely purge the counter to remove air, as is required when starting operation. A small amount of air will probably enter with the sample but it is relatively quickly flushed, and a slight positive pressure excludes air during the counting period. A glass bubble tube assists in regulating gas flow.

Although theoretically attractive for very high efficiency beta counting, there are a number of practical problems with windowless gas flow counting. The sample has to be completely dried and must be free from either solvent or trapped air which would affect the counting atmosphere. Difficulty has also been experienced with the build-up of a positive charge on samples, as they are poor conductors. Not only does this result in a loss of counting efficiency but can lead to sample particles jumping onto the wall of the chamber or the electrode. A partial cure for this is to use a metal planchet in contact with the cathode, while on occasion samples have been mixed with a trace of graphite as a conductor. The possibility of reduced counting efficiency can be checked if each sample is counted for several minutes, the count being noted at the end of every minute to establish if there is a decrease in count rate.

Gas flow G-M counters have been adopted for radiochromatogram scanning/detection, in conjunction with a rate-meter and a flat-bed chart recorder. This procedure is especially efficient for the radiochromatograms having ^{3}H and ^{14}C. A typical example is shown in Fig. 4.7.

Gas flow counters operating in the Geiger region can usually be operated also in the proportional region. For this it is usually necessary to change the straight anode wire to one in the shape of a loop, and also to use a gas having more quenching action than that use for G-M counting. This may be argon and methane, 90%:10%.

A useful development has been the Tracerlab CD series of gas flow counters which can be operated either as G-M or proportional detectors simply by changing the gas and the operational voltage. These detectors have a stainless steel loop centre wire and take the form of end-window counters, but the windows are exceptionally thin, $< 150\ \mu g/cm^2$ for aluminized Mono-Mol windows and $900\ \mu g/cm^2$ for aluminized mylar. They clearly offer the possibility of a counting assembly with the advantages of gas-flow G-M and proportional counting, without some of the disadvantages hitherto experienced. Helium/isobutane/butadiene is used for counting in the G-M mode and

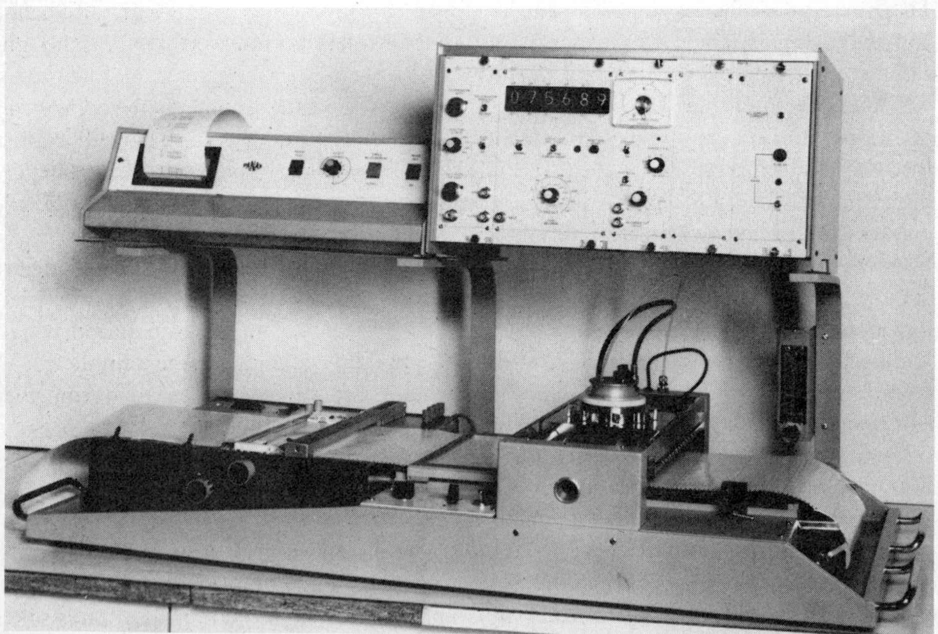

FIG. 4.7 Radiochromatogram scanner incorporating a windowless gas flow G-M counter, utilizing a 98%:2% argon/propane mixture. The assembly includes a flat-bed chart recorder, scaler, rate-meter and printout.

argon/methane or pure methane for proportional operation. This type of detector is now available in fully automatic counting systems.

Anticoincidence Low Level G-M Counting

The purpose of the anticoincidence unit has already been described (page 49). G-M counters employing an anticoincidence circuit are available in which the main counter is surrounded by either a ring of secondary guard counters or else by a specially constructed circular G-M tube. Then a pulse from the surrounding guard tube or tubes which coincides with a pulse from the main counter is not registered, through the operation of the anticoincidence unit, as indicated in Fig. 4.2.

By this means the background count due to external cosmic radiation, instrument noise and natural radioactivity in the assembly can be reduced to an extremely low level (0.1–1 c.p.m.) when the counter is contained in a thick lead castle.

Antico counting is particularly useful for assaying very low activity beta samples, with amounts of radioactive material giving "background level" counts of only a few disintegrations a minute.

Proportional Counters

Proportional counting is a very effective means of counting soft beta radiation, although it is comparatively little employed in biological and agricultural studies. There are probably several reasons for this. The equipment and its handling are somewhat sophisticated, and except where there are regular numbers of ^{14}C or ^{3}H gas samples to be assayed, there are usually easier means of counting available. End-window G-M counting for example is much less trouble if the samples are sufficiently active, while liquid phosphor scintillation counting is usually the method of choice for biochemical samples. Secondly, the strong point of proportional counting is its high efficiency for alpha emitters and the ability to distinguish between alpha and beta radiation. This, however, is of virtually no interest in biological applications.

Apart from beta counting efficiency, a further characteristic of the proportional counter is its ability to accept high count rates, as the "dead time" after each pulse is very small. Thus the resolving time is only 1–10 μsec, compared with two to three hundred microseconds for G-M counters. In practice this means that while G-M counting rates are limited to 15–20,000 c.p.m., in the proportional region count rates can be as much as 200,000 c.p.m. Additionally the background is relatively low.

Gas counting, that is the counting of samples in gaseous form, is especially effective with proportional counters. They are much less sensitive to changes in gas mixture, e.g. from the injection of $^{14}CO_2$ samples into the counting gas, than are gas counters operating in the Geiger region. For this reason proportional counters can be usefully linked up with gas chromatography equipment for the analysis of radioactive components of complex organic substances. Counting efficiency for ^{14}C in gaseous form is as high as 95% (4π). With anticoincidence circuitry, background counts can be reduced as low as 2–3 c.p.m.

Proportional counters are similar to G-M tubes, but the gas they contain must have greater quenching effect, and the anode is in the form of a wire loop. Usually argon and methane, 90%:10%, or argon and carbon dioxide is the mixture used. That is, a simple gas which encourages high amplification and a more complex gas to ensure stability of operation. The purpose of having the anode in the form of a wire loop, instead of a straight concentrically-mounted wire as in G-M tubes, is to increase the potential gradient near the anode and correspondingly decrease it towards the cathode. The objective is to control the avalanche of electrons by ensuring that they do not achieve sufficient energy to produce secondary ion-pairs until they are very close to the anode, and hence have little opportunity of colliding with atoms in the relatively short distance that they have to traverse.

Proportional counters may be made as end-window tubes as in the case of G-M tubes, but in the majority of cases the counters are of the gas-flow type in which the sample, either in solid or gaseous form, is placed inside the counter. The newer ultra thin-window proportional or G-M gas flow detectors have already been described in connection with G-M gas flow counting. Scaling assemblies for proportional counting must have a linear amplifier to increase the pulse size for scaling and also a discriminator to resolve the pulses both from radiation source and from instrument noise.

In the proportional region there are separate alpha and beta thresholds, the former being reached at a much lower voltage. There are corresponding alpha and beta plateaux, analogous to the Geiger plateau but very much shorter and not nearly so clearly defined, although the alpha plateau is relatively well defined because of the mono-energetic nature of the alpha particle. Operational voltage is variable, but may be in the range from 500–1800 volts, depending on the counter. Counting of either alpha or beta particles is carried out with the potential adjusted to the centre of the appropriate plateau, or at least where there is the minimum slope.

Proportional counting of samples from biological experiments is straightforward, as these usually contain only beta-emitting isotopes, alpha-emitters having no place in normal biological tracer work. In radiochemical work alpha and beta emitters may have to be assayed in the same sample. Advantage is then taken of the proportional nature of the amplification that leads to an alpha particle giving a larger pulse than a beta particle, Fig. 4.3. Thus the alpha contribution to the activity can be assayed by setting the discriminator to reject small pulses but to allow large pulses to pass through and be registered. The voltage potential is then raised till the counter operates on the beta plateau. At this setting both beta and alpha particles are registered because it is not possible to discriminate against the alpha particles. The observed count is therefore the sum of the beta and alpha pulses, the beta contribution to the count being found by difference.

Neutron Detectors

Neutron detectors are also based on the principle of gas ionization, although the method is indirect. Neutron detectors are made essentially like G-M tubes which have been lined internally with an elemental boron or a boron carbide coating, or else contain BF_3 as a gas. The operating theory of such tubes is based on the following reaction:

$$^{10}_{5}B + ^{1}_{0}n \rightarrow ^{7}_{3}Li + \underset{\alpha\text{-particle}}{^{4}_{2}He} + 2.5\,MeV$$

whereby if a slow neutron hits the nucleus of a boron-10 atom, then an alpha particle is ejected. This is a typical (n,α) reaction (page 23). The alpha particle induces ionization with the formation of primary ion-pairs and is counted in the usual way. As the reaction is especially dependent on boron-10, because of its high neutron cross section, and this isotope is only present to an extent of 19% in natural boron, the boron compound used is enriched with excess boron-10 isotope. Enrichment may be as high as 96% for BF_3 gas filled tubes. Tubes are made relatively large to achieve high efficiency.

Such detectors are normally operated in the proportional region as this gives excellent discrimination against radiations except alpha. Additionally, operation in the proportional region permits very high counting rates. These detectors are primarily for thermal and slow neutrons and to determine fast neutrons with such a counter requires that they be slowed down (moderated) before it is possible to count them. Moderation can be effectively done with a layer of paraffin wax around the counter.

Neutron detectors have no role in tracer studies, but they have an important application in studies of soil moisture content with the neutron moisture meter (page 33). In this device a fast neutron source emits neutrons which are scattered and moderated by the hydrogen atoms of water. The slow neutrons which result are subsequently detected and counted.

Detectors specifically for fast neutrons have hydrogen or methane as the counter gas. A property of fast neutrons is that they collide elastically with hydrogen nuclei (protons). Therefore, recoil protons generated in the counter tube by the incident neutron flux have enough energy to ionize the gas, giving rise to pulses which are counted.

Helium-3 detectors are now available. They are based on the reaction:

$$^{3}_{2}He + ^{1}_{0}n \rightarrow ^{1}_{1}p + ^{3}_{1}H + 764\,keV$$

These tubes have high filling pressures (circa 10 atm) and are much more efficient in counting thermal neutrons than are BF_3 detectors; they are consequently of especial value in neutron moisture meter equipment. They are however more sensitive to gamma-radiation than BF_3 tubes.

Less commonly, slow neutrons may also be detected and counted with a scintillation counter (see next section). With this method an element may be added to the scintillation material with which neutrons will interact to give particles which will produce scintillations. This may be done by incorporating a boron compound in a sodium iodide (thallium activated) phosphor. Alternatively, use may be made of the reaction:

$$^{6}_{3}Li + ^{1}_{0}n \longrightarrow ^{4}_{2}He + ^{3}_{1}H + 4.8\,MeV$$

utilizing a scintillation crystal of europium activated lithium-6 iodide to detect the neutrons. The count rate of scintillations systems for neutrons is much lower than that for gas ionization counters.

SCINTILLATION COUNTERS

Principles of Scintillation Systems

A zinc sulphide screen which emitted luminescent flashes when exposed to incident radiation particles and photons was one of the earliest methods used to detect and measure radiation intensity.

If certain materials are exposed to gamma photons or particulate radiation then they emit *scintillations* or flashes of light. The scintillations are produced by a complex process involving the production of an excited (higher energy) state of the atoms of the material. When the orbital electrons of these atoms become deexcited the excess energy is then given off in an infinitely small time as a flash of light (scintillation).

Substances which emit such flashes of light or scintillations are known as *scintillators, phosphors* or *fluors* and may be either solids or liquids.

The scintillations produced in the transparent scintillator by incident radiation are ''collected'' by the photocathode of a type of photocell known as a *photomultiplier*. A photomultiplier is an evacuated glass tube containing a series of *dynodes*, which are positive electrodes of increasing potential. The scintillations received by the cathode cause it to emit electrons by a photoelectric effect and the purpose of the dynodes is to successively multiply these electrons. This multiplication, or amplification is of the order of a million, and the resulting pulse can be linearly amplified and registered on a scaler or rate-meter.

Whereas a G-M tube may detect as few as 1 per cent of the gamma photons which may enter, a scintillation crystal of sodium iodide can detect well over 50% of them, primarily because of the greater density. It is therefore the method of choice for gamma counting, particularly as the resolving time is only a few microseconds, count rates up to 100,000 c.p.m. being possible. Each gamma photon or particle intercepted by the scintillator produces a scintillation, these light photons giving rise to a proportionate stream of photoelectrons from the cathode. It follows that the number of photoelectrons emitted, and hence also the final output of the photomultiplier tube, are proportional to the original incident γ-photon energy. This fact is of particular importance when gamma spectrometry is considered (page 106).

Scintillation counting is not however confined to gamma-emitters. With a suitable choice of crystal, beta-emitters can also be counted efficiently, while *liquid scintillation* counting in which the sample to be counted is mixed with a liquid scintillator is now the method of choice for low energy beta-emitters such as tritium or ^{14}C.

The basic scaling assembly for a scintillation counter includes a very stable high voltage source, a linear amplifier and a discriminator. Extremely stable high voltage is required because the photomultiplier amplification is particularly dependent on voltage. The discriminator (bias) makes it possible to prevent pulses below a certain height from being counted, thus reducing thermal and electronic noise from the counter.

The background count of scintillation detectors is nevertheless higher than for G-M counters, ranging from 50–500 c.p.m., about 100–150 c.p.m. being fairly normal, because efficiency for cosmic radiation is high.

Scintillation counters may be operated in the region of 600–1300 volts, although the majority of photomultiplier (PM) tubes are customarily operated between 850 and 1000 volts. Unlike G-M and proportional gas counters, scintillation detectors do not have a clearly defined operational voltage. As has already been noted, the size of the pulse induced by a radiation photon or particle which enters a scintillator is proportional to its dissipated energy. Further, if the high voltage is systematically increased then linear amplification results in the proportional amplification of all pulses. Thus the largest pulses due to the most energetic radiation are detected first, but with increasing voltage and hence increased amplification the smaller pulses reflecting the less energetic and scattered radiation and photomultiplier noise are also detected.

If then a plot is made of counts as a function of voltage a curve will be obtained somewhat analogous to that obtained with a G-M counter. However, this *integral spectrum* as it is called, is essentially different, inasmuch as it is characteristic of the specific isotope being counted rather than of the detector. These characteristic spectra form the basis of scintillation spectrometry (page 106). Such an integral spectrum shows a series of steps or irregular plateaux which (Fig. 4.8) reflect the main *photopeak* due to the primary radiation, with secondary steps or regions due to scattered radiation (*Compton region*) and photomultiplier noise.

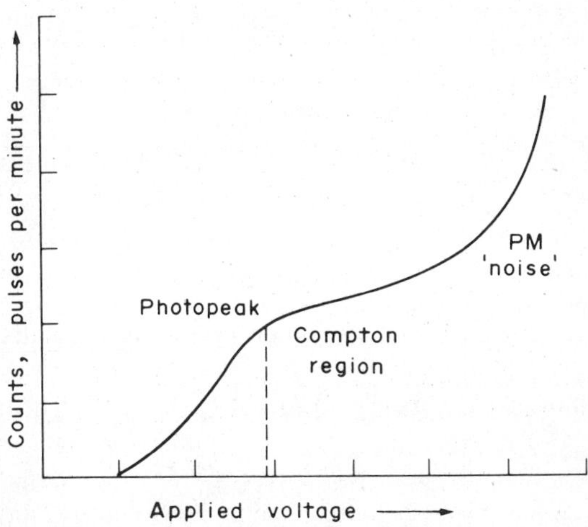

FIG. 4.8 A generalized integral spectrum of a γ-emitter as determined by scintillation counting.

Count rate rises sharply when a voltage is reached where noise pulses predominate, and if this voltage is maintained then the PM tube will soon be destroyed.

The slope and length of a scintillation "plateau" are dependent on the strength of the source, the slope increasing and length decreasing with a low-strength source, so that near background there is no effective "plateau". The optimum operational voltage for a scintillation detector is therefore defined theoretically both by the sample count/background count ratio, and especially in the case of gamma-emitters by the energy of the isotope being counted. In theory then, the voltage on the PM should be adjusted each time to suit both the isotope being counted and the activity of the samples. This was originally done, but changing the voltage on the PM is now considered to be bad practice, both reducing the life of the tube and causing instability. In practical counting it is now customary to maintain the PM at a constant voltage as determined and set by the manufacturer of the particular apparatus.

The problem of achieving as high a sample count-background count ratio as possible, especially for low activity samples, is now met by pulse height analysis and increasing the gain of the system by greater amplification of the pulses from the PM rather than by increasing PM operational voltage. This is considered under Radioassay, Chapter 5.

Solid or Well-crystal Scintillation Counting

A number of solids possess properties which make them suitable for use as scintillators or phosphors. In practical use the type of phosphor employed depends on the particular type of radiation that is to be detected. This is because of different characteristics, particularly in regard to size of light pulse, the decay time of the light pulse and the density and transparency of the phosphor. Organic phosphors such as naphthalene were originally used, but it was found that certain inorganic crystals could also give scintillations provided that they were not pure. Such inorganic crystals have to be treated with a small amount of an *activator*. Plastic, polystyrene or polyvinyl toluene-based scintillators are also available and offer the advantage of being obtainable in large pieces which an be made into any shape.

The solid scintillators commonly used for detecting the different types of radiation are as follows:

For Gamma Photons

A thallium activated (0.1%) sodium iodide crystal is normally used for gamma counting. It has a high light output and the high density gives it considerable stopping power for gamma photons, and hence makes it a very efficient detector. The light pulse has a comparatively long decay time in comparison with organic phosphors. Crystals used are normally from 2.5 × 2.5 cm to 7.5 × 7.5 cm. The larger the size the greater the efficiency, but also more expensive. Sodium iodide crystals are hygroscopic and are therefore encapsulated in an air tight metal mount with a glass window. Cesium iodide (thallium activated) crystals may also be used.

Where the volume of the sample is relatively large, sodium iodide crystals of the "well" type are useful.

For β-particle Counting

Either an anthracene or naphtalene crystal is usually used for beta counting, the former being preferred because of the more intense light flashes. These organic scintillators emit smaller light pulses than thallium activated sodium iodide crystals, but the decay time of the light pulse is shorter thus permitting higher count rates. Plastic, polystyrene-based and trans-stilbene scintillators are also used for beta counting. The transparent material can be used as a plane surface piece for normal counting or can be of well-type, in which a hole is drilled out to receive a sample tube or micro-beaker. For the counting of beta emitters in liquid or gas samples, special thin glass or plastic cells coated with anthracene are available. These cells may be of either continuous flow or non-flow type. Organic and plastic scintillators are much more efficient for beta counting than are sodium iodide crystals.

For α-particle Counting

α-detectors normally consist of a 5 cm diameter zinc sulphide (10 mg/cm^2) screen on a perspex disc or some similar transparent material. A trace of silver is used as an activator.

For Neutrons

Neutrons may be detected using scintillators of stilbene, lithium iodide, or lithium silicate glass activated with cerium.

Well counters have already been referred to. They are used for counting either a liquid or a powder in a container of about 2–20 ml capacity. The advantage of the well counter is that it has improved geometry due to the sample being almost surrounded by the scintillator, whether crystal or plastic. This gives high counting efficiency, but one important consideration in the use of these counters is the size of the sample. The volume of sample and standard must be standardized in order to keep the geometry constant. The optimum volume can be found by comparing count as a function of volume, activity being kept constant.

Liquid Scintillation Counting

It has been found that certain organic substances when dissolved in an aromatic solvent such as toluene, xylene or dioxan, also show scintillation properties. Accordingly it is possible to dissolve or mix an active sample into a liquid phosphor contained in a thin-wall glass vial or jar. The sample can then be counted by placing the container in optical contact with the end of a photomultiplier tube in a light-tight counting chamber. Liquid scintillation counting is primarily used for low energy beta-emitters such as ^{14}C and ^{3}H. Due to the low penetrating power of beta-particles and the self absorption of such samples, solid scintillators are relatively inefficient. The intimate contact of liquid scintillator and sample, in which every radioactive molecule is surrounded by molecules of scintillator, is much more efficient.

Although liquid scintillators achieve high *overall* efficiency through the intimate contact of the molecules of active sample and scintillator they are inherently less efficient scintillators than are crystals. The majority of liquid systems are now based on 2.5-diphenyloxazole (PPO) as the primary phosphor, although p-terphenyl, which suffers from poor solubility, was originally widely used. Butyl-PBD is coming into use as an improved alternative to PPO.

In addition to the primary phosphor, a secondary scintillator has usually been added as a wavelength shifter, the most commonly used secondary phosphors being POPOP (1.4-di-2(5-phenyloxazolyl)-benzene), and dimethyl-POPOP. The purpose of this has been to increase the wavelength of the light emitted from about 370 mμ to about 420 mμ, because the photocathodes of the older photomultipliers were primarily sensitive to wavelengths about 400 mμ. The effect of the secondary scintillator is to absorb the light from the primary phosphor and then re-emit light of longer wavelength. Secondary scintillators are not now so important for this purpose, as the newer PM photocathodes have a much higher sensitivity for light of shorter wavelengths.

The solvent used in the scintillation mixture must be able to transfer its excitation energy (derived from the emission of β-particles) through the solution till it reaches a molecule of the scintillator. Additionally, the solvent must be miscible with the sample, and ideally there should be no chemical reaction between the sample and the solvent. In practice, about only three solvents, toluene, xylene and dioxan, meet these requirements. Of these solvents, toluene has the highest scintillation efficiency, but unfortunately scintillation systems based entirely or mainly on toluene (or xylene) are not suitable for aqueous samples. Addition of methyl alcohol or ethanol to toluene based mixtures makes it possible to mix-in samples containing small amounts of water but the counting efficiency is greatly reduced. This inability to accept aqueous samples is a severe limitation for much biological work.

Dioxan is the solvent of choice for aqueous samples, although its basic scintillation efficiency is only about 70% that of toluene. However, its efficiency can be greatly improved by the addition of naphthalene as a secondary scintillator, it being also comparatively insensitive to quenching by water. Water-containing dioxan based mixtures have a tendency to freeze when counted under refrigeration, but this can be counteracted by including cellosolve (2-ethoxyethanol) as an anti-freeze. This is the general theory underlying mixtures such as the dioxan-based one (Bruno and Christian) ([7]) given below.

A survey of the literature on liquid scintillation counting will reveal a very large number of scintillation mixtures. In general they may have 0.2–1% of PPO, p-terphenyl or butyl-PBD as primary scintillator and about 0.02–0.05% POPOP as secondary scintillator. The scintillators are carried in a solvent system consisting of pure toluene, xylene or dioxan, or mixtures such as toluene + ethanol, toluene + dioxan, toluene + methanol + phenethylamine, xylene + dioxan, in various proportions. Solvent mixtures with dioxan invariably contain naphthalene to an extent of 2.5–12%. It would be pointless to give the formulas for a large number of scintillation mixtures, but the non-specialist will find that the two mixtures given below, a much-used toluene mixture for non-aqueous samples and a highly successful dioxan based mixture for aqueous samples, will meet most counting requirements. Such a dioxan + naphthalene mixture will accept aqueous samples up to about 25–30% water content in ethanol, without too great a loss in counting efficiency. Very many proprietary ready prepared scintillation mixtures are also available, both for general and special purposes.

Special purpose mixtures and systems are considered in Chapter 5.

Toluene-based scintillation mixture for non-aqueous samples:

PPO	4 g
POPOP	100 mg
Toluene to	1 litre

Dioxan-based scintillation mixture (Bruno and Christian,[7]) for aqueous samples:

PPO	10 g
POPOP	500 mg
Naphthalene	50 g
Cellosolve	166 ml
Dioxan to	1 litre

A major potential source of error in liquid scintillation counting is the phenomenon known as "*quenching*", that is processes which lead to a decrease in the amount of light reaching the photocathode of the PM.

The light-absorbing properties of some biological materials cause the so-called *colour-quenching*, due to chromogens in the sample reducing the amount of light reaching the photomultiplier and hence lowering counting efficiency. *Chemical quenching* may also occur when reaction between the sample and the scintillation mixture takes place, so that the energy of the beta particles is absorbed by the solvent as heat rather than resulting in light flashes. A significant quenching effect is normally overcome in practice by the use of an internal standard and a correction curve, by external standards, or by the method known as "channels ratio". This is considered later in connection with Radioassay.

Detector Systems for Scintillation Counting

A scintillation detector consists of a crystal or other phosphor coupled with its mount to a PM tube with a silicone oil light-couple. The oil light-couple is necessary to obtain an intimate connection of the phosphor with the photomultiplier to ensure the efficient passage of very small light photons. The oil must have a refractive index similar to the glass of the tube. As of course it is light sensitive, the whole counter assembly is held in a light-tight metal counter-tube or support-tube, one end of which accommodates the multi-pin photomultiplier tube base.

Such a detector may be mounted upright, i.e., scintillator uppermost, for "end-on" or "well" counting of gamma samples. Strong beta emitters may also be counted in a well-type counter. Alternatively the detector may be inverted for so-called "windowless" counting of beta emitters, in which the sample on a planchet is placed beneath the scintillator. Simple arrangements of this sort are shown in Fig. 4.1.

FIG. 4.9 A relatively inexpensive approach to liquid scintillation counting. Specification includes twin ambient temperature PM tubes with coincidence circuitry. With two scalers it can be operated in twin channel mode (Nuclear Enterprises).

In the case of liquid scintillation counting the vial containing the sample and scintillator is placed in direct optical contact with the end of the PM tube. In order to protect the PM tube from the light during sample changing, a light-tight automatic shutter device is always built-in to these instruments to prevent accidental exposure which would greatly increase the background count.

Individual PM tubes of nominally identical manufacture differ in the degree of ''noise'' that they produce, due to normal manufacturing tolerance. It is therefore usual to select ''low noise'' tubes for the more demanding scintillation counting applications, such as liquid scintillation counting of ^{3}H and ^{14}C.

A constant temperature is also important for stable photomultiplier performance, and in practice for the liquid scintillation counting of low activity beta samples this has meant refrigeration. Thus the detector may be put in an ordinary refrigerator, or cooling coils incorporated in the lead shielding, or refrigeration units built-in to the larger automatic equipment. With the older type of PM tube there is a substantial reduction in thermal or ''dark'' noise when the operating temperature is reduced from 15–80°C to 5°C, but little improvement is gained by working at still lower temperatures, and 0–5°C can be regarded as a practical working temperature range for these tubes.

Nowadays, the more recently developed PM tubes such as the Beckman-RCA tube can be operated with high efficiency at ambient temperatures, and refrigeration is no longer necessary for this type of tube, exceptionally in very hot conditions where refrigeration may be used to keep the ambient temperature at about 15°C. A constant operating temperature still remains important however.

By comparison, reduced operating temperatures have not usually been necessary for gamma counting with a scintillation crystal. This is because the gamma photons are of higher energy and the amount of light produced in a sodium iodide crystal is much greater than that produced by a weak beta emitter in a scintillation liquid. In the latter case the pulses produced in the PM only just exceed the thermal noise pulses of the PM, and thus every effort must be made to keep thermal noise to a minimum for liquid scintillation counting.

Although a basic scaling assembly can certainly be used for the scintillation counting in integral mode of energetic gamma emitters, actual practice favours much more sophisticated arrangements. Thus both for solid and liquid scintillation counting, equipment with pulse height analysis facilities has become the general rule. By means of variable lower and upper discriminators it is possible to accept only pulses which pass through the ''window'' of the discriminator. Thus noise pulses with amplitudes less than or greater than the window are rejected.

Such a scintillation spectrometer system makes it possible to count a γ-emitter on its appropriate photopeak, or to characterize the radiation from a sample and thereby identify a specific isotope in a mixture e.g. in connection with activation analysis (page 186). Equally, for the liquid scintillation counting of ^{3}H and ^{14}C it ensures the maximum possible differentiation between background count and sample count, and with two or three ''channels'' available it is practicable to count the two isotopes simultaneously or to use the method of channels-ratio for quench calibration (page 115).

Often a more sophisticated detector assembly is employed for liquid scintillation counting because of the low intensity of the scintillations from low-energy beta-emitters. Thus coincidence circuitry (page 49) is now normal, with two PM tubes facing each other either side of the sample. In this case only when a pulse is received by both tubes simultaneously is it recorded. As noise pulses are random they tend to be rejected, as the chances of a noise pulse from one photomultiplier arriving at the coincidence unit simultaneously with a noise pulse from the other photomultiplier is relatively small. Coincidence circuitry can almost eliminate thermal noise from the PM.

The technical difficulties, yet the obvious potential of liquid scintillation counting for biological and medical research have stimulated much development of this equipment. There are now complex automatic instruments (Fig. 4.10) with capacity for several hundred samples; with programming of instrument operation; low count sample reject; automatic background subtraction; automatic quench calibration by external standard and channels ratio; counter systems which automatically adjust to optimum counting conditions for each sample; simultaneous assay of two or more isotopes; automatic data processing and print-out of the results in various forms; the list of automatic options is endless.

This type of equipment, and the comparable automatic gamma counting systems are expensive, with the cheapest automatic counter costing about $20,000 while it is easy to spend twice this sum. The technical ingenuity and nucleonic excellence of many of these automatic instruments should not obscure the fact that for small laboratories with only a small number of samples to process at irregular intervals, it is possible to set up an effective manually operated single sample liquid scintillation counting system for quite a small expenditure (Fig. 4.9).

CERENKOV COUNTING

Liquid scintillation counters can also be used for Cerenkov counting. If a charged particle passing through a clear liquid medium exceeds the speed of light then a flash of light is emitted. The threshold for this effect is 0.26 MeV, although the efficiency increases markedly with the energy of the β particle. Water can be used successfully as the liquid, and the method is proving very valuable for ^{32}P counting where scintillation equipment is available.

The technique is considered further in Chapter 5.

SEMI-CONDUCTOR DETECTORS

Certain crystals which are usually poor conductors of electricity become conductors when exposed to ionizing radiation. If such a crystal is placed between two electrodes to which a potential is applied, then an ionizing particle will induce a pulse which can be recorded. This is the basis on which semi-conductor detectors function. They are therefore primary ionization detectors but having the advantage of solid state detectors with a high radiation absorption coefficient.

Semi-conductor detectors are mainly based on germanium or silicon crystals "doped" with impurity atoms. Silicon based detectors are stable at ambient temperatures, but germanium detectors have to be cooled in liquid air to achieve stable operation. They are therefore much less flexible for biological purposes and this discussion will primarily refer to silicon detectors. The germanium-based detectors are used for very high resolution γ-spectrometry.

It is not possible here to go into detail concerning the nature and theory of semi-conductor materials and the construction from these of devices such as diodes, resistors and transistors. However, the non-specialist wishing to understand more of the background of semi-conductors can find very readable accounts in references 11 and 14 and a more technical approach in references 10, 12, and 13.

Principles and Properties of Semi-conductor Detectors

When free outer shell electrons move from atom to atom an electric current is formed. Now, if a crystal of silicon or germanium is taken and doped with "impurity" atoms, deficiency or excess of outer shell electrons can be artificially achieved according to whether the impurity atom has a valency electron less or more than the semi-conductor atom. Where the absence of an electron in the impurity atom attracts an electron from the crystal atom a "hole" is said to be left, creating a positive charge and the crystal being then known as "P-type" (acceptor). When an excess electron from the impurity atom constitutes a negative charge it is termed "N-type" (donor).

The movement of electrons or holes (i.e. positive charges due to the absence of electrons) forms an electric current, the current being conveyed primarily by electrons in "N-type" material and mainly by holes in the "P-type".

Consider a semi-conductor detector as a reversibly biased diode, consisting of a

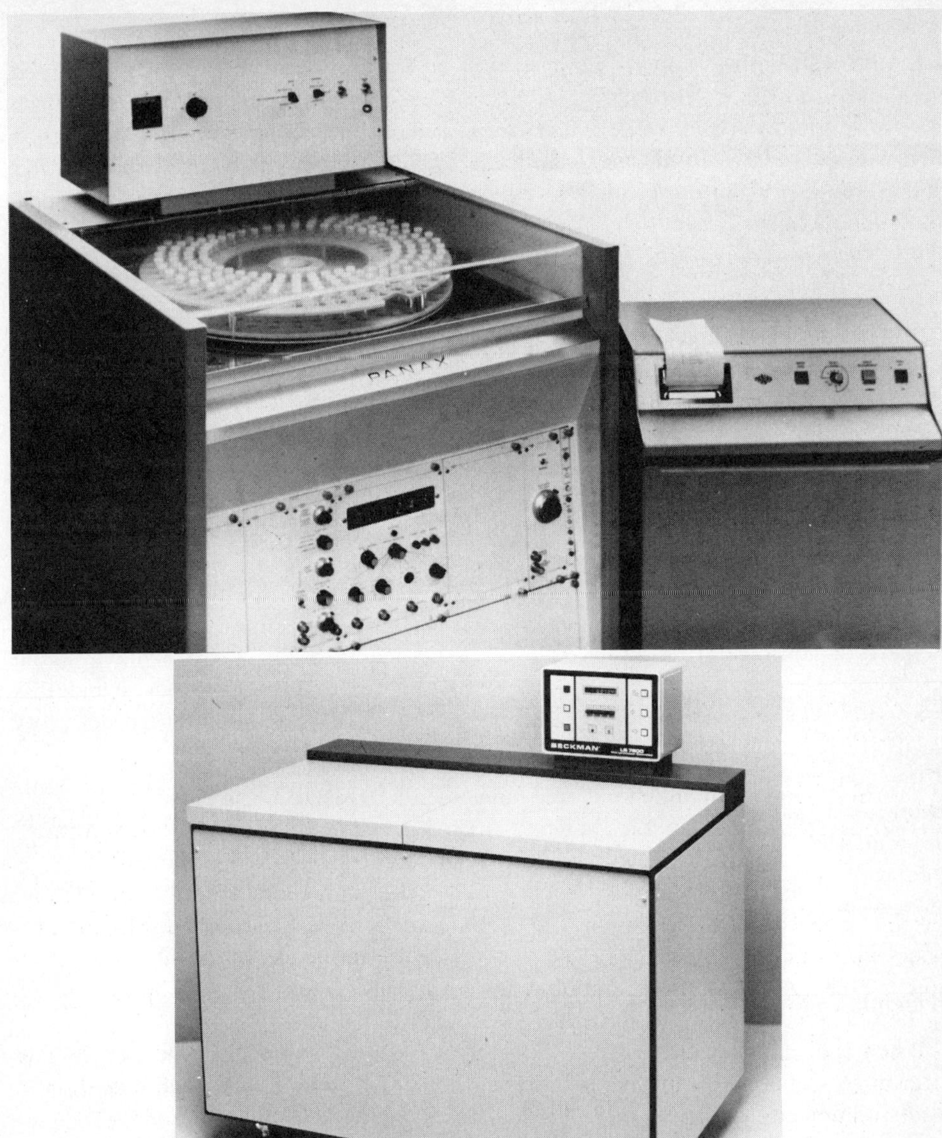

FIG. 4.10 Automatic scintillation counting systems. *Top:* Automatic gamma well-counting system with capacity for 160 samples, 7.5 × 7.5 cm sodium iodide crystal detector enclosed in low background shielding. Instrumentation includes provision for counting three isotopes simultaneously, automatic low count sample rejection and background subtraction. Printout of data. *Bottom:* Liquid scintillation system with twin ambient temperature photomultiplier tubes and coincidence circuitry. Two or three channel counting systems with programmed operation. Pre-set discriminator and gain plug-in-modules for specific isotopes, and provision for quench calibration by external standard and channels-ratio. Varied types of data output; printer shown.

FIG. 4.11 Part of the sample changer compartment of Beckman automatic liquid scintillation counter showing serpentine belt changer. There are standard samples in the belt near the photo electrically operated sample elevator (centre).

junction of "P-type" and "N-type" material areas or regions. Such a diode allows current to flow primarily in one direction and negative electrons will move towards the P-region and positive "holes" move across to the N-region as they are "filled" by electrons from adjoining atoms, the holes then left vacant being filled, and so on.

When an electric potential is applied to such a detector a depleted region is formed in the detector where there will be hardly any free charge-carrier remaining. If a charged particle then enters this region it will cause ionization, with the production of electron/hole pairs in direct proportion to the radiation energy absorbed. The charges are drawn to the corresponding electrodes under the influence of the strong electric field, where after collection, amplification and shaping the resulting pulse can be counted. The pulse height is directly proportional to the energy.

Semi-conductor materials of the donor or N-type are frequently produced by the addition of antimony or arsenic as impurity elements, while indium, gallium, boron or aluminium may be used for the production of P-type acceptor material.

There are a number of types of semi-conductor detectors, of which those of interest to us include *lithium drifted detectors, surface barrier detectors*, and the *diffused junction diode* type.

The lithium drifted type of detector is used for γ-detection and γ-ray spectrometry as it can be made sufficiently large. For these detectors P-type material has a layer of lithium deposited on the surface and diffused in, to form an N-P junction, i.e., between the lithium-rich and the lithium-deficient side.

Surface barrier and diffused junction diode type detectors will detect β-particles, but as they are very thin they have scarcely any sensitivity for γ-radiation. For the surface barrier detectors a section of N- or P- type material is rigorously cleaned and has an extremely thin layer of gold deposited on one side. In this manner a rectifying contact is formed. In the case of diffused junction diodes, an extremely thin section or wafer of P-type material is bombarded at high temperature with donor type atoms which diffuse in, forming a layer of N-type material. A rectifying contact is then formed by the N-P junction.

Like scintillation counters, semi-conductor detectors are sensitive to light, so that in use they must be covered with a thin (less than 1 mg/cm^2) light shield.

Application of Semi-conductor Detectors

The advantage of semi-conductor detectors for certain biological studies depends on their characteristics of high efficiency and high density with a low γ-background sensitivity yet with what is in effect a thin "window." Moreover, they may be made of a size and geometry to suit any special application, and in particular they may be made extremely small in diameters of a few millimetres.

Such characteristics are especially suited to transport studies as the small size of the probe gives good localization, the background count is low and the sensitivity permits the determination of even ^{14}C and ^{45}Ca *in vivo*. Mixed β- and γ-emitting isotopes can be used and easily distinguished by the fact that β-particle detectors are so thin that they have little sensitivity for γ-rays.

In the case of transport studies involving stems or roots the use of two or more detectors positioned at intervals will enable the measurement of transport velocity. For animal studies the small size of the detectors opens up the possibility of implantation. G-M detectors have in the past been used for transport studies, but the semi-conductor detector gives a much more clear-cut indication of when the tracer reaches a given point, and is vastly more sensitive. Other biological applications requiring probe sensitivity and good localization of detection will readily suggest themselves.

The use of germanium-based detectors for γ-spectrometry has already been mentioned.

REFERENCES FOR FURTHER READING

General

1. CHASE, G. D. and Rabinowitz, J. L. *Principles of Radioisotope Methodology*, 3rd Ed. Burgess Publishing Co. (1968).
2. GLASSTONE, S. *Sourcebook of Atomic Energy*, 2nd Ed. Van Nostrand, pp. 641 (1960).
3. PRICE, W. J. *Nuclear Radiation Detection*. 2nd Ed. McGraw Hill, New York (1964).
4. SHARP, J. *Nuclear Radiation Detectors*. 2nd Ed. Wiley, New York (1964).
5. SNELL, A. H. Ed. *Nuclear Instruments and Their Uses*. Vol. I. Wiley, New York (1962).

(The above references give good accounts of principles of operation, but catalogues and technical literature of the various manufacturers of radioisotope counting equipment are useful sources of current technical information and developments.)

Scintillation and Cerenkov Counting

6. BIRKS, J. B. *Theory and Practice of Scintillation Counting*. Pergamon Press (1964).
7. BRUNO, G. A. and Christian, J. E. Determination of ^{14}C in aqueous bicarbonate solutions by liquid scintillation counting techniques. Application to biological fluids. *Anal. Chem.* **33**, 1216 (1961).
8. PENG, C. T. Liquid Scintillation and Cerenkov Counting. *In: Radiochemical Methods in Analysis* (D. I. Coomber, Ed.). Plenum Press, New York, 79–149 (1975).
9. WANG, C. H. and Willis, D. L. *Radiotracer Methodology in Biological Sciences*. Prentice-Hall Inc. (1965).
(See more extensive references in Chapter 5.)

Semi-Conductor Detectors

10. BERTOLINI, G. and Coche, A. Eds. *Semi-conductor Detectors*. North Holland Pub., Amsterdam (1960).
11. CHEDD, G. *The Half-Way Elements*. Aldus, London, p. 88 (1969).
12. DEARNALEY, G. and Northrop, D. C. *Semi-conductor Counters for Nuclear Radiation*. Wiley, New York (1963).
13. STROBEL, H. A. Chemical Instrumentation: A systematic approach to instrumental analysis. *Semi-Conductors*, Chapter IV. Addison-Wesley, Reading, Mass. (1973).
14. WORCESTER, R. *Electronics*. Hamlyn, London, p. 159 (1969).
15. MELLONI, M., Rechenmann, R. V., Ringoet, A., Van de Geijn, S. C. and Wilthagen, R. J. C. In depth localization of beta-emitting isotopes with a semiconductor detector spectrometer assembly. *Nuclear Instruments and Methods*. **74**, 101–8 (1969).

CHAPTER 5

Radioassay

SAMPLE PREPARATION

In general, biological material which has been exposed in some way to a radioisotope requires a certain amount of preparation, sometimes quite a lot, before it is in a form suitable for satisfactory "counting" or radioassay. We are concerned here primarily with biological samples, as soil is rarely counted for radioisotope content except when investigating ion or soil-water movement. In the same way, samples of radioactive fertilizers are easily brought into solution or extracted for counting. Basic preparation often follows classical methods ([1,4,5,7]).

To avoid sampling problems with plants it is desirable to take the whole of the experimental material as a single sample e.g. portions of excised roots in ion-uptake experiments; whole leaves or leaf discs; portions of plant stem; etc. However, the amount of material may be too great to consider it as a single sample, as in the case of material from field experiments or from many pot experiments in the greenhouse. For these samples it may be necessary to grind the plant material, to permit sub-sampling in the normal way.

Grinding Radioactive Plant Material

When it is necessary to grind radioactive samples of plant material a face mask must always be worn to prevent any inhalation of dust. If the amount of material is small it is possible to place a small mill, such as the Wiley* micro-mill or the Glen Creston** mill in a glove box. There should be a slight negative pressure and the outgoing air should be filtered. Both mills have certain advantages. The Wiley micro-mill takes the smallest sample, but is not self-cleaning and does not return all the sample. The Glen Creston mill requires a slightly larger sample but is self-cleaning, almost all the sample is recovered and if the special plastic tubes are used which are a push-fit on the outlet, then very little dust is produced.

In the case of rather active samples then a small vibratory mill should be used, of the type in which the sample is placed in a steel or aluminium screw top container with a steel ball. All the sample is retained during the grinding operation and there is no dust produced except during transfer.

Sometimes relatively large plant samples containing ^{32}P have to be milled e.g. from field experiments. In this case activity is usually rather low and the major problem is to count it, so open grinding is often possible if a dust shield is constructed around as much of the mill as possible and connected by ducting to a powerful suction fan. Either the large Wiley mill or a Christy and Norris*** mill can be used, the latter having the advantage of being self-cleaning. The dust must be retained until the activity has decayed to background.

Sample Preparation for G-M and Solid Scintillation Counting

There are a number of common methods of preparing plant material for counting, and their use depends on particular circumstances: the isotope being counted; the size of sample; the activity of the sample; and frequently the equipment available.

Plant material may sometimes be counted directly if it is in a standard form e.g. excised root samples of constant weight from ion-uptake experiments. This is only possible if the sample size is small and we are working with a fairly energetic isotope such as ^{32}P, ^{22}Na, ^{90}Sr or ^{86}Rb for instance.

Sub-samples of finely ground plant material taken from bulk samples may also be counted directly if the activity is fairly high. The finely ground material may be placed in a deep planchet using a standard technique to ensure constant sample size, the excess material being ''struck'' off the top. Actual weighing is usually unnecessary. A better method, but more time consuming, is to compress the finely ground plant material into ''discs'' or ''briquettes''. For this purpose the briquette is compressed at 10,000 lb/in^2 for 1 minute, about 2 g sample being needed for a disc 2 cm in diameter.

Both these procedures depend on the principle of counting a sample which is in excess of ''infinite thickness''. The advantage lies in the rapidity and simplicity of sample preparation, but problems of self-absorption arise with soft beta-emitters. The preparation of standard samples may also prove troublesome but can usually be overcome by adding aliquots of standard radioisotope solution to known amounts of inactive plant material of the same type, using acetone as a carrier and spreader. The standard radioactive plant samples are then very carefully mixed and dried under an infra-red lamp prepared in the same form and volume as the samples to be determined.

In general the preparation of samples of negligible density is to be preferred. A thin ''film'' of active material is counted in a planchet (usually about 25 mm diameter). This is commonly the method used with end-window G-M counting. The active material is pipetted into the planchet in solution form i.e. it has already been wet digested, ashed or extracted by some other means. The solution is then evaporated, preferably from above by means of an infra-red lamp, otherwise on a temperature-controlled electric hot plate.

Alternatively for elements which are not volatile on ashing, small samples of root, stem and leaf material may be placed in a stainless steel planchet, preferably of the deep well type and ashed at a low temperature in the furnace. The ash is then taken up in a drop of dilute nitric acid and the liquid evaporated as previously.

When preparing a sample like this, every effort should be made to get a uniform ''film'' of active material and the use of a small amount of detergent as a ''spreader'' is sometimes useful. The crystallization of solutions must be avoided as this will give a variable film thickness.

Plant material may be brought into solution either by wet digestion or by ashing in the furnace and taking up in dilute acid, either for subsequent counting in liquid form or for sub-sampling, planchetting and evaporating. The method chosen will depend on the usual chemical considerations for each element, summarized in Table 5.1.

TABLE 5.1

Preparation of plant material

Preparation	Element
Wet digestion	
HNO_3 + $HClO_4$ + H_2SO_4 or H_2SO_4 + H_2O_2 + $LiSO_4$	Al, Ca, Co, Cu, Fe, Li, Mg, Mn, Mo, P, K, Rb, Si, Na, Sr
HNO_3 + $HClO_4$	S
acid permanganate (or Schöniger combustion)	I
Ashing in the furnace	
normal: 500–650°C	Al, Ca, Cu, Fe, Mg, Mn, Mo, K, Si, Na, Sr
low temperature: 450°C	P, Zn
ash with alkali: $Ca(OH)_2$ or CaO	B, Cl
Schöniger combustion	H_2O, CO_2, SO_2

Where it is possible to make a choice between dry ashing and wet digestion of biological material, the former may sometimes be chosen as there is less risk of spills from broken beakers etc. It is also especially convenient for large samples. It is important to avoid losses due to initial overheating of the furnace giving rise to volatilization e.g. of phosphorus and zinc. Heat to 220–260°C initially for 2 hours, then raise the temperature to 550–650°C for completion of the ashing.

Wet digestion may be carried out using the normal nitric-perchloric-sulphuric (12: 3: 5) acid mixture, with the addition of extra nitric acid if necessary to prevent drying out before the digestion is completed. Heating of such digests is best carried out on a sandbath, or on an electric hot plate, all operations being carried out in a fume hood. The final volume of solution should normally be less than 10 ml. Where nitrogen is also to be determined use the sulphuric acid/H_2O_2/$LiSO_4$ digest mixture of Parkinson and Alan (page 155). This method is also well adapted for use with block digestors such as Technicon.

Liquid samples may then be counted by a number of means. Aliquots may be taken and evaporated on planchets as previously described for end-window G-M counting. A better method, particularly where the volume of solution is quite large, 4–10 ml, is to use a liquid G-M counter such as the Mullard MX-124/H or the 20th Century Electronics M6H, in which the radioactive solution is placed inside the tube. The tube must be properly washed and decontaminated between samples, and it is best to work with two or three tubes, cleaning one while the other is being counted. A major advantage of liquid counting is the improved and uniform geometry.

Liquid counting is particularly useful for handling large sub-samples (4–10 grams) of plant material from ^{32}P fertilizer field experiments and from pot experiments. In this case the large plant samples are ashed in the furnace overnight, the ash being taken up in dilute nitric or hydrochloric acid, placed in the liquid counter and made up to constant volume.

Total phosphorus may then be determined colorimetrically on an aliquot, using ammonium molybdate or ammonium vanadate. The latter is usually preferred because of its stability, though the former is more sensitive.

G-M tubes of the dipping type which are placed in radioactive solutions for counting are also available, but are seldom used.

Alternatively, liquid samples may be placed in glass or polystyrene sample vials or tubes and counted by scintillation counter. The vials may either be placed end-on to the counter, or hard B-emitters counted by Cerenkov technique.

Samples such as those containing ^{45}Ca may require more elaborate preparation because of the low energy of the β. In this case a sample of appropriate size, containing 0.2–5.0 mg Ca, may be ashed or wet-digested. The calcium is then precipitated from the dilute acid extract as the oxalate, using the normal procedures for semi-micro oxalate precipitation as described below.

The precipitate is filtered off using some form of sintered glass filter stick or crucible, so that the precipitate is collected on a 25 mm filter paper disc. There are demountable glass and stainless steel filter assemblies available for the purpose e.g. Tracerlab. These discs are then counted, the amount of precipitate being such that the counting is usually carried out on "infinitely thick" samples. It will be necessary to prepare standards in exactly the same manner.

If total calcium is to be determined then the precipitate is subsequently dissolved and washed off the filter paper with hot, 60–80°C 2 N sulphuric acid. Calcium is then determined either by permanganate titration, by complexometric titration using EDTA with calcein as indicator using standard procedures, or by atomic absorption.

Semi-Micro oxalate precipitation for ^{45}Ca

1. Take 10 or 20 ml of digest (containing 0.2–5.0 mg Ca) and place in a 250 ml beaker.
2. Add 5 ml 1:4 HCl.
3. Heat solution to boiling point.
4. Add 10 ml of hot 4% ammonium oxalate.
5. Add a small drop of 0.5% methyl red solutions.
6. Almost neutralize with concentrated ammonium hydroxide (until solution is faint pink).

7. Digest at 90–100°C for 20 minutes.
8. If no precipitate apparent after 10 minutes digestion add 10 ml of 95% ethanol and continue digestion another 20 minutes. If the solution shows signs of drying up add the minimal amount of distilled water from a wash bottle.
9. Allow to stand until cold—convenient to leave overnight.
10. Filter with great care, a little at a time, carefully washing down the walls of the beaker and the filter.

Procedures for ^{14}C samples now centre around liquid scintillation counting, the preparation of samples for this technique being described principally in the next section. In the past and where liquid scintillation counting is not available, $^{14}CO_2$ was released from the sample by the Van Slyke-Folch (9) or persulphate wet oxidation procedure, the $^{14}CO_2$ being absorbed in sodium hydroxide solution and subsequently precipitated as $Ca^{14}CO_3$. Alternatively, dry combustion is possible and can also be used for tritium determination from large samples (e.g. [2]).

In these procedures some form of standard glass joint, e.g. Quickfit, apparatus is used. A system is assembled consisting of a reaction flask containing the sample, plus inlet funnel for the oxidation mixture, connected to a $^{14}CO_2$ absorbing trap or flask. Provision is made either to flush the system with N_2 or CO_2-free air or else a connection is provided to permit evacuating and sealing off the system before commencing the reaction. Typical macro- and micro-apparatus are shown in Fig. 5.1. Commercial Van Slyke apparatus measuring the gas volume manometrically have also been available (e.g. Thomas, Philadelphia).

> *Van Slyke mixture (12.5 g CrO_3, 83.5 g H_3PO_4 S.G. 1.7 and 166.6 ml of fuming H_2SO_4)* is introduced to the sample in excess and heated gently to complete the reaction. Heating should continue at 100°C. If a vacuum system is used then 30–45 minutes should be allowed for the absorbent to come into equilibrium. When N_2 gas is passed through the system, continue the gas sweeping for 10–15 minutes after final addition of acid and apparent completion of the reaction. The CO_2 is collected in 0.1 N NaOH. The NaOH solution is removed and the $Ca^{14}CO_3$ is precipitated by the addition of saturated $CaCl_2$ solution or alternatively $Ba^{14}CO_3$ is precipitated by the addition of 1 M $BaCl_2$—NH_4Cl solution. Filter the precipitate as described previously.

In micro-scale determinations the CO_2 may be directly absorbed in saturated $Ba(OH)_2$. It is important that the amount of $CaCO_3$ or $BaCO_3$ precipitate should be about the same for each sample to avoid varying self absorption effects. About 0.5 g $BaCO_3$ will probably be found suitable, and the sample size to give this may be found by trial.

The Van Slyke-Folch procedure is well established but the persulphate method has a distinct advantage for samples with high water content, as the former method cannot be used effectively for these samples, though it is useful for oxidizing soil organic matter. Water is used as a solvent in the persulphate method.

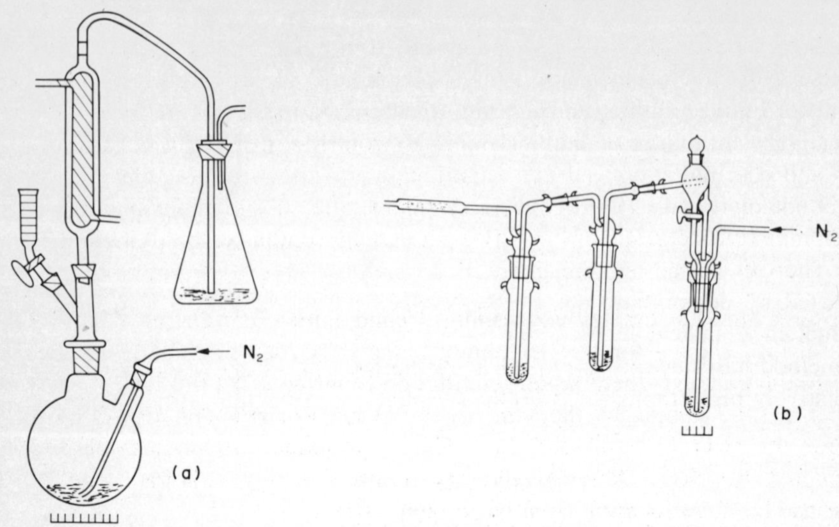

FIG. 5.1 Van Slyke-Folch wet oxidation apparatus, (a) macro, (b) micro-scale.
Features: inlet funnel, reaction flask, CO_2 absorbing trap(s), N_2 flushing of system.

Persulphate oxidation mixture (about 0.2–0.4 g $K_2S_2O_6$, 1–2 ml 5% $AgNO_3$, and 20–30 ml H_2O per sample) is added to the sample in the reaction flask. Unlike the Van Slyke procedure no reaction will take place at room temperature, and gentle heat is applied to start the reaction. Sweep N_2 gas, as in the previous method. Slowly bring the temperature up to 70°C and maintain for about 20–25 minutes. Then raise the temperature to 100°C and maintain until the solution becomes clear. Then simmer for a further 10–15 minutes, continuing to sweep with N_2 gas to remove all $^{14}CO_2$. Collect the $^{14}CO_2$ and precipitate as $BaCO_3$ as described previously.

Dry combustion. The sample is burned with CuO, or the gases passed over CuO, as in the classical combustion train for the determination of formula composition. The CO_2 evolved is collected quantitatively in 0.1 N NaOH and precipitated as $BaCO_3$ as described above. For tritium determinations the water is collected in a glass trap immersed in liquid nitrogen or a dry ice-carbon tetrachloride-chloroform freezing mixture. The typical method of Peets *et al.*([16]) oxidizes the sample at a temperature of 750°C with an oxygen flow of 30 ml/min, completing the combustion by passing the gases over a catalyst section at 950°C.

Macro-combustion procedures, such as by Peets *et al*, can handle samples up to 2 g, which is their chief advantage, but micro procedures using the typical Pregl ([8]) apparatus are quite practical. For these, about 0.75 g CuO is required for every 10–25 mg of sample. As both Van Slyke-Folch and Pregl micro procedures can provide a quantitative determination of CO_2 and hence C, the specific activity of the ^{14}C can be readily determined.

In all the above procedures for $^{14}CO_2$, instead of absorbing the CO_2 in NaOH and precipitation as $BaCO_3$, the gas may be absorbed in ethanolamine, phenethylamine, or hyamine hydroxide in methanol. This is particularly appropriate for micro procedures and for liquid scintillation counting, as discussed in the next section.

Alternatively an aliquot of the 0.1 N NaOH containing $^{14}CO_2$, say 1 or 2 ml, is pipetted into a scintillation vial and scintillation mixture added so that the NaOH sample is not more than 10% of the total volume, i.e. for 1 ml sample add 10 ml scintillation mixture (e.g. 4 g of PPO, 100 mg POPOP, 650 ml toluene, 350 ml Triton X-100). The CO_2 in the remaining NaOH or $Ba(OH)_2$ absorbtion solution can then be estimated by determining the excess NaOH by titrating with 0.1 N HCl using phenolphtalein as indicator.

This method has much to commend it for the non-specialist, when the amount of ^{14}C is relatively high, though it is not as sensitive as the Schöniger technique:

Typical Calculation

If volume of absorption NaOH was 15 ml and 1 ml was withdrawn for counting, 14 ml will remain. Determine the blank titre, *B* ml, on 14 ml of 0.1 N NaOH. But the sample titre is *S* ml, so the mequ CO_2 in a 14 ml sample is $0.1 \times (B - S)$, or in the original 15 ml, $0.1 \times (B - S) \times \frac{15}{14}$. But the GEW of *C* is 6, so the total mg C absorbed is $6 \times 0.1(B - S) \times \frac{15}{14}$ mg, and the 1 ml which was counted contains 1/15 this quantity.

Sample Preparation for Liquid Scintillation Counting

Preparation of samples for counting by liquid scintillation is extremely varied, in keeping with the wide diversity of materials and radioisotopes that must be handled. The preparation is closely linked to the requirements and limitations of the scintillation mixture, and the weaker the β energy, i.e. with ^{14}C and ^{3}H, the more closely these limitations must be respected ([11,17,55]).

Scintillation mixtures have been discussed in Chapter 4, which should be referred to, but essentially we are limited by: relatively few solvents; the inability of scintillation mixtures to take more than 25–30% H_2O without severely reduced counting efficiency; the need to keep colour—and chemical—quenching to a minimum ([14,17]).

Basically, preparation methods fall into three groups: samples with relatively energetic β-emitters; solids, suspensions and emulsions; and total combustion procedures for ^{14}C and ^{3}H.

Samples with Relatively Energetic β-emitters

Samples may be prepared by dry ashing or wet digestion as appropriate and described in the previous section. The ash is extracted with the minimum of 1.0 N acetic acid, and the extract placed in a counting vial with 2 or 3 times its volume of a dioxan based scintillation fluid. Alternatively an emulsion may be prepared using the surfactant Triton X-100 which ([19]) emulsifies polar sample solutions. Use the toluene based scintillation mixture formula given in the last section.

The use of ethylene glycol monomethyl ether (EGME or Cellosolve) should be noted here in connection with general sample preparation. It has the unique capacity of enabling the incorporation of larger amounts of salt in dioxan based scintillation mixtures, and more importantly of water in toluene mixtures. It can replace dioxan or toluene on about a 1:1 basis.

Hyamine hydroxide may also be used for extraction of ash etc. The powerful caustic nature of this quaternary ammonium compound makes it in addition a useful solubilizer of biological tissues, blood, and gels of all sorts, dissolving as much as 20 mg/ml. It is used as a molar solution in methanol, digesting at 55–70°C. This offers a different, if rather expensive, approach to the preparation of small organic samples, though a disadvantage of hyamine hydroxide is its strong chemical quenching. Use with a toluene scintillation mixture.

Radioisotopes which may be handled in these ways include ^{45}Ca, ^{36}Cl, ^{35}S, ^{28}Mg, ^{63}Ni and ^{33}P.

Suspensions and Emulsions

To save preparation time and work, it may be possible to prepare suspensions of finely divided solids, small pieces of filter paper or emulsions of non-miscible liquids in the scintillation mixture. A further advantage is the absence of chemical quenching, assuming the sample does not contain soluble quenching products. Although usually not practicable for ^{3}H because of self absorbtion losses, the method is suitable for most other β-emitters including ^{14}C. It is particularly useful for counting "spots" scraped from thin layer chromatograms, and for "spots" or areas cut from paper chromatograms. In general, the position of the paper in the vial is not critical ([10,11]).

An extension of this technique is to "spot", absorb and dry on pieces of filter paper, solutions of compounds which are normally insoluble in the scintillation mixture. Similarly suspensions of bacteria, algae, spores, etc. may be collected on filter paper, dried and placed in a scintillation mixture for counting. These samples may often be recovered for other purposes. Inevitably, self absorbtion reduces counting efficiency. A problem with the "unconventional" methods is the preparation of standards, as it is important that the standards are both prepared and presented to the counter in identical manner to the samples.

Very finely ground or powdered samples may be successfully prepared for counting in emulsion form with the aid of very finely divided SiO_2, available under trade names "Cabosil" or "Aerosil", or else by use of "Thixcin", a castor-oil derivative. About 2–5% solution of the latter in the scintillation mixture gives a thixotropic gel, pourable after shaking, and this gels after adding the solid sample, at about 0.5–1.0 g per 20 ml scintillation mixture.

Silica suspensions may often take greater amounts of sample, as much as 2 g or so being shaken into a gel with 3–5% SiO_2 in 20 ml of scintillator. Preparation of gels may be done directly in the scintillation vial, are stable long enough for counting and are transparent to scintillations, even for ^{3}H. Insta-gel is a commercially prepared mixture (Packard).

The diverse materials that can be handled in emulsions include nucleic acids, lipids, protein solutions, carbohydrate solutions, amino acids, fatty acids, enzymes, and various biological fluids, etc.

Reference has already been made to the use of Triton X-100 (Rohm and Haas), a mixture of polyethoxy alkylphenols, in the preparation of emulsions of polar samples, and quite high proportions of aqueous solutions can be emulsified. The nature and composition of the sample influences the amount of sample that can be handled, and also the proportion of Triton required. This is a matter for trial under the specific experimental conditions, but about 40% H_2O is the maximum that can be effectively emulsified using a toluene based scintillation mixture plus Triton in the proportion of 2:1 or 2½:1. When Triton is added to a scintillation mixture the amount of PPO and POPOP scintillators must be increased to maintain the same proportions ([19]).

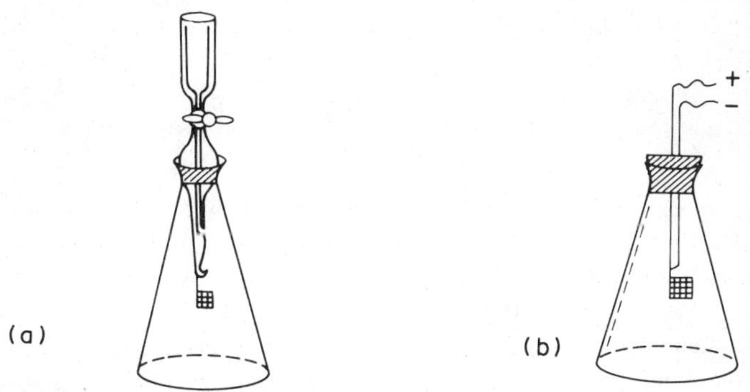

FIG. 5.2 Schöniger combustion flasks with platinum sample baskets
(a) Thomas-Ogg pattern with funnel head to permit introduction of inflammable scintillation mixture after combustion.
(b) simple pattern with rubber stopper carrying a wire resistance igniter.

Clear liquids are formed with low levels of water but at higher concentrations of water the emulsion tends to be opaque with a cloudy appearance which may or may not disappear on brief standing. The method is very suitable for ^{3}H counting, but if possible high concentrations of water should be avoided as the counting efficiency is reduced and also may not be reproducible.

Total Combustion Procedures for ^{14}C and ^{3}H

For low energy β-emitters such as ^{14}C, ^{3}H, ^{35}S and ^{131}I, particularly when sample activities are low, then complete oxidation of the sample with absorbtion and counting of the oxidation products is the method of choice. This procedure also eliminates the major causes of quenching.

Wet and dry combustion procedures suitable for ^{3}H and ^{14}C have already been described, and these methods may handle samples of 2–3 g dry weight. However, some variation of the Schöniger technique is now more frequently used. This consists of combusting the sample in an atmosphere of pure oxygen and absorbing $^{14}CO_2$ or

$^{35}SO_2$ in phenethylamine, hyamine hydroxide or ethanolamine. In the case of ^3H_2O the flask is cooled in liquid nitrogen or a freezing mixture, and the water condenses on the walls, subsequently being taken up in the scintillation mixture which is introduced ([18]).

Schöniger flasks (Fig. 5.2) consist merely of heavy walled conical Erlenmeyer pattern, or round flasks of 500 ml–2000 ml capacity, fitted with a stopper to which is attached a platinum wire spiral or basket as the sample carrier.

Flasks may be fitted with side arms or funnel heads fitted with taps to permit the introduction (after combustion) of inflammable absorbing or scintillation mixtures without breaking the seal.

To carry out a combustion the sample, often wrapped in dialysis tubing (visking membrane) or low ash paper, is placed in the basket, the flask flushed completely with pure oxygen and the stopper firmly inserted. The combustion is usually started by means of an infra red lamp or a projector lamp. Alternatively, if combustion is difficult a wire resistance heating device may be used as in Fig. 5.2(b). To avoid possible accidents a wire mesh screen should be placed around the flask before combustion.

After combustion the flask is allowed to cool and the absorbing-scintillation solution run in from the funnel head, the stop-cock closed, and the flask swirled around and allowed to stand for 15 minutes to complete absorbtion. The solution is then transferred to a scintillation vial for counting.

A suitable phenethylamine based $^{14}CO_2$ absorbing-counting mixture consists of phenethylamine : methanol : toluene in the proportions of 1:1:2 plus 4–5 g PPO and 250 mg POPOP per litre. A similar ethanolamine based mixture can be prepared from ethanolamine, ethylene glycol monomethyl ether (EGME) and toluene, in the proportions of 1:10:10 plus the same amounts of scintillators.

Sample size should not be more than 300 mg for normal flasks. For very small samples not exceeding 3 or 4 mg direct in-vial oxygen combustion is also practicable, absorbing the $^{14}CO_2$ in 1 ml of equal parts of phenethylamine in cellosolve (EGME), placed in the vial before combustion. Scintillation mixture is added after combustion ([12]).

Oxygen flask combustion provides a very effective preparation method for ^3H and ^{14}C, and is the method of choice for low activity work, but in its basic form is rather time consuming for large numbers of samples. This can be overcome with the Kaartinen automatic apparatus available commercially (e.g. Packard Instrument Co.), in which the ^3H_2O is condensed directly into the scintillator vial. Apart from making possible a greater throughput of samples the automatic apparatus has some positive technical advantages, including almost unlimited sample size for tritium, and about 500 mg samples for ^{14}C containing about 20 millimoles of C; the elimination of possible oxygen quenching; and the separate collection of ^3H and ^{14}C. This latter feature permits separate counting of ^3H and ^{14}C under optimal conditions. This is however suitable mainly for laboratories with large numbers of samples to process. Inexpensive ^{14}C and ^3H methyl methacrylate calibration tablets are available to check the day to day operation and recovery efficiency of such apparatus ([13]).

The refinement of ^{14}C determination techniques often makes the estimation of total

C for specific activity determination more troublesome than the radioisotope, especially when the sample size is small. For very small samples it may be essential to resort to such micro-techniques as involve Pregl, Dumas or Conway-diffusion ([3]) methods. However, when the number of samples justifies it there are available automatic elemental (C,H,N,O) analyzers (e.g. Perkin-Elmer) which can handle samples of about 1–3 mg.

THEORY AND PRACTICE OF COUNTING

Choice of Counting Method

In these days of well-equipped laboratories it may seem anachronistic to say that choice of counting method depends on the equipment available. Nevertheless, it is as well to recall that a great deal of useful and important work has been carried out using a G-M-tube in a simple lead castle. This assembly is very adaptable.

The choice of detector depends primarily on the nature and energy of the radiation. Table 6.1 gives the energy data for a number of tracer isotopes frequently used in biology and soil studies, and should be read in conjunction with this section. Detection and counting of gamma-emitting isotopes presents few problems; the main difficulties arise with weak beta-emitters such as ^{45}Ca, ^{14}C, ^{35}S and ^{3}H.

The most suitable radioisotopes for G-M counting are hard beta-emitters such as ^{32}P, ^{89}Sr and ^{90}Sr. Alternatively, Cerenkov counting may be conveniently used, but requires the presence of liquid scintillation counting equipment.

^{45}Ca and ^{35}S can present problems, due to the weaker β rays and self absorbtion of the sample. Usually separation and precipitation as calcium oxalate and barium sulphate respectively has been the answer, together with thin window-G-M or windowless scintillation counting using organic or plastic scintillators. Liquid scintillation counting is now frequently used for these and ^{35}Cl with appropriate sample preparation.

In general, gamma emitters are best counted by solid scintillation counting, as the efficiency of G-M detectors for gamma rays is not very great, as they tend to pass through without causing ionization. Nevertheless, the special gamma G-M tubes can if necessary be successfully used, e.g. for ^{54}Mn and ^{55}Fe.

A number of radioisotopes are both beta and gamma emitters, and can conveniently be counted with a G-M detector; they include ^{131}I, ^{28}Mg, ^{99}Mo, ^{42}K, ^{86}Rb, ^{22}Na, ^{24}Na and ^{64}Zn. Scintillation counting will be found more sensitive for ^{131}I, while ^{42}K, ^{24}Na and ^{86}Rb can also be determined by Cerenkov counting.

When gamma spectrometry is carried out either in connection with activation analysis or in qualitative or quantitative determinations of mixtures of isotopes, then high efficiency scintillation counting will always be adopted, because it provides an output pulse proportional to the energy input. More recently the greater availability of semiconductor detectors has made them the detectors of choice, due to their high resolution.

Few people nowadays would contemplate any extensive work with ^{14}C or ^{3}H without having liquid scintillation counting facilities, even if only of a basic nature such as shown in Fig. 4.9. In the past ([7]) ^{14}C has often been transformed into $^{14}CO_2$ and counted directly as a gas, or else precipitated as a carbonate and determined either by thin window G-M tube, or "windowless" scintillation counting, or by means of a gas flow detector. Gas flow detectors are sensitive (over 20 times more sensitive than end window counting) and of high efficiency when working properly, but they can be troublesome. Their use is now mostly confined to radiochromatogram scanners where their sensitivity can detect ^{14}C-labelled spots on chromatograms.

Apart from the inherent sensitivity of liquid scintillation counting for ^{14}C and ^{3}H, a major factor in the almost universal move to this technique was the troublesome nature of the preparation of the gas samples which require individual preparation on a rather complex line by skilled personnel, with little prospect of developing full automation. Nevertheless, internal gas counters operating in the proportional region are inherently very efficient (95%). There is therefore a limited and specialized use of such counters in radiocarbon dating studies and for tritium environmental studies, where the counts are extremely low, the samples relatively few, and the counting times long.

It will be apparent that there is no one "correct" way of counting any particular isotope. Usually the investigator has more than one possibility at his disposal, within the limitations imposed by radiation energy, the nature and size of the sample, and the amount of activity anticipated.

When counting a sample by any means a number of factors relating to geometry, background, counting statistics, etc. have to be taken into account. These are considered in the next sections.

Geometry

The "geometry", or geometry factor, G, of a counter is used loosely to describe its counting efficiency or the general arrangement of a sample or source in relation to the detector. The geometry factor is really a measure of the proportion of rays reaching the detector from the source and primarily depends on the distance of the source from the detector and the radius of the detector aperture ([25,27,36]).

For ionization detectors the geometry may be defined more formally as the solid angle at the source subtended by the aperture of the detector, divided by the solid angle, 4π. If we assume a sphere of radiation surrounding a source this may then be expressed as the area of the sphere enclosed by the aperture of the detector in relation to the total area of the sphere, as in Fig. 5.3.

As radiation is given off in all directions it is clear that with a counting system comprising only one detector, the maximum geometry factor that can be achieved is 50%, for a small source that is close to the detector. Such counting arrangements are known as 2π. It is possible to obtain 4π counting, either by having two detectors opposite to each other (e.g. as in some radiochromatogram scanners) or for labelled gas samples which are actually inside an ionization chamber. For the latter, 95% efficiency may be achieved.

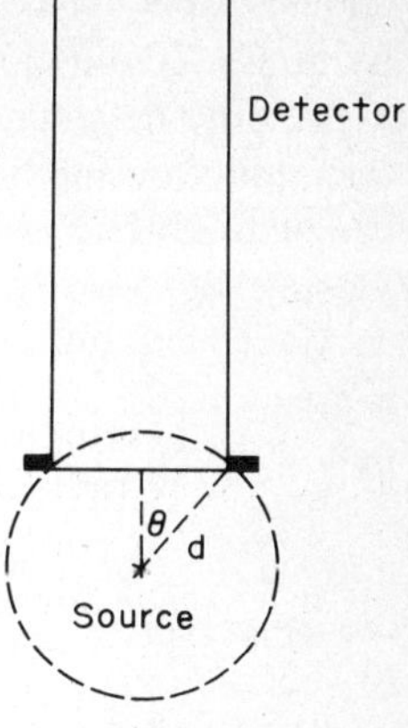

FIG. 5.3

If S_w = area of the sphere enclosed by the aperture
$= 2\pi d^2(1 - \cos\theta)$
S = total area of the sphere
$= 4\pi d^2$
d = distance from source to the edge of the aperture
= radius of the sphere

then,

$$G = \frac{S_w}{4\pi d^2}$$

$$= \frac{2\pi d^2(1 - \cos\theta)}{4\pi d^2}$$

In practical terms such a calculation is seldom made, but it is the basis of our understanding of the topic.

Efficiency is determined by means of a calibrated standard, which is a source of known activity in terms of disintegrations per second, d.p.s.

$$\text{The efficiency, } E = \frac{\text{count rate of sample, c.p.s.}}{\text{disintegration rate of sample, d.p.s.}} \qquad (2)$$

The geometry of an ionization counter is affected mainly by the distance of the source from the detector but to some extent by ''forescattering'' of β-particles by air between the source and the detector, by ''side scattering'' from the support of the detector and sample holder and by ''backscattering'' from the planchet and sample holder. In addition there are ''self-absorbtion'' effects within the sample itself.

We speak of ''good geometry'' when the arrangement of sample (source) in relation to the detector tube results in the greatest part of the radiation encountering the detector with the minimum of scattering effects. Poor geometry is when a substantial part of the radiation reaching the detector is due to scattered radiation. Scattered radiation results when the sample is too far from the detector, or sometimes if an absorber is incorrectly inserted to reduce radiation reaching the detector. This is the case when the absorber is too close to the detector or the shielding of the detector is inadequate for good collimation of the beam, and internally scattered radiation from the absorber will contribute to the count. The shielding of the detector should always have the objective of achieving narrow beam collimation of the radiation from the sample.

Nowadays the widespread use of automatic sample changers, with the geometry determined by the manufacturers, has tended to make considerations of geometry theoretical for most users. Previously, holders for end window counters often had shelves with a definite geometrical relationship, so that any shelf had half the geometry of that above it. It was common practice to count the more active samples on shelves further from the detector, applying a ''shelf factor'' to equalize the resulting counts from different shelves. This is a thoroughly bad practice: all samples should be counted under identical conditions, and if wide variations in activity make this impossible then separate standards of appropriate similar activity should be prepared, rather than subject the counts to an unsatisfactory correction factor.

Background Correction

A certain proportion of every count will be due to extraneous radiation, such as from cosmic radiation, and the general background radiation of the laboratory, and electronic "noise" of the equipment. This is called the *background*, and a correction must be made for it. Good handling techniques and clean working should keep the general laboratory background radiation quite low, but inevitably much work, especially where a number of workers are using different isotopes, causes the general background to rise. It is instructive to determine the background count immediately before and after a training course for 15–20 people!

As background counts are relatively low, the "mean background count" must be determined over a fairly long period for the counts to be statistically sound. All the normal experimental conditions should be observed except for the absence of a radioactive source e.g. an empty planchet or sample tube should be placed in the detector. Background counts are typically 15–25 c.p.m. for an average type of lead castle with a GM-tube, and about 1–2 c.p.m. for low level counting systems with thick lead shielding and anticoincidence circuitry. Scintillation counters give background counts of the order of 100–200 c.p.m. or even higher.

An unusually high background is likely to indicate a loose cable connection, or a nearby source of contamination.

The background count should always be immediately subtracted from every observed sample count to obtain the true activity before carrying out any further necessary calculations.

Resolving Time and Instrument Count Losses

Resolving time, or dead time, is the minimum time necessary for a counting tube or system to recover between events for each event to be recorded. Thus if two particles arrive at a counter tube in a time which is less than the resolving time then only one particle can be recorded, and for a series of such events the total count will be less than the true count. The greater the count then the greater the loss (Fig. 5.4). This is known as *coincidence loss* and especially for higher count rates it may be necessary to make a correction.

The coincidence loss for a counting system can be determined by preparing a series of dilution standards and counting each in turn, plotting a graph, c.p.m. versus activity. The theoretical curve can also be plotted by taking the count of the most dilute standard and calculating by simple proportion what the counts of the more active standards should be, assuming no counting loss, as in Fig. 5.4. A factor for correcting sample

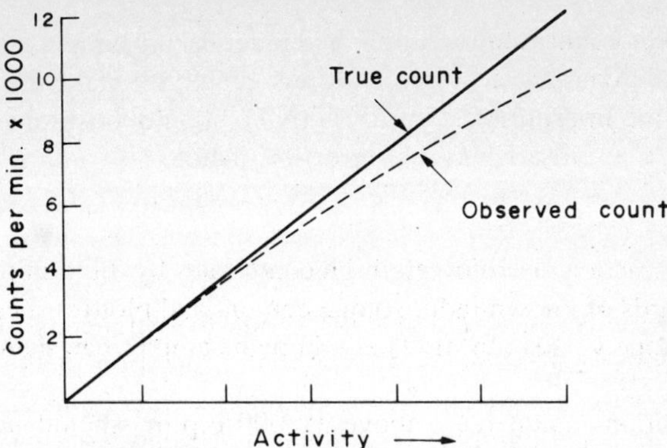

FIG. 5.4 Coincidence loss of G-M counters affects the true count rate.

counts can then be determined for any observed counting rate by dividing the true (calculated) counting rate by the observed counting rate, e.g. if the observed count is 5000 c.p.m. and the true count 5075 c.p.m. then a sample count of between 4500–5500 c.p.m. would have to be multiplied by a factor of 1.015 to obtain the true count.

In practical terms correction for coincidence loss is usually only required with G-M tubes which can have a relatively long dead time of 300 μsec, though some modern high voltage tubes have dead times as little as 15 μsec. Seven with the longer dead time tubes correction is usually not necessary for count rates below 2000 c.p.m. For a dead time of 300 μsec the coincidence loss is about 0.5% per 1000 counts. Instrument count losses are negligible with modern instruments.

An approximate correction for dead time can be found mathematically, when the dead time is known ([34]):

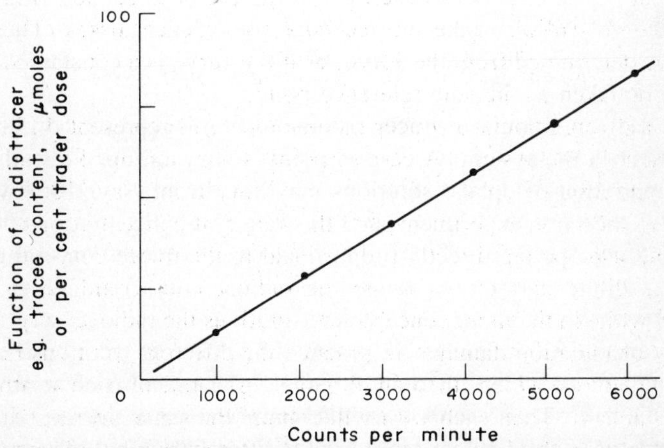

FIG. 5.5 A typical calibration curve. A function of the radiotracer versus observed counts.

If R_1 = observed count rate
R_2 = corrected count rate
and T = dead time in minutes
then

$$R_2 = \frac{R_1}{1 - R_1 T}$$

It is much better practice to avoid very high count rates by dilution or by preparation of a series of standards of known radioisotope content and plotting a calibration curve, amount of radioisotope versus c.p.m. The unknown samples can then be interpolated from the curve.

Even with correction, count rates above 10,000 c.p.m. should not be used with G-M tubes. Count rate has comparatively little effect in proportional counting. Dead times are only 1 or 2 μsec and count rates up to 200,000 c.p.m. can be employed with coincidence correction unnecessary below 100,000 c.p.m. Similarly, scintillation detectors will accept count rates of the order of 100,000 c.p.m.

Standards

Preparation of adequate standards is a critical part of all radioassay. The standard must be as close as possible in all respects to the form in which the samples are prepared. In particular the thickness, density, form, and mass should be the same, while the volume of liquid in liquid counters must always be the same, both for standards and samples. Sample holders must be identical. All standards should be prepared in duplicate or triplicate.

It is highly desirable that the counts of the samples and the standard should not be widely different, and when samples do vary widely in their activity it will be necessary to prepare standards of corresponding activity. The best procedure is to prepare a series of standards of increasing radiotracer and tracee content and plot a standard curve against observed counts as described in the previous section (Fig. 5.5). Such a calibration curve will also take into account coincidence losses. Then, either the samples can be determined from the curve, or if the curve is a constant slope then one standard may be taken as the sole reference point.

Sometimes a given amount of tracer radioisotope will represent different amounts of tracee element in the system. A case in point is with ion uptake studies. Here the range in concentration of uptake solutions may vary from, say, 10^{-3} M to 10^{-5} or even 10^{-6} M in the same experiment, and this range of concentration is far too great to have the radioisotope tag directly proportional to the tracee concentration. This is because in the dilute part of the range the radioactivity could be inadequate for measurement, while in the more concentrated solutions the radioactivity might present the possibility of radiation damage. In practice the different solutions have the same amount of radioisotope added, then an aliquot, e.g. 1 ml, of each treatment solution is taken as a standard. Then each 1 ml will contain the same amount of radioactivity but for each treatment the 1 ml will represent a different number of μmoles of tracee.

Standards must obviously be prepared from the same batch of radioisotope as is used for the experiment and it is usually convenient to make them when the test solutions or materials are being initially prepared. As the radioactive decay of the standards and the experiment samples will be the same, all that is necessary to take decay into account when subsequently counting is to count both samples and standards at the same time. Additionally a reference standard should be counted periodically during sample counting as a check on the counter and instrument.

With radioisotopes with very short half-lives, or if counting is spread over a long period, it is often more convenient to record the time that the count is made and then correct for decay, rather than continually count the standards. This can be done very conveniently using the graphical method developed by Curtiss ([24]).

We have seen (page 7) that the relationship of the half-life of a radioisotope to the decay constant is given by

$$\lambda = \frac{0.693}{t_{1/2}} \tag{4}$$

and that due to the exponential nature of radioactive decay a plot on semi-log paper of log activity versus time on a linear scale will give a straight line. Similarly, a straight line is given if the decay factor, f ($f = e^{-\lambda t}$), is plotted against the fractional half-period $\frac{t}{t_{1/2}}$ (Fig. 5.6). $\frac{t}{t_{1/2}}$ represents the time elapsed between one measurement of a radioisotope and a subsequent measurement, expressed as a fraction of the half-life. Obviously both t and $t_{1/2}$ must be expressed in the same units, whether minutes, hours or days.

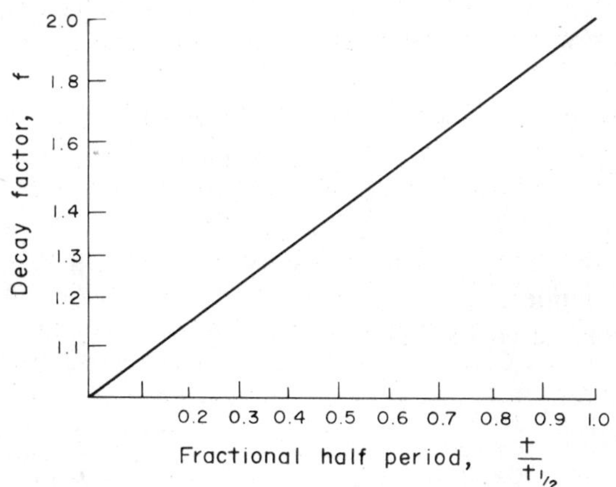

FIG. 5.6 Graphical determination of radioisotope decay (24).

As the fractional half-period can be easily calculated, then f can be determined from the graph, but note that when the elapsed times exceed the half-life time of any radioisotope, then a decay factor of 2 must be added for every half-life which has passed. Then if the observed count rate is N at an elapsed time t, the original count rate, N_o, will be given by fN:

Example

If in an experiment one of a number of samples containing ^{42}K ($t_{1/2}$ = 12.4 hr) was counted and gave an observed count of 2985 c.p.m., what would the count have been 8 hours previously when the first sample was counted? The fractional half-period is $\frac{8}{12.4} = 0.645$, and so from the graph f = 1.57. Then the count rate at zero time, N_o, would have been 2985 × 1.57 = 4686 c.p.m.

In most biological work knowledge of the exact amount of radioactivity present is seldom required. The radioisotope is merely a tracer and it is usually immaterial when, for example, in labelling a test substance with a nominal 10 μCi of activity it is in fact actually 8 or 12 μCi. In most cases it is sufficient to calculate the aliquot, required for labelling with a certain activity, from the data supplied by the manufacturer with the stock solution at the time of shipment.

Nevertheless, all radioisotope laboratories need a certain number of reference standard sources. These are required: to check the activity of new shipments, if there is any doubt about the data supplied or if it is necessary to check the amount of activity lost due to shipping delays, customs clearance, etc.; to check regularly the operation of counters and instrumentation, as low counting rates can result from deteriorating counter tubes or electronic components; for use in "ratio" counting when samples are counted in relation to a long-lived standard, so that each batch of samples over a period is automatically corrected for long term counter instability; to determine absolute amounts of activity in test substances and samples on the relatively few occasions it is necessary; and to establish the position of energy peaks when carrying out spectrometry.

Reference standards are either primary or secondary. The international primary standard is of radium chloride and additional standards are established by national agencies such as the National Bureau of Standards (U.S.) or the National Physical Laboratory (U.K.), using rigorous procedures. These need not concern us here, but primary standards are used to prepare secondary standards which are readily available from radionuclide suppliers.

The most commonly used sources are:

α	β	γ
^{241}Americium	^{210}Bismuth	^{133}Barium
^{210}Polonium	^{45}Calcium	^{109}Cadmium
	^{14}Carbon	^{137}Cesium
	^{137}Cesium	^{60}Cobalt
	^{36}Chlorine	^{129}Iodine
	^{57}Cobalt	^{54}Manganese
	^{60}Cobalt	^{22}Sodium
	^{3}Hydrogen	
	^{147}Promethium	
	^{90}Strontium	
	^{99}Technetium	

Reference sources are available an encapsulated discs, point sources on rods, or as liquids, with activities ranging from 0.1–10 μCi. The sample of unknown activity must as far as possible be prepared in the same form as the reference source. For liquid scintillation counting quenched and unquenched ^{3}H and ^{14}C standards can be obtained in scintillation vials. For general use it is frequently convenient to buy sets of calibrated reference sources.

Due to the relatively short half-lives of the very frequently used ^{131}I and ^{32}P, simulated standards of the isotopes can be obtained for checking counter operation, or some suppliers have actual standards available at several specified times during the year. The checking of the day to day efficiency of liquid counters can be conveniently carried out with the naturally occurring ^{40}K radioisotope, using a 10N solution of potassium carbonate or potassium acetate. For carbon dating, as in some organic matter studies, contemporary standards of oxalic acid are available.

High intensity gamma reference sources with activities from 1–100 mCi are used to calibrate survey meters. They are supplied with special storage shields and safe handling facilities.

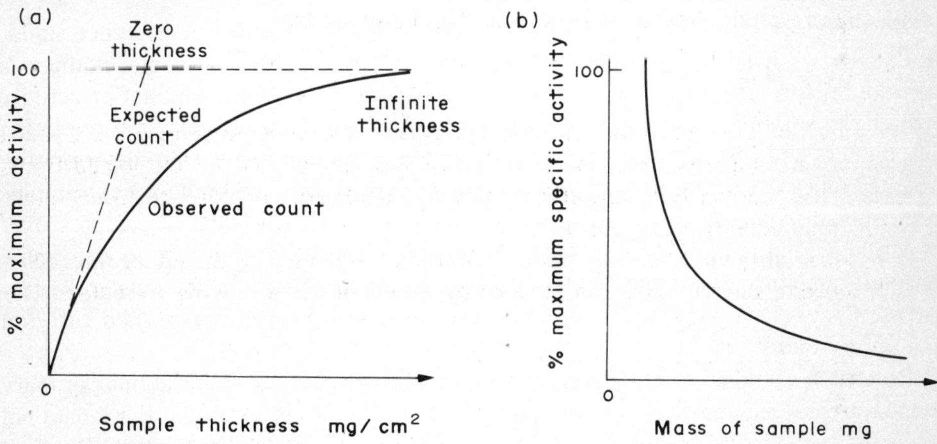

FIG. 5.7 (a) Relationship of sample thickness to observed activity. (b) Correction curve for sample mass e.g. for $Ba^{14}CO_3$ or $^{45}CaCO_3$.

Self-Absorbtion Effects

The radiation emitted from a solid is attenuated and scattered by the mass of the sample itself, and the thicker the sample the more the radiation from the lowest layers is absorbed. Thus if increasing amounts of sample are placed in a planchet and counted, it will be found that the count rate at first rises and then achieves a constant plateau value (Fig. 5.7 a) as the radiation from the bottom layers is totally absorbed ([28]).

Self-absorbtion effects are especially relevant in the case of β-emitters because β-particles have a defined range and are readily absorbed by quite small amounts of sample. This is particularly important in the preparation of solid samples of such weak β-emitters as ^{14}C, ^{45}Ca and ^{35}S when thickness of sample becomes critical. Nowadays the problem is increasingly avoided through liquid scintillation counting.

Self absorbtion can be countered in five ways:

(i) by counting an ''*infinitely thin*'' sample where self-absorbtion effects are negligible, but in practice this may not be feasible, as this is as little as 0.5 mg/cm^2 for $CaCO_3$;

(ii) by preparing all samples so that they are counted at ''*infinite thickness*'', with corresponding standards;

(iii) by preparing a set of *standards with the same specific activity but increasing thickness* of material (Fig. 5.7 b). This means that the observed specific activity, S_0^* (cpm/mg) $= \frac{R_1}{M}$ where R_1 is the observed count rate and M is the mass (mg) of the sample. It is then possible to relate the observed count rate of a sample at any thickness to the ''normalized'' specific activity S_n^* that would be given by an infinitely thin sample as,

$$S_n^* = S_0^* \frac{100}{\%\ \text{of max. sp. act.}} \tag{5}$$

The ''normalized'' total activity R_2 will then be given by (6)

$$R_2 = M\, S_n^*$$

When a calibration curve for self absorbtion has been prepared (Fig. 5.7 b), then each unknown sample is weighed, its observed specific activity S_o determined from the graph, its ''normalized'' specific activity S_n calculated, and from this the ''normalized'' total activity R_2 (c.p.m.).

(iv) by *correcting to ''zero'' or ''infinite thickness''* having determined the thickness of each separate sample. This can be done by means of the following formulae ([32]):

$$\frac{N}{N_o} = \frac{1 - e^{-\mu T}}{\mu T} \tag{7}$$

and

$$\frac{N}{N_\infty} = 1 - e^{-\mu T} \tag{8}$$

where N = observed activity

N_o = activity at zero thickness

N_∞ = activity at infinite thickness

T = sample thickness (mg/cm^2)

μ = self-absorbtion coefficient (cm^2/mg), e = 2.718

(v) by *preparing standard thickness samples*, having approximately the same weight of material in each sample. This can be readily undertaken for routine analyses where the amount of $CaCO_3$ for instance is readily known. It will clearly be more troublesome with unknown samples, but previous analysis for the total stable tracee isotope can make it possible to select suitable sized aliquots which will give standard samples.

Theoretically this last method is the best approach to the problem of self absorbtion. The other correction methods are subject to such uncertainties as are inherent in counting infinitely thick samples i.e. lack of uniformity in the sample and decreased sensitivity as parts of each sample are counted inefficiently, while very thin samples may suffer from self-scattering of β-rays and difficulty in obtaining a uniform spread. Additionally, the variation in thickness of samples may alter the distance of the surface of the sample from the detector, which could be critical with low activity and short distances.

Counting Statistics

Radioactive decay is a random process and therefore there are statistical fluctuations in the count rate as recorded by a detector. These fluctuations must be taken into account, especially with extremely low-level counting, as the ultimate accuracy of radioisotope determination is limited by this factor ([23,26,29,30,31]).

In general the accuracy depends on the total number of counts recorded. Radioactive decay follows the Poisson distribution law and the standard deviation, σ, is equal to the square root of the number of counts. Now the standard deviation is the amount a single count may vary from the mean, so for counts of 1000 and 10,000 for example, σ is 32 and 100 respectively, or put another way: 1000 counts give a standard deviation of 3% and 10,000 counts a deviation of 1% ([35]).

Normally the sample deviation is expressed in terms of count rate. Then, if N is the total number of counts recorded in time t (minutes), the deviation is

$$\sigma = \frac{\sqrt{N}}{t} \tag{9}$$

As an example, if after subtracting the background count 6400 counts were recorded in 5 minutes then the deviation is $\frac{\sqrt{6400}}{5} = 16$ c.p.m, and the true count rate is 1280 ± 16 c.p.m.

It is necessary to know the probability of the counting error being greater or less than a particular value, and the standard deviation is the basic "error" used to define confidence levels. A more or less rigorous confidence level can be adopted. A useful parameter of the error is the relative error, r, which is equivalent to the number of standard deviations in the error, and relates to the probability, P, by the following formula:

$$\begin{aligned} P &= r\sigma \\ &= r\sqrt{N} \end{aligned} \tag{10}$$

The frequently used values of r and the corresponding probabilities are ([35]):

Relative error r	Probability P	Equivalent to:
0.6745	0.500	50% chance of an error exceeding this
1.0000	0.317	"One sigma level", 31.73% chance of an error greater than this
1.64	0.100	10% chance of an error exceeding this
1.96	0.05*	5% chance of an error exceeding this
2.576	0.01**	1% chance of an error exceeding this

*0.05 and **0.01 significance levels respectively.

The 0.1 P level (1.64 $\sqrt{N}$), sometimes known as the "reliable error", is most often used for reporting count data, though $2 \times \sqrt{N}$ will give values statistically significant at the 0.05 level.

So far the situations considered have been essentially where the count is large compared with the background. In such circumstances it is sufficient to estimate the standard deviation by taking the square root of the total counts. But when the sample count is little more than the background count it becomes necessary to consider the deviation of the background and sample counts separately. Then the standard deviation is:

$$\sigma = \sqrt{\text{Sum of squares of separate standard deviations}}$$

$$= \sqrt{\frac{N_s}{t_s^2} + \frac{N_b}{t_b^2}} \qquad (11)$$

where N_s and N_b are the counts recorded for sample + background and the background alone respectively.
t_s and t_b are the respective count times.

Therefore the true sample count rate is

$$\frac{N_s}{t_s} - \frac{N_b}{t_b} \pm \sqrt{\frac{N_s}{t_s^2} + \frac{N_b}{t_b^2}} \qquad (12)$$

A further refinement when counting low activity samples is to choose a counting time which gives the most efficient distribution between sample activity and the background, because it can be demonstrated that counting error will be at its lowest for one particular time distribution.

Where A_s = total activity of sample and background (c.p.m.)
A_b = activity of background (c.p.m.)
t_s = time to count total activity
t_b = time to count background activity

then

$$\frac{t_s}{t_b} = \sqrt{\frac{A_s}{A_b}} \tag{13}$$

In practical use, values for A_s and A_b would be approximately determined from short periods of counting, and the ratio t_s/t_b determined. If this ratio were, say 8, then the sample should be counted for eight times longer than the time taken to count the background, if 10, then ten times longer, etc. In general one seldom uses this procedure except when the sample count is little greater than background.

In practice, fluctuations in count rate are not solely due to the random nature of radioactive decay, known as natural uncertainty. Variation in count rate may also be due to the positioning of the sample in relation to the detector, operator technique, counter efficiency, electronic fluctuations etc., which is usually called technical uncertainty. Natural and technical uncertainty add up geometrically, so that

$$\sigma^2_{\text{tot}} = \sigma^2_{\text{nat}} + \sigma^2_{\text{tech}} \tag{14}$$

With samples where the sample count is large compared with the background there is usually little need to consider the distribution of natural and technical uncertainty, but with low activity samples it assumes greater importance. It has already been seen that for practical purposes the natural standard deviation of a single counting of a sample is given by

$$\sigma_{\text{nat}} = \sqrt{N} \tag{15}$$

If however such a sample is counted repeatedly in the same position then the total standard deviation of the count would be determined following normal distribution law, as

$$\sigma_{\text{tot}} = \sqrt{\frac{\Sigma\,(N - N)}{n - 1}} \tag{16}$$

where N = count of each separate count
$\bar{N}$ = the average value of N
n = number of repeated counts

In effect this will give a result little different from that calculated by the preceding equation (9). But if the sample position is altered between counts, or a number of separate but replicated (say 20–29) samples are counted successively, then it is to be expected that a larger value of σ_{tot} will be obtained. This additional variation is due to technical uncertainty, which may be estimated by substituting in

$$\sigma^2_{\text{tech}} = \sigma^2_{\text{tot}} - \sigma^2_{\text{nat}} \tag{17}$$

As a working rule, sufficient counts should be accumulated so that the percentage natural standard deviation is about one half or a third of the percentage technical standard deviation. In practice this is usually not a problem.

Counting Samples with More Than One Isotope

Samples may frequently contain more than one isotope, because certain pairs of isotopes may usefully be used together in tracer experiments, or in the case of activation analysis many radionuclides will be present other than the one or more that must be counted.

The isotopes ^{13}C, ^{14}C and ^{3}H are very frequently used as double labels in biochemistry and metabolism studies, ^{15}N and ^{32}P are commonly used in fertilizer experiments, and any one of a number of combinations of ^{32}P, ^{42}K, ^{45}Ca, ^{54}Mn, ^{55}Fe, ^{22}Na and ^{14}C in plant nutrition and physiology work. Double-label experiments permit obtaining more data from the same experimental material, and determining interaction and relationships relative to uptake, transport, synthesis and degradation. The isotope labels may be applied to the system separately but simultaneously, or different parts of the same molecule may be labelled, making it possible to follow different behaviour of the active groups.

Clearly there is no problem in determining a mixture of one or more stable isotopes with one radioisotope because the stable isotope(s) will be determined by mass spectrometry or emission spectrometry and the radioisotope by any of the usual counting techniques. Where there is more than one radioisotope in the mixture there are a number of possibilities open to the investigator.

(i) By *chemical separation*: Separation of the labelled isotopes of different elements by any of the normal classical techniques is possible. Separation on ion exchange columns is also a method that has been extensively used. In general, every effort is made to avoid tedious chemical separation procedures because they are the negation of the simplicity of radiotracer counting techniques. However, especially in activation analysis (Chapter 8) it is sometimes necessary to resort to these methods, but usually other procedures can be used.

Any chemical separation process must take into account the potentially harmful contamination effects of radiotracers and suitable working precautions and clean working techniques must be adopted.

(ii) By *different decay*: The total activity of the sample is the sum of the separate activities at a given time of the radioisotopes present. Therefore, when two radioisotopes, whose half lives vary by at least a factor of four, are present in a mixture then it is possible to determine them by counting the samples at successive times.

If the total activity is then plotted on semi-log paper, log counts versus time, the curve obtained will be a composite of the two activities (Fig. 5.8). To obtain the contribution of the long-lived isotope to the total activity, the lower, linear, portion of the curve is extrapolated back to zero time as in line b, Fig. 5.8. The activity of the long half-life component is then subtracted from the total activity to obtain the activity of the short half-life radioisotope. When this is done for a number of points on the time axis it is then possible to construct line a. This method may also be used to check the purity of a radioisotope sample, or to determine the half-life and hence identity of a possible contaminant nuclide, provided the half-life is not too long.

An alternative method is to count the sample containing the two radioisotopes, and

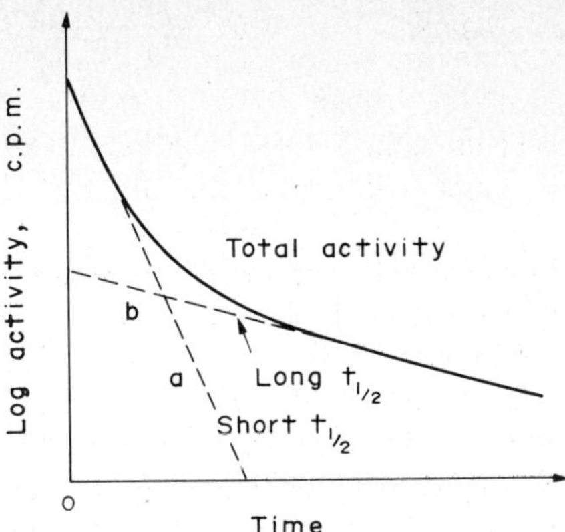

FIG. 5.8 Resolution of the decay curve of two radioisotopes present in a sample.

then count it again after the elapse of a period of time. The following equation can then be applied:

$$A_o + B_o = N_o \tag{18}$$

and

$$\frac{B_o}{A_o} = \frac{F_A - \dfrac{N}{N_o}}{\dfrac{N}{N_o} - F_B} \tag{19}$$

if N_o = count rate of the mixture at zero time
N = count rate of the mixture at time t
A_o = count rate of isotope A at zero time
B_o = count rate of isotope B at zero time
F_A = fraction of isotope A remaining at time t
F_B = fraction of isotope B remaining at time t

F_A and F_B can be found from a decay curve on semi-log paper or by calculation from the decay equation:

$$t_{1/2} = \frac{0.3t}{\log\left(\dfrac{N_o}{N}\right)} \tag{20}$$

Substituting in equation (19) permits using equation (18) to calculate what proportion of the count rate at zero time, N_o, was due to isotope A and what due to isotope B.

(iii) By *different detection systems*: The efficiency of detection and counting of α-, β- and γ-radiation depends on the detector that is used. As some detectors may be extremely insensitive or inefficient for a particular radiation or radiation energy, then it may be possible to count one radioisotope in a sample with negligible contribution to the total count from the second isotope.

The classic example of this is with ^{55}Fe and ^{59}Fe in the same mixture, when the X-ray of ^{55}Fe is determined by solid scintillation counting with a sodium iodide crystal, and the β-rays from ^{59}Fe either by G-M counter or possibly with liquid scintillation.

Cerenkov counting permits the detection of ^{32}P in mixtures with ^{33}P, ^{45}Ca, ^{35}S and ^{14}C without any contribution to the total count from the low energy β-emitters. Even with such radioisotopes as ^{65}Zn, ^{54}Mn and ^{22}Na the efficiency of detection by Cerenkov counting is so low as to permit counting of ^{32}P with very low interference.

It is possible with a proportional counter to count α- and β-emitting radionuclides in the same sample by varying the operational voltage as described in Chapter 4 (page 60). Thus the α-emitter is counted at the low voltage "alpha plateau" setting, then the voltage is raised so that the counter is operating on the "beta plateau" so that the sum of both beta and alpha pulses is recorded, the beta contribution to the total count being derived by subtraction.

(iv) By *differential absorbtion of radiation*: If two β-emitting radioisotopes have different energies then it is possible to separate the individual contributions to the total count by G-M counting with and without aluminium absorbers. The optimum condition for this procedure is when the radiation from one of the radioisotopes can be completely absorbed. In the past ^{45}Ca and ^{32}P were most commonly separated by this means, but Cerenkov counting now offers an easier approach to counting ^{32}P without interference from ^{45}Ca.

Equations (18) and (19) developed for differential decay can similarly be applied to differential absorbtion, the symbols then representing:

N_o = count without absorber
N = count with absorber
A_o = count rate of isotope A without absorber
B_o = count rate of isotope B without absorber
F_A = fraction of isotope A counted with absorber
F_B = fraction of isotope B counted with absorber

In practice, F_A and F_B will have to be determined empirically using the individual isotopes A and B. In the instance of ^{45}Ca and ^{32}P an absorber of 55 mg/cm will absorb all the ^{45}Ca β-rays but reduce the count rate of ^{32}P by about 67%. Therefore, in this case, $F_{Ca} = 0$ and $F_P = 0.67$ ([22]).

(v) By *spectrometry*: Spectrometry is one of the most effective means of counting a mixture of two or more isotopes if the equipment is available. Individual radioisotopes, both β- and γ-emitters, can be identified by the characteristic radiation energies and spectra. For β-emitters the energy spectrum is continuous, as β-particles are emitted with energies between zero and the characteristic maximum energy. With γ-emitters the γ-rays given out have a distinct energy, or energies in the case of those radioisotopes which emit more than one γ-ray.

In either case, when counting with scintillation or proportional counters the pulses recorded are directly proportional to the energy dispersed in the detector. Therefore, as the equipment can be adjusted to select and count the appropriate part of the pulse height spectrum, then theoretically each individual radioisotope can be separated and counted. In practice there are limitations due to interference, especially in the case of β-emitters, because of their continuous energy spectrum. Thus although a high energy β-emitter can be counted largely free from interference by a low energy β-emitter, the count of the latter will almost always have a component from the high energy emitter. Separating γ-emitters is somewhat easier, especially if the γ-energies are widely different.

Further discussion on pulse height spectrometry for counting mixtures of radionuclides is given in the sections on gamma spectrometry (page 106), liquid scintillation counting (page 117), and activation analysis (page 188).

GEIGER-MÜLLER AND PROPORTIONAL COUNTING

This is basically straightforward and has largely been considered in other sections. Thus the principles and operational characteristics of G-M and proportional counters have been dealt with in Chapter 4, while very many of the topics discussed in the previous sections on sample preparation (page 78) and on the theory and practice of counting (page 87) are directly relevant to G-M counting. In particular, geometry, resolving time and self-absorbtion effects have to be taken into careful consideration.

After switching on the scaler and allowing it to warm up, the high voltage is turned up to the appropriate operational voltage for the particular G-M tube. This is usually marked on the tube by the manufacturer, and when a counter assembly with lead castle etc. is put together, the correct working voltage should be marked on the side of the sample changer or scaler. If for any reason the operational voltage is not known then the Geiger plateau must be determined as described in Chapter 4. Excessive voltage will destroy a G-M tube and should be avoided by careful use of the high voltage controls. The high voltage should always be turned down before switching the instrument off.

Having set the instrument to the correct high voltage, the background count should then be determined over a significant period, preferably at least 30 minutes, though on subsequent occasions a quick check over a shorter period is frequently all that is necessary. If the equipment is provided with an automatic background subtraction device this should then be set appropriately. A number of samples should then be chosen at random and given a quick count in order to select an appropriate counting period. The instrument should then be set to the required "pre-set time" or "pre-set count", and systematic counting of the samples can begin. It is desirable to occasionally recount "background" or a standard sample, to check on instrument operation.

As with most electronic equipment, it is best to leave the equipment switched on continuously until any given series of samples has been counted. This usually ensures the best stability and the least change in operating characteristics, and also results in reduced stress on the instrument due to minimizing the shock voltage loadings which occur at switch-on.

SOLID SCINTILLATION COUNTING AND γ-SPECTROMETRY

The general principles of scintillation counting have been considered in Chapter 4 (pages 64–67), which should be read in conjunction, in particular the section on solid or well-crystal counting [38,44].

Although typically used for gamma counting with a sodium iodide crystal, solid scintillation techniques may also be used for hard β-particle counting using a crystal of anthracene, naphthalene or plastic. For β-emitters either "well" or "windowless" arrangement of the crystal gives the greatest efficiency, as illustrated in Fig. 4.1.

With gamma emitters it is common to have the crystal arranged in an upright or "end-on" position as there are no problems of absorbtion by the planchet or sample vial. Well counters may also be used for liquid samples and give high counting efficiency. With both β- and γ-emitters it is important that all sample tubes contain exactly the same amount, and that this should be "normalized" i.e. the volume giving the maximum efficiency should be determined. As little as 1 mm height difference can affect efficiency by 1%.

For exceptionally large liquid samples "Marinelli" beakers made of polythene, which fit closely over the crystal, may be used [41]. There is usually an optimum volume for such beakers, above which efficiency declines greatly.

Basic γ-counting in Integral Mode

It was shown in Fig. 4.8 for a scintillation detector, that if a plot is made of counts as a function of voltage then a characteristic integral spectrum is obtained for the radioisotope being counted. This spectrum comprises the main photopeak due to the primary radiation, a secondary region due to Compton scattered radiation, and in addition background pulses which are due to electronic noise. There is no significant "plateau" as for a G-M counter.

Even a very simple scaling system for scintillation counting usually has a discriminator and most have two, the purpose of which is to reject pulses below or above a certain level, thus improving the peak to background-noise ratio. For counting in integral mode with only one discriminator, the bias voltage is set at an appropriate level to reject pulses due to dark noise from the PM tubes, as these are smaller than the pulses due to the gamma photons. Typical lower-level discriminator (bias) voltage is about 5 volts, but higher in older equipment and less in the newest instruments. If a sample is then counted the scaler will record all pulses exceeding this lower discriminator setting.

Such a procedure is quite suitable for counting samples containing only one isotope of relatively high activity, but the background will necessarily be high. However, in some circumstances counting in such a broad band including the Compton region gives very stable results. To count on integral mode when the scaler has two discriminators the upper level discriminator should be adjusted to its maximum setting, or on some instruments it can simply be put out of action by switching to "integral".

The working high voltage (HV or HT) will usually be that recommended by the manufacturers of the PM tube, but if it is not known then an integral spectrum can be plotted using a fairly active standard of the given isotope. The high voltage chosen for counting samples should then correspond to that of the photopeak as in Fig. 4.8.

An appropriate gain (amplifier) setting can be determined by placing a low activity standard in the detector, adjusting the high voltage on the PM tube to a relatively low operational level, say 600–750 volts, and turning up the gain until counts begin to register on the scaler. Alternatively some instruments have an "attenuation" switch rather than a "gain", and with this the height of the pulses is decreased. Attenuation controls may be "continuous" or "discontinuous", and often both are used. In the case of the latter, then attenuation may be altered by factors of two, so that for example, if the setting is "3" then the attenuation is $2^3 = 8$.

With the high voltage and amplifier settings adjusted to the optimum, and a suitable pre-set time or count chosen, the counter is then set up for operation. The background count should be determined before any samples are counted, and thereafter periodically as a check.

It will usually be desirable to reduce the background count beyond what is normal when counting in integral mode. This can be done with scalers possessing two discriminators, and these also make it possible to count separately more than one gamma emitter in the same sample, providing their energies are sufficiently different.

A reduced background can be obtained by selecting a "channel" so that the lower level discriminator is just below the photopeak, at the same time setting a "window" i.e. the space between lower and upper discriminators, that is as wide as the photopeak at its base. Although the count rate of the sample will be lower without the contribution of the Compton pulses, the background count should also be much lower than when counting in integral mode and thus there will be an overall improvement in sample count/background ratio.

When there is more than one gamma emitting radioisotope in the sample it becomes necessary to choose channels which will reduce as far as possible the mutual interference. This leads on to a consideration of gamma-spectrometry.

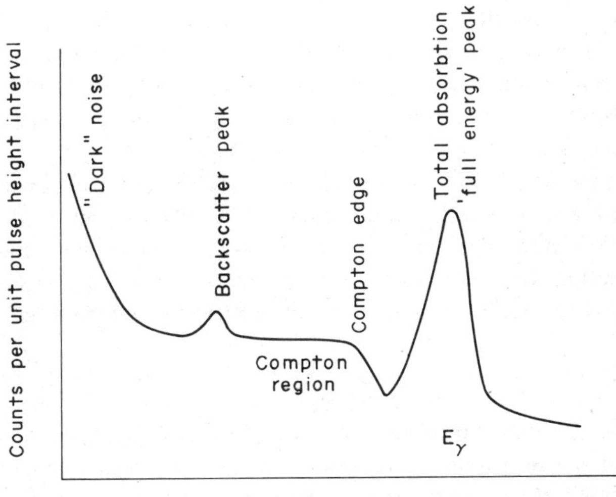

FIG. 5.9 Diagrammatic representation of the components of a pulse height spectrum of monoenergetic γ-rays.

Gamma-Spectrometry

The characteristic ways in which gamma-rays react with matter were considered in Chapter 1. Thus, an incident particle of energy E (MeV) impinges on a scintillation crystal and loses its energy by ionization and excitation, through the photoelectric effect, Compton scattering and pair production. Now, the pulse produced by the scintillation detector is proportional to the γ-energy absorbed by the crystal, but not all the energy is always absorbed by the crystal. Moreover, the proportion of energy lost through each of the three processes is different; therefore, it follows that the pulses from the PM tube will not be of uniform size or height. A plot of the number of pulses of a certain height against the pulse height is known as a "pulse height spectrum" and is characteristic for each radionuclide.

A γ-spectrum is extremely complex, as shown diagrammatically in Fig. 5.9. The "total absorbtion" or "full energy" peak, $E\gamma$, results from interactions in which a gamma-ray loses all its energy e.g. photoelectric effect, or Compton scattering followed by a photoelectric interaction. Overlying the Compton distribution are pulses due to 180° or other back-scattering of gamma rays which had left the crystal and were deflected back into the crystal by the PM tube or its base, or the shielding. Background pulses due to the dark noise of the PM tube are of importance at the low pulse height end of the spectrum. Back-scattering is reduced by the use of as large a crystal as possible.

In addition to the main photopeak and possible back-scatter peaks, there is the possibility of so-called "escape-peaks" when the gamma ray is of sufficient energy ($E > 1.02$ MeV) to induce pair production. If one of the 0.511 MeV annihilation gamma rays escapes from the crystal it results in a "single escape" peak at E = 0.511 MeV, and if both annihilation gamma rays escape a "double escape" peak is produced at E = 1.02 MeV.

There may also be X-rays arising from the interaction of γ-rays with the lead castle. This can be largely eliminated by lining the lead with material of low atomic number such as cadmium or plastic. With radionuclides also having very energetic β-emission it is desirable to have an absorber of low atomic number (bremsstrahlung!) to prevent any contribution to the pulse coming from β-particles. Table 5.2 gives a summary of the major events influencing a pulse height spectrum.

The problem therefore in gamma spectrometry is to separate out the meaningful total absorbtion, $E\gamma$, photopeaks from extraneous incidents, and to identify by their energy and position in the spectrum the photopeaks of "unknown" nuclides. As the area under the total absorbtion peak is proportional to the activity of the radionuclide present, quantitative as well as qualitative determinations are possible.

Pulse Height Analysis

Pulses from the photomultipliers are sorted according to their pulse size, by means of two discriminators, and stored in an electronic memory. Following integration and scanning through different energy levels, the read-out will give the γ-ray spectrum of the radionuclide(s) present in the sample.

TABLE 5.2
Summary of major events influencing the origin of peaks in a γ-ray pulse height spectrum (Simplified from Birks) ([38]).

Origin	Energy of peak	Name
Photoelectric effect	E	Photopeak
Single Compton scattering	T_{cm} to O	Single Compton distribution
Multiple Compton scattering	E to O	Multiple Compton distribution
Compton 180° scattering	$T_{cm} = \frac{E}{1+(1/2\alpha)}$	Compton edge
External Compton 180° scattering	$E'' = \frac{E}{1+2\alpha}$	Back-scatter peak
Pair production with escape of one annihilation γ-ray	$E - m_0c^2$	Single escape peak
Pair production with escape of two annihilation γ-rays	$E - 2m_0c^2$	Double escape peak

where, E = full energy; E'' = energy of back-scattered γ-rays; m_0c^2 = energy equivalent to the resting mass of an electron = 0.51 MeV;

$$\alpha = \frac{E_\gamma}{m_0\ c^2}$$

Pulses are fed into lower and upper discriminators, the setting of which can be continuously varied. The lower discriminator, when set for a pulse height E, rejects any pulse smaller than E. When the window, or channel, between the discriminators is set at a value of ΔE it will reject all pulses which are greater than $E + \Delta E$. An anti-concidence unit receives pulses from both discriminators, rejecting coincident pulses but passing on non-coincident ones to a scaler. By this means only those pulses with an amplitude between E and $E + \Delta E$ pass through the analyzer to the scaler.

When the lower level setting E (the base line) is adjusted to 0.0 volts and the window ΔE to a small part of the total range of the lower discriminator range, say 1% (N.B. Linear amplification may produce an output pulse range up to 100 volts for old equipment and as little as 5 volts for modern instruments). Then a count is taken in the channel from 0–1; next the window, with the same setting, is moved up the energy range and a count is made in the channel from 1–2. In the same manner successive counts are taken in the channels from 2–3, 3–4, 4–5 etc. until the whole range has been scanned from channels 0–1 to 99–100. One hundred pulse height intervals (channels) has been taken as an example, but we could count in half the number of channels with twice the window (50 channels) or twice the channels with half the window (200 channels). The setting is a compromise. When the count from each channel is plotted against channel position (i.e. pulse height interval) a pulse height spectrum is obtained. The scanning may be done manually, adjusting from channel to channel, but most instruments nowadays have a continuous automatic scanning device and the pulses are fed into a ratemeter, the spectrum being recorded on a chart recorder, as a print-out, or fed into a desk-top computer.

Such a spectrometer, in which successive counts are taken in channel after channel, one at a time, is known as a "single channel" analyzer. Instruments are available, known as "multi-channel" analyzers, in which a large number of channels can be counted and recorded simultaneously (Fig. 5.10). Multi-channel analyzers of several hundred channels are available for use with regular sodium iodide scintillation crystals and of many thousand channels for use with germanium based semiconductor detectors. The latter are essential for achieving the necessary high resolution for such sophisticated spectrometers. Multi-channel analyzers are necessary for such applications as activation analysis, when it is essential to achieve rapid determination of the complicated pulse height spectra resulting from the many radionuclides likely to be present.

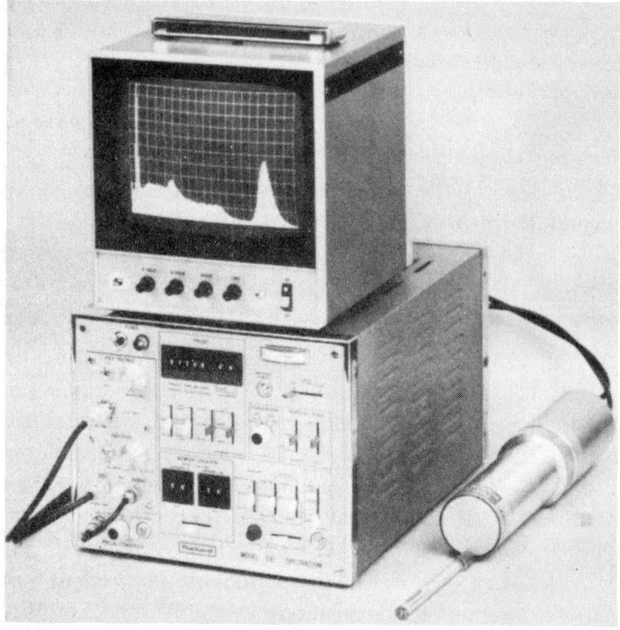

FIG. 5.10 Multichannel analyzer, providing up to 800 channels and two independent memory groups.

Resolution and Efficiency

The resolution of the detection system reflects the extent to which monoenergetic radiation produces pulses of the same height. The peaks in the spectrum are subject to broadening by statistical fluctuation in several processes, such as light production in the crystal, and electron multiplication by the dynodes of the PM tube. However, it is desirable that the peaks should be as narrow as possible so that γ-photons of different energies can be separated from each other, especially when the energies are relatively close.

The resolution of a detector system is determined as the width of the photopeak of a γ-emitter when measured at half the maximum height of the photopeak. This is normally stated as "full width at half maximum" (FWHM) expressed as a percentage of the mean pulse height of the photopeak. FWHM for a Na(T1) crystal is usually 7–10%, but for germanium based semi-conductor detectors may be as low as 0.5%.

As resolution also depends on the γ-energy it is usual practice to use the 0.66 MeV γ-photons emitted by ^{137}Cs as a standard reference.

Efficiency and energy calibration are of considerable importance in critical γ-spectrometry, but will only be briefly mentioned here, and references 37 and 38 can be consulted. The parameters involved include the *geometry factor* (gV), defined for a point source as the fractional solid angle subtended by the crystal face from the source; the *incident intrinsic efficiency* (ε_E), which is the fraction of monoenergetic radiation of photon energy E, interacting at the crystal face to give a scintillation; *source intrinsic efficiency* ($\Omega\varepsilon_E$), which is the fraction of the *total* monoenergetic radiation emitted by the source at energy E, interacting to give a scintillation; *incident full energy peak efficiency* ($\varepsilon_E R_E$) otherwise known as the peak-to-total ratio, where R_E is the ratio of the area of the total absorbtion peak to the area of the total spectrum of a monoenergetic photon source of energy E; and finally the *source full energy peak efficiency* ($\Omega\varepsilon_E R_E$).

Calculation of point source intrinsic efficiencies is possible knowing the crystal dimensions, distance of source from crystal and the linear absorbtion coefficient for E, and are also available from tables ([37]). Table 5.3 gives a limited number of radionuclides especially useful for full energy peak efficiency calibration. Determination of the peak-to-total ratio is complex, because of the effect on the total area of the gamma spectrum by back-scattered peaks, electronic noise, and possibly characteristic X-rays. Their contribution must be subtracted.

TABLE 5.3
Radionuclides suitable for peak efficiency calibration

Radionuclide	E_γ(keV)	$t_{1/2}$
^{210}Pb	46.5	21 y
^{109}Cd	88.0	453 d
^{203}Hg	279.1	47 d
^{7}Be	477.6	53.1 d
^{137}Co	661.6	30.0 y
^{54}Mn	834.8	303 d
^{65}Zn	1115.4	245 d
^{22}Na	1274.5	2.60 y
^{88}Y	898.0 (93.5%) 1836.1 (92.5%)	104 d

Instrument Operation and Calibration

No two spectrometers are the same, and detailed operation of them has to be learned individually, but there are certain features and procedures in common.

There are potentiometers for lower level (threshold) and window settings, and usually a choice of differential or integral measuring possibilities. The threshold potentiometer is usually divided into 1000 divisions, the window potentiometer divided similarly but sometimes with fewer divisions. The amplifier will have coarse and fine gain or attenuation controls, both of which may be continuous, but frequently the coarse gain is altered in discrete steps e.g. 1, 2, 4, 8, 13, 32 and 64.

If we have a threshold potentiometer divided into 1000 units, then this corresponds to 100% of the total range, and 600 units to 60% range and 450 units to 45% of the range and so on. Where the window potentiometer also has 1000 divisions then a window or channel of 1% of the total range is set by adjusting it to "10".

It is necessary to calibrate gamma-ray energy against channel number for the particular detection system and control settings. Radioisotopes which are convenient for this purpose, as they have distinct full energy peaks, conveniently long half-lives, and taken together cover a large part of the useful energy spectrum, are: ^{241}Am (0.06 MeV), ^{22}Na (0.511 MeV), ^{137}Cs (0.662 MeV), ^{54}Mn (0.835 MeV) and ^{60}Co (1.17, 1.33 MeV).

Calibration of a multichannel spectrometer can then be done by successively recording the pulse height spectra of three or four of the radionuclides given above. Then the channel numbers, corresponding to the maximum counting rate of all the full energy peaks, are abstracted from the spectra. When a plot is made of channel number against peak energy it will give a calibration curve establishing the relationship between gamma photon energy and channel number.

This can be used for the identification of an unknown nuclide, such as a contaminant, or in a mixture as in activation analysis. The energy of the "unknown" is determined from the position of its photopeak. As the energy is known it is then possible to identify the unknown radionuclide by consulting tables of gamma energies of different radioisotopes. Where practicable, determination of the half-life will confirm identification.

Alternatively, the instrument may be calibrated just on the 0.662 MeV energy of ^{137}Cs. This method is not so accurate but is especially applicable to single channel spectrometers where running a complete spectrum is a good deal more tedious than on multi-channel instruments. In this case the gain of the spectrometer would be adjusted so as to locate the 0.66 MeV photopeak of ^{137}Cs at a pulse height corresponding to 66% of the total range of the lower discriminator. Because the pulse heights are proportional to the energy of the γ-photons the position of other γ energies can be easily determined.

To set up a single channel spectrometer in this manner the gain controls are set to about the middle of the range with a very wide window, and the base line (lower discriminator) voltage adjusted so that it is 66% of the full range. With a ^{137}Cs standard in the counter, the high voltage is then gradually increased until counts begin to appear on the scaler (alternatively, increase the HV direct to the working level when known). The window is then reduced to the desired opening, say 1%, and the fine gain and high voltage are adjusted so as to give the maximum count rate. When correctly set up neither the high voltage, gain, or window controls should be altered, the subsequent analysis of spectra being carried out entirely by means of the lower discriminator control.

Set up in this manner, the range corresponds to 0–1 MeV, and with a channel of

1% the pulse height interval is effectively 0–01 MeV. If the coarse gain is reduced by a factor of 2, then the range would be increased to 0–2 MeV.

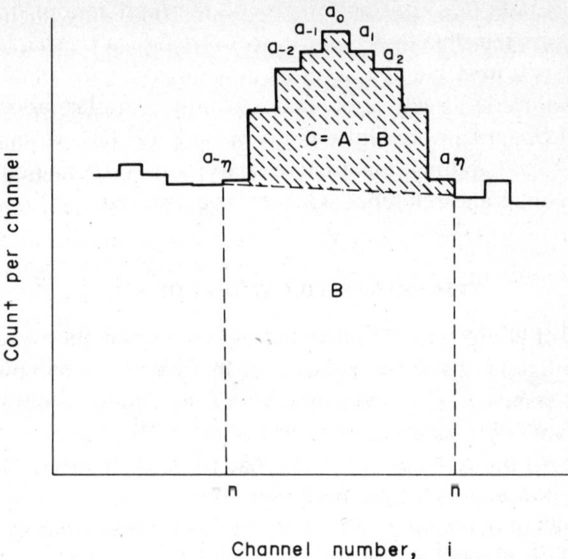

FIG. 5.11 Diagram of a full energy peak illustrating a method for determining the total area, A, and the Compton continuum, B.

Quantitative Determinations

The identity of the element(s) is established by determining the energies corresponding to the photopeaks, and then comparing them against known standards and/or checking them with data tables. The area under the appropriate full energy photopeak, minus background, is proportional to the amount of element present, and this can be quantified by comparison with known standard samples. In effect it is comparing the peak-to-total ratio R_E (the "photofraction") of the unknown with the R_E of the standard.

In the simplest manner it can be done by straight-forward measurement, as described on page 187 of the chapter on activation analysis and shown in Fig. 8.2. However, this is only suitable for very simple cases of a single peak, or at the most, two widely separated peaks. More sophisticated handling of the data is required for the output from multi-channel analyzers. Reference is made to Fig. 5.11. If A is the area of the full energy peak including background, and B is the area of peak due to the Compton distribution continuum, then the nett area C, due solely to major pulses from the nuclide, is given by $A - B$. The total area A may then be calculated by

$$A = a_o + \sum_{i=1}^{n} a_i + \sum_{i=1}^{-n} a_i \qquad (21)$$

when a_o represents the highest count rate recorded for any channel forming the peak. The area of the peak, B, due to Compton distribution, is given by $(n + \frac{1}{2})(a_n + a_{-n})$.

As the area C relates in a constant manner to the total area of the photopeak, then comparison of C of the unknown sample to C of a known standard sample of the radionuclide makes it possible to determine the amount of the unknown. The method is particularly applicable to activation analysis, and a similar procedure, differing in calculation detail but not in principle, is given on page 188. Computer programmes have been written for handling quantitative analysis of γ-spectra but they are beyond the scope of this book and reference 43 may be consulted.

LIQUID SCINTILLATION COUNTING

The principles of liquid scintillation counting, instrumentation and the requirements for scintillation mixtures have been discussed in Chapter 4. Sample preparation and more specialized scintillation mixtures have been discussed earlier in this chapter, so this section is primarily concerned with operation ([46,55]).

In practical terms the two major factors that cause difficulty in liquid scintillation counting are the low energy of the beta rays from ^{14}C and ^{3}H, especially the latter, and the phenomena of quenching. While ^{14}C has a maximum energy of 155 keV, quite small in comparison to hard β-emitters, the maximum energy of ^{3}H is only 19 keV. As this is the *maximum* energy it follows that many of the pulses resulting from the disintegration of ^{3}H are scarcely greater than thermal noise pulses originating from the PM tube. Except in very basic equipment this problem is now overcome by having two PM tubes in coincidence circuitry. Thus a noise pulse originating in one PM tube will usually not be coincident with a noise pulse from the other tube, but each flash of light in the scintillation vial will usually give rise to a pulse in both PM tubes, the pulses being summed before registering. By this means the interference of noise pulses is almost entirely overcome.

No universal solution has been found to the problem of quenching. Quenching results from certain substances either inhibiting the transfer of radiation energy to the scintillator or else resulting in the absorbtion of the emitted light. Phosphorescence and chemiluminescence may also interfere. Particularly bad quenchers are acids, dissolved gases especially oxygen, alcohol, and any chemical reaction which results in a coloured solution. The use of internal or external standards and the procedure known as channels ratio counting are the methods adopted at the present time to overcome quenching ([47,49,54,59,60]).

Liquid scintillation counting is basically dependent on pulse height analysis, as in γ-spectrometry, especially as it is often necessary to count more than one radioisotope in the same sample. With β-emitters there are no clear cut mono-energetic peaks, but rather a continuous spectrum of energies (see Fig. 1.5) with a certain maximum energy. Pulse height resolution is rather poor. Counting is therefore usually carried out in somewhat broader channels than in the case of γ-spectrometry; moreover there is no need of a scanning facility, as in practice the radionuclides in any sample are always known. Frequently though, liquid scintillation spectrometry systems also incorporate facilities for gamma pulse height analysis.

Instruments have at least one set of upper and lower discriminators, and usually all but the basic instruments have three or more of such counting channels each with independent gain control. This makes it possible to count samples with ^{14}C + ^{3}H and ^{32}P + ^{14}C etc. as well as performing channels ratio counting. As samples with more than two radioisotopes are very uncommon three channels serve most purposes, but some instruments are provided with extra channels specifically for using external standards or for automatic efficiency determination.

Some automatic instruments can be programmed to process both single and double-labelled, or intermixed samples in the one counting run. Others have plug-in modules which pre-select the optimum window for the particular isotope of interest.

Efficiency

The counting efficiency of any counter is defined as

$$E = \frac{\text{count rate of sample (c.p.s.)}}{\text{disintegration rate in sample (d.p.s.)}} \times 100 \qquad (22)$$

Inevitably, as not all the disintegrations result in pulses being recorded, the counting efficiency of any system is always less than 100%.

In liquid scintillation counting, when a β-ray loses all its energy to the scintillation mixture only about 2% of the energy results in fluorescent light, and it has been found that it takes about 1 keV of β-energy to produce seven fluorescence photons. This means that with a quantum efficiency for the photocathode of 25% then 1 keV of β-energy results in less than 2 photoelectrons being dislodged from the photocathode, *assuming no loss*. In fact there is likely to be a loss of 15–20% of photons between the scintillator and the photocathode.

As it is clearly essential to have one photoelectron in each PM tube to trigger the coincidence unit, this means that any β-particle must have an energy of about 1.5 keV to be recorded. The real problem is with ^{3}H, where many of the β-particles have energies less than 1.5 keV and therefore are not registered. Thus the counting efficiency of ^{3}H is inevitably much lower than for other β-emitters of higher E_{max}. Counting efficiencies of 50–60% may be obtained for ^{3}H under the best circumstances, compared with 85–95% for ^{14}C and almost 100% for ^{36}Cl.

In comparing counting systems or particular counting arrangements, it is common to use a Figure of Merit derived from $\frac{E_2}{B}$, that is the square of the counting efficiency divided by background. This should be as high as possible.

Single Labelled Samples

Counting a single β-emitter and determining its pulse height spectrum. This assumes the most simple case without quench correction.

The lower level discriminator of one of the channels is set to a low value, say not more than 10% of the range, or sometimes a preset low setting is provided, and a window of maximum width is chosen by suitable adjustment of the upper level discriminator. A known standard of ^{14}C or ^{3}H is placed in the detector, and counting commenced with the gain set at its lowest value. The sample is then counted at different, increasing gain values.

The lower discriminator will eliminate many noise pulses, but when the gain is very low many pulses arising from radionuclide disintegration will also remain under the lower level. As the gain is increased the count in the channel will rise as more pulses exceed the lower level. The count rate will continue to rise with increased gain until a maximum value is reached, because beyond a certain increase in gain the pulses become so large that they are rejected by the upper discriminator and the count rate decreases.

A plot should be made of count rate against gain and from this is determined the gain setting which gives maximum count rate. This is known as the ''balance point'' position, and gives very stable counting. It should be observed that the gain setting for balance point position will vary with the scintillation mixture used, as some are more efficient than others.

To determine the pulse height spectrum of ^{14}C, ^{3}H or other β-emitter, a sample should be placed in the counter, and the gain adjusted to balance point position. The lower level discriminator is set at a low value and a channel width of about 5% is selected (some instruments offer a push button facility of e.g. 2%, 4% and 8% channel widths). The sample is then counted at successive increasing values of the lower discriminator. In many instruments the 5% window will be maintained automatically, but in others it will be necessary each time to adjust the upper discriminator the same amount as the lower one.

When the scan has been completed the count rate of the different channels is plotted against pulse height, which should be expressed as the setting of the lower level discriminator plus half the channel. Due to the continuous nature of β-spectra relatively few measurements are actually necessary to establish the form of the spectrum providing the maximum is carefully determined.

When a pulse height spectrum has been established for any β-emitter it is then possible to establish the optimum window setting and channel position for any counting situation. This is particularly important when there are two isotopes in the same sample and it is desired to count each with the minimum interference from the other. For these cases some experimentation may be necessary, although some instruments have pre-set controls for common situations, such as ^{14}C + ^{3}H and ^{32}P + ^{14}C. In general, an optimum setting will involve a channel width approximately equivalent to the width of the peak at half the maximum height, with the maximum pulse height towards the centre of the window. When the pulse height spectrum shows considerable ''skew'' e.g. ^{32}P (see Fig. 1.5), it may be inappropriate for the maximum pulse height to be centred in the window.

Optimum channel selection is particularly important for measurements of low activity samples, in order to reduce the background as much as possible. With these samples it may prove effective to reduce the window width. On the other hand, when measuring samples of high activity, that is high in comparison with background, integral counting can be carried out with only a lower level discriminator, the upper level being either switched off or adjusted to maximum value.

Quench Correction by Internal Standard

This method is the most simple procedure available; it can be used for samples of very low activity, and has the advantage that the counting efficiency is determined directly, without the need for a correction curve. There are a number of severe disadvantages: the need for counting twice; the method is very time consuming for routine work; once the internal standard has been added the sample cannot be re-counted; the possibility of pipetting errors in adding the μl quantities of standard ([56, 57]).

The procedure involves first counting the sample in the normal manner, then an additional small amount of *the same isotope of known activity* is added and the sample is counted again. The volume added must be very small, 50 or 100 μl, so as not to alter the efficiency of counting, i.e. the quenching effect, that is being estimated. For ^{14}C and comparatively energetic β-emitters about 20–50,000 d.p.m. should be added and for ^{3}H about twice this amount of activity. The standard may be hexadecane or toluene—^{3}H or ^{14}C. Benzoic acid—^{14}C and $^{3}H_2O$ should be avoided because of their own severe quenching effect.

Calculation is simple. Suppose a sample to have a count rate of 3500 c.p.m. Internal standard amounting to 25,000 d.p.m. is added, and the new count rate is found to be 22,000 c.p.m. Clearly the increase in count rate of 22,000 − 3500 = 18,500 c.p.m. comes from the internal standard. Therefore, the counting efficiency of the standard is given by:

$$E = \frac{18{,}500\ \text{c.p.m.}}{25{,}000\ \text{d.p.m.}} \times 100 = 74\%$$

But the sample was also counted with 74% efficiency so it must have an absolute activity of

$$\frac{3500}{0.74} = 4730\ \text{d.p.m.}$$

Quench Correction by Channels Ratio

The advantages of the channels ratio method are that samples require only one counting, assuming instruments with two or three channels; nothing is added to the samples so that they may be subsequently counted again or used for other purposes; it permits automatic quench correction. The principle disadvantages are that samples with low activity require long counting times to give a sufficient degree of accuracy, and that a calibration curve must be constructed and regularly checked (47,48,49,51,52,.57).

The method is based on the fact that the effect of quenching is to decrease the energies in the β-spectrum and effectively to shift the spectrum to lower pulse height, to the "left" as spectra are normally plotted. If therefore two channels are taken next to each other, as A and B in Fig. 5.12a, then channel A especially reflects the effect of quenching and the ratio of the count rates of the two channels will indicate the degree of quenching. Thus the ratio of counts in channel B divided by counts in channel A decreases with increased quenching.

A calibration curve, as in Fig. 5.12b, is constructed by adding increasing amounts of a quenching agent such as benzoic acid, chloroform, or methyl orange (for colour quenching) to a series of samples of known activity. Amounts suitable for this purpose are up to 1 ml of chloroform or benzoic acid per scintillation vial, or up to 2 ml of methyl orange (0.05 g/l). Then the ratio of B/A is plotted against the counting efficiency as determined in channels A + B. It is a useful feature of the channels ratio method that the calibration curve determined for one quenching agent is generally applicable, with the possible exception of colour quenchers.

Appropriate activity for the standards is about 0.01 μCi of ^{14}C and 0.05 μCi of ^{3}H.

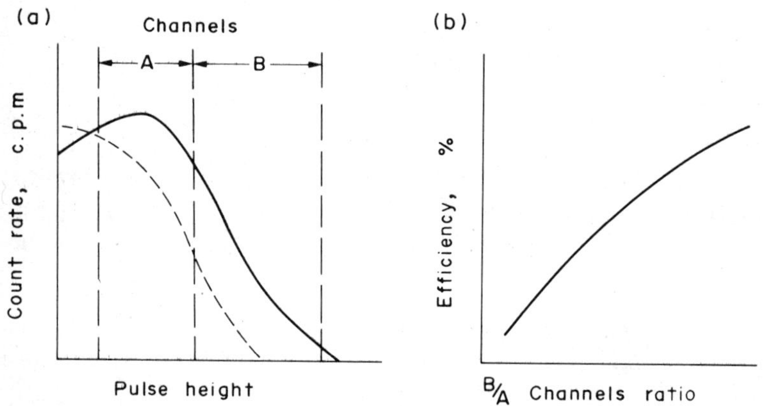

FIG. 5.12 (a) Influence of quenching on shifting the β-spectrum in liquid scintillation counting. (b) A quench correction curve established by "channels ratio".

Careful selection of the channels can result in the calibration curve being a straight line, enabling a constant factor to be established for multiplying the channels ratio to give the counting efficiency.

Quench Correction by External Standard

This method is becoming the normal procedure for the more sophisticated automatic instruments, and although single channel counting can be used, it is usually combined with channels ratio counting. The major advantages of the method are that it can be made automatic, is suitable for routine use and for samples with low activity, while nothing is added to the sample. The major drawback is that the samples have to be counted twice, and as in the previous method a calibration curve must be constructed and checked ([57,60]).

The principle of the method is that a γ-source is placed in close proximity to the scintillation vial when it is in the counting position. Now the γ-photons produce electrons by means of Compton scattering in the scintillation mixture which is principally of low atomic number elements. But the continuous energy spectrum of the Compton electrons is of very similar form to a β-energy spectrum, and moreover the influence of quenching on the spectra is similar. The effect of quenching on the pulse height spectrum of the standard can be readily determined, and a calibration curve is made which establishes the relationship between counting efficiency and the channels ratio of the external standard as influenced by increasing amounts of quenching agent.

In practice, the external standard is counted together with the sample and then the sample is counted without the external standard. The two figures obtained are then used to derive an external standard nett ratio figure, which indicates the amount of quenching occurring in the sample.

The external standard is commonly ^{226}Ra but ^{241}Am and ^{137}Cs are also used. If only a single channel is used for counting, then the calibration curve will be affected by the decay of the gamma source, but not if channels ratio counting is adopted. An overriding requirement of the system is that each time the external standard source is reproduceably and accurately positioned near to the sample vial.

Double Labelled Samples

A basic difficulty in counting samples with more than one β-emitting isotope is that with the continuous energy distribution of β-emitters there will always be an overlap of pulse height spectra at the low pulse height end of the scale. Usually there will only be two β-isotopes in a sample and the greater the difference in β-energy, say four-fold, then the better the results obtained ([48,49,51,60]).

Consider a sample containing a high energy and a low energy β-emitter, such as ^{14}C and ^{3}H, the most frequently used double label. If the lower level of an upper counting channel is adjusted so that very few pulses from the low energy β-emitter are counted in the upper channel then the high energy β-emitter can normally be counted without a significant contribution from the soft β. However, this usually results in some loss of counting efficiency as the counting channel that is used will not be optimum.

In the case of the low energy β-emitter such good separation cannot be achieved because it is inevitable that in the lower channel pulses are counted which come from both β-emitters. The contribution of the hard β-emitter can be reduced to some extent by lowering the upper level of the lower channel but which will leave the counting efficiency of the low energy β-emitter relatively unchanged.

Further separation of two β-emitting isotopes and the determination of their absolute activities, is carried out by the use of standards and application of equations (28) and (29) derived below.

Achieving good instrumental separation of ^{14}C and ^{3}H really begins at the planning stage of the experiment, because if there is too great a proportion of the relatively high energy β-emitter ^{14}C in the sample it will be harder to get an accurate count of the much softer ^{3}H. Therefore, considerably higher specific activities of ^{3}H should be used compared with ^{14}C.

Counting Two β-emitters Using a Single Channel

It is necessary to count each β-emitter in the sample separately, one after the other, adjusting the single channel for each isotope by appropriate setting of gain and discriminators. Then, using standard samples of known absolute activity, the counting efficiencies of the two isotopes at *both* channel settings must be determined i.e. four separate counting efficiencies.

With the counts of the two isotopes in the sample known and the counting efficiencies determined it is then possible to calculate the absolute activities of the samples, where

N_A = c.p.m. in channel A
N_B = c.p.m. in channel B
D_l = d.p.m. (absolute activity) of *l*ower energy β-emitter
D_h = d.p.m. (absolute activity) of *h*igher energy β-emitter
E_{lA} = counting efficiency of lower energy β, channel A
E_{lB} = counting efficiency of lower energy β, channel B
E_{hA} = counting efficiency of higher energy β, channel A
E_{hB} = counting efficiency of higher energy β, channel B

then,
$$N_A = E_{lA}D_l + E_{hA}D_h \tag{23}$$

and,
$$N_A = E_{lB}D_l + E_{hB}D_h \tag{24}$$

solving,
$$D_l = \frac{N_A E_{hB} - N_B E_{hA}}{E_{lA}E_{hB} - E_{lB}E_{hA}} \tag{25}$$

and
$$D_h = \frac{N_B E_{lA} - N_A E_{lB}}{E_{lA}E_{hB} - E_{lB}E_{hA}} \tag{26}$$

If the upper channel can be arranged so that the count of the lower energy β in this channel is negligible compared to the high energy β,

then
$$E_{lB} = 0 \tag{27}$$

and equations (25) and (26) will become,

$$D_l = \frac{N_A E_{hB} - N_b E_{hA}}{E_A E_{hB}} = \frac{N_A - D_h E_{hA}}{E_{lA}} \tag{28}$$

$$D_h = \frac{N_B}{E_{hB}} \tag{29}$$

Counting Two β-emitters Using Two (or More) Channels

When using two channels to simultaneously count two β-emitters of differing energy, each channel must be adjusted to count one isotope as efficiently as possible with as little interference from the other as can be achieved. Consider the specific and very common case of a sample with ^3H and ^{14}C. The channel for the lower energy β should be set first.

A ^3H standard of known and fairly high activity is taken and placed in the counter. Then channel 'A' is adjusted to the balance point position with the lower level discriminator adjusted as low as possible, and with as wide a window as practicable. It is then useful, but not essential, to determine a pulse height spectrum as described for a single channel spectrometer in order to check the settings. With the channel now set at an optimum for ^3H the upper level discriminator should be decreased so that the counting efficiency for ^3H drops about 2–3% in absolute terms. The effect of this will be to disproportionately reduce the potential counts due to ^{14}C in the tritium channel.

The ^{14}C channel 'B' may now be adjusted. Keep the ^3H standard in the counter and set the upper discriminator to its maximum, and the lower level to a normal low position. Then steadily lower the gain (or increase the attenuation if appropriate) until the counting efficiency for ^3H in the ^{14}C channel 'B' has fallen to 1–2%. This will permit the counting of ^{14}C with very little interference from ^3H.

With the instrument now set up, the background count should be determined for each channel, and standards of known activity should be used to determine the counting efficiencies of ^3H and ^{14}C in their respective channels.

Double labelled $^3H/^{14}C$ samples may then be counted and if desired the absolute activities can be calculated by means of the equations given in the previous section.

On a number of sophisticated instruments, the selection of appropriate $^3H/^{14}C$ channels has been reduced to a push-button operation, but the principle is the same.

Quench Correction in Double Labelled Samples

All the quench correction methods previously described, internal standard, channels ratio and external standard, can be used. In present practice the use of an external standard with channels ratio counting is now almost universally used. With modern instrumentation this method of quench correction can be made entirely automatic, only a single counting is required and no additions to the samples.

With only very basic equipment available it may be necessary to use the internal standard method. In this case following the counting of both isotopes in appropriate channels a known amount of standard of the isotope with the lower energy is added. Counts are then made again in both channel settings. Finally a known amount of the higher energy β-emitter is added and a third count is made in both channels. Then using the procedure previously described for the internal standard technique, the efficiencies are calculated for both isotopes in both channels.

This method is clearly very tedious and to be avoided. Moreover, it can never be very accurate for the lower energy β-emitter, ^3H for instance, even if a 5–10 fold excess of activity is added as standard.

In order to carry out quench correction on double labelled samples by means of an external standard and channels ratio counting it is necessary to have an instrument with three or four channels. When there are just three channels, as two channels are essential for counting the two isotopes, there is only one channel available for the external standard. This can be overcome by using the ^3H channel as the second channel for the channels ratio estimation, giving the ratio $\frac{\text{'C'}}{\text{'A'}}$ where channel 'C' is used for the external standard. With four channels this problem does not arise.

If a double labelled $^3H/^{14}C$ sample is to be counted, it will be necessary to determine three quenching calibration curves for later efficiency estimations. These comprise the external standard ratio versus: ^3H efficiency in ^3H counting channel 'A'; ^{14}C efficiency in ^{14}C channel 'B'; and ^{14}C efficiency in the ^3H channel 'A', as illustrated in Fig. 5.13.

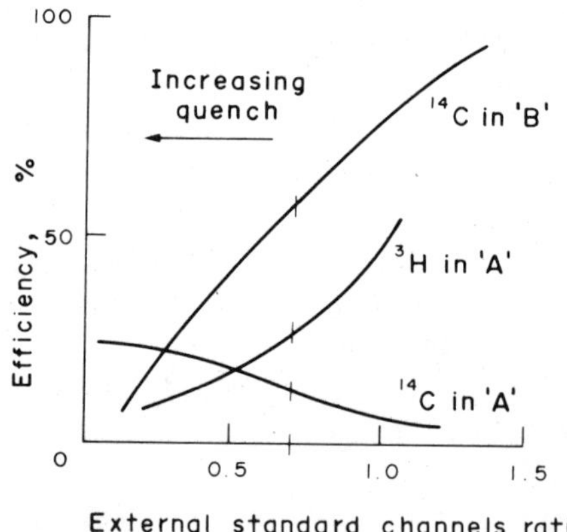

FIG. 5.13 Effect of increasing quench on efficiency versus external standard channels ratio.

The correction curves are constructed essentially as already described for channels ratio counting (page 115), with increasing amounts of quenching agent being added to a series of five or six standard samples of the same known activity. Duplicate unquenched standards of known absolute activity are also prepared for each isotope.

First count in channel 'A' a series of ^3H quenched and unquenched standards prepared in this way. The standards of known absolute activity make it possible subsequently to calculate the counting efficiency. Additionally, during the same measurement also determine the channels ratio of the external standard for each sample. Then plot on graph paper the counting efficiency of the ^3H standards in channel 'A' versus the external standard channels ratio.

In the case of ^{14}C, go through the same procedure of counting quenched and unquenched standards in *both* channels 'B' and 'A', simultaneously determining the channels ratio of the external standard for each sample. Likewise calculate and make a plot of efficiency versus ratio for ^{14}C in channel 'B'. A similar plot should be made of efficiency versus external standard ratio for ^{14}C in channel 'A', because it gives a measure of the 'overspill' of ^{14}C from channel 'B' to 'A', and its contribution to the ^3H count.

An unknown $^3H/^{14}C$ sample may then be determined by counting in channels 'A' and 'B', at the same time obtaining the external standard channels ratio. From this data a logical series of steps permits the calculation of the absolute activities of the ^{14}C and ^3H.

(i) From the appropriate calibration curve and the determined channels ratio of the external standard, obtain the counting efficiency of ^{14}C in channel 'B'. From this counting calculate the absolute activity (d.p.m.) of the ^{14}C in the unknown.

(ii) From the calibration curve of the ^{14}C in channel 'A' obtain the counting efficiency of ^{14}C in this channel and with this calculate from the already determined absolute activity of the ^{14}C in the sample the nett count rate (c.p.m.) due to ^{14}C in channel 'A'.

(iii) From the total nett count rate in channel 'A' subtract the nett count rate due to ^{14}C in this channel, and the difference is the nett count rate of ^3H in the unknown.

(iv) With the ^3H calibration curve and knowing the channels ratio of the external standard of the sample determine the counting efficiency of ^3H in channel 'A'. Then calculate the absolute activity of the ^3H in the sample, using the nett activity due to ^3H in channel 'A' (step iii) and the counting efficiency.

Example. Assuming a four channel instrument, the basic data is as follows:

Count rate, ^3H in channel 'A'	=	3390 c.p.m.
Count rate, ^{14}C in channel 'B'	=	984 c.p.m.
External standard, preset count	=	10,000 c.p.m.
External standard	=	7200 c.p.m.
∴ external standard channels ratio	=	0.72

Taking the efficiency calibration curves as in Fig. 5.13, and a channels ratio of 0.72 we determine and calculate:

(i) ^{14}C efficiency in ^{14}C channel 'B' = 56.0%

$$\therefore {}^{14}C \text{ absolute activity} = \frac{984}{0.56} = 1757 \text{ d.p.m.}$$

(ii) ^{14}C efficiency in ^{3}H channel 'A' = 11.6%

∴ ^{14}C nett count rate in channel 'A' $= \frac{1757}{0.116} = 204$ c.p.m.

(iii) ∴ nett count rate of ^{3}H in the unknown is

$$3390 - 204 = 3186 \text{ c.p.m.}$$

(iv) ^{3}H efficiency in ^{3}H channel 'A' = 25%

∴ ^{3}H absolute activity $= \frac{3186}{0.25} = 12{,}744$ d.p.m.

The activity in the sample is therefore:

^{14}C = 1757 d.p.m.

^{3}H = 12,744 d.p.m.

CERENKOV COUNTING

Background

Cerenkov counting was only briefly referred to in Chapter 4. The flash of light that is emitted when a charged particle of not less than 0.26 MeV passes through a clear liquid is known as the Cerenkov effect. It can be seen as a ''blue'' light in swimming pool reactors and was originally used for the study of high energy particles.

If a radioisotope of an energy higher than 0.26 MeV is present in a clear solution such as water it has been found that the Cerenkov light can be detected by the PM tubes of a liquid scintillation counter. Scintillation fluid is not necessary and counting efficiencies vary between 10–50%, sometimes more ([50,53,55]).

The Cerenkov effect is due to moving charged particles forming an electromagnetic field around themselves, and if their velocity exceeds the speed of light in the liquid constructive interference results in an electromagnetic ''shock wave'' which is the Cerenkov light. The angle which the emitted light makes in relation to the direction of movement of the particle is known as the Cerenkov angle, Θ, and has a maximum of 41.3° in water. There is a relationship between the Cerenkov angle, the velocity of the particle and the refractive index of the liquid, and this has some importance in relation to counting efficiency.

There is a very considerable theoretical and mathematical background to the Cerenkov effect but it is not really necessary for a working understanding of practical counting and will not be gone into here.

The efficiency is much less than for normal scintillation counting. This is because both the number of photons emitted is much smaller with the Cerenkov effect than in liquid scintillation counting, and because for a certain number of Cerenkov photons the pulse produced is much smaller than for the same number of scintillator/fluorescent photons, as PM tubes are better able to convert this light to photoelectrons.

Thus with coincidence circuitry the counting efficiency of a β-emitter with a maximum energy of 1 MeV will be as low as 5–7%. This is because the *average* energy is much less, about 0.3 MeV, barely above the threshold value for the Cerenkov effect. The more energetic the β-emitter the higher the efficiency; a β with a maximum energy of 2 MeV will have a counting efficiency of about 35%.

Practical Application

The advantages of Cerenkov counting are relatively straightforward sample preparation, the sample is uncontaminated and can be used subsequently for other purposes, there is no expense for scintillation mixture, it is suitable for certain γ-emitters, and there are no problems due to chemical quenching. It may be used for solids as finely divided samples in water, providing sample self-absorbtion is not too great. The main disadvantages are that the counting efficiency is much lower than for liquid scintillation counting and the method is only suitable for β-emitters having a maximum energy above a certain level ([45,55,58]).

Normally the channel setting used for ^{3}H liquid scintillation counting is also used for Cerenkov counting because the output pulses of the photomultiplier are comparable to those produced by ^{3}H. Plastic vials are usually to be preferred, partly because glass absorbs Cerenkov light of low wave lengths and also because due to the Cerenkov light being given off at an angle, plastic vials ensure better distribution between the two PM tubes. Similarly it has been established beyond doubt ([58]) that where it is possible to switch off one of the PM tubes greatly improved counting efficiency is obtained. This is because the coincidence circuitry may eliminate events producing enough photons to give at least two electrons, but due to the Cerenkov angle the photons may have only contacted one PM tube.

The amount of liquid in the sample vial greatly affects the sensitivity and efficiency of counting. It is usually desirable to have the vial as full as possible to achieve maximum sensitivity. Although chemical quenching is not a problem, colour quenching can be, particularly from yellowish colours as Cerenkov light is essentially in the ultraviolet and blue end of the spectrum.

The use of fluorescent compounds as wavelength shifters has sometimes been tried, with a view to increasing the wave length of that part of the Cerenkov light which is below 400 nm, and hence reduce its absorbtion by the glass of the vial or PM tube or the water medium. This does improve counting efficiency but it takes away from the inherent, and attractive, simplicity of the procedure. Moreover, there may be trouble from chemical quenching.

Cerenkov counting is especially suitable for ^{32}P, ^{36}Cl, ^{89}Sr and ^{90}Sr, while such γ-emitters as ^{42}K, ^{24}Na and ^{86}Rb may also be effectively counted. High γ-activities may give problems in regard to background, when used in sample changers not primarily designed for this purpose.

REFERENCES FOR FURTHER READING

Sample Preparation—General

1. Association of Official Analytical Chemists. *Methods of Analysis*. 12th Ed. Washington, D.C. (1975).
2. CALVIN, M., Heidelberger, C., Reid, J. C., Tolbert, B. M. and Yankwich, P. F. *Isotopic Carbon, Techniques in its Measurement and Chemical Manipulation*. Wiley, New York (1949) [Methods for ^{14}C before liquid scintillation counting available].
3. CONWAY, E. J. *Microdiffusion Analysis and Volumetric Error*. 3rd Ed., Crosby Lockwood, London (1950).
4. JACKSON, M. L. *Soil Chemical Analysis*, Prentice-Hall (1958).
5. O'BRIEN, R. D. Nitric acid digestion of tissues for liquid scintillation counting. *Anal. Biochem.* **7**, 251 (1964).
6. PAECH, K. and Tracey, M. V. *Modern Methods of Plant Analysis*. Volume I. Springer-Verlag, Berlin-Heidelberg (1956).
7. PIPER, C. S. *Plant and Soil Analysis*. Interscience New York (1950).
8. PREGL, F. and Roth H. *Quantitative Organische Mikroanalyse*. Springer, Vienna (1958). (Also Pregl and Grant J. Blackiston, Philadelphia 1951).
9. VAN SLYKE, D. D., Plazin, J. and Weisiger, R. Reagents for the Van Slyke-Folch Wet Carbon Combustion. *J. Biol. Chem.* **191**, 299–304 (1951).

Sample Preparation—Liquid Scintillation

10. DAVIES, J. W. and Cocking, E. C. Liquid scintillation counting of ^{14}C and ^{3}H samples using glass filter or filter paper discs. *Biochem. Biophys. Acta* **115**, 511 (1966).
11. FOX, B. W. Sample preparation for scintillation counting. *Lab. Pract.* **17**, 595 (1968) [solids].
12. GUPTA, G. N. A simple in-vial combustion method for assay of ^{3}H, ^{14}C and ^{35}S in biological, biochemical and organic materials. *Anal. Chem.* **38**, 1356 (1966).
13. KAARTINEN, N. A new oxidation method for the preparation of liquid scintillation samples. Packard Instrument Co. *Tech. Bull.* **18** (1969).
14. KOWALSKI, E. *et al.* Criteria for the selection of solutes in liquid scintillation counting: new efficient solutes with high solubitlity. *Int. J. Appl. Rad. Isot.* **18**, 307 (1967).
15. MOORE, R. B. *et al.* An apparatus and a method using phenethylamine for liquid scintillation counting of $^{14}CO_2$ obtained by wet oxidation of biological materials. *Anal. Biochem.* **24**, 545 (1968).
16. PEETS, E. A., Florini, J. A. and Buske, D. A. *Anal. Chem.* **32**, 1465–68 (1960).
17. ROBERTS, W. A. Preparation of liquid scintillation mixtures for the measurement of ^{14}C and ^{3}H samples. *Lab. Pract.* **17**, 703 (1968).
18. SCHÖNIGER, W. Die mickroanalytische Schnellbestimmung von Halogenen und Schwefel in organischen Verbindungen. Microchimica Acta, *Heft* **6**, 869–76 (1956).
19. TURNER, J. C. Tritium counting with the Triton X-100 scintillant. *Int. J. Appl. Rad. Isot.* **20**, 499 (1969).
20. WEYMAN, A. K. *et al.* Collection of $^{14}CO_2$ for scintillation counting by a modification of the Van Slyke procedure. *Anal. Biochem.* **19**, 441 (1967) [direct absorbtion of $^{14}CO_2$ in vial].
21. WILLIAMS, R. H. Liquid scintillation counting of tritium in water with Triton emulsion systems. *Int. J. Appl. Rad. Isot.* **19**, 377 (1968).

Counting Theory

22. COMAR, C. L., Hansard, S. L., Hood, S. L., Plumlee, M. P. and Barrentine, B. F. Use of calcium-45 in biological studies. *Nucleonics* **8**, 19–31 (1951).
23. COOK, G. B. and Duncan, J. F. *Modern Radiochemical Practice*. Oxford (1956).
24. CURTISS, L. F. Nat. Bur. Stands. U.S., Circ. 473 (1948).
25. DUNN, W. L. and Gardner, R. P. The determination of intrinsic gamma ray detection efficiencies for

cylindrical Geiger-Müller tubes by Monte Carlo methods. *Nucl. Instrum. Methods* **103**, 373–84 (1972).
26. ELMORE, W. C. Statistics of counting. *Nucleonics* **6**, 26–34 (1950).
27. GARDNER, R. P. and Verghese, K. On the solid angle subtended by a circular disc. *Nucl. Instrum. Methods* **93**, 163 (1971).
28. GRAF, W. L., Comar, C. L. and Whitney, I. B. Relative sensitivities of windowless end-window counters. *Nucleonics* **9**, 22–27 (1951).
29. JARRETT, A. A. *Statistical Methods used in the Measurement of Radioactivity*. AECU-262 (Mon P-126) ORNL, USA.
30. JOYCE, W. B. Origin of counting statistics in simple nucleonic instruments. *Am. J. Phys.* **37**, 489–94 (1969).
31. KUYPER, A. C. The statistics of radioactivity measurements. *J. Chem. Ed.* **36**, 128–32 (1959).
32. LIBBY, W. F. *Anal. Chem.* **19**, 2 (1947).
33. LOEVINGEN, R. and Berman, M. Efficiency criteria in radioactivity counting. *Nucleonics* **9**, 26–29 (1951).
34. SKINNER, S. M. *Phys. Rev.* **48**, 438 (1935).
35. SNEDECOR, G. W. *Statistical Methods Applied to Experiments in Agriculture and Biology*. 5th Ed. Iowa State College (1956) [general statistics].
36. VERGHESE, K., Gardner, R. P. and Felder, R. M. Solid angle subtended by a circular cylinder. *Nucl. Instrum. Methods* **101**, 391–93 (1972).

Solid Scintillation Counting by Gamma Spectrometry

37. ADAMS, F. and Dams, R. *Applied Gamma Ray Spectrometry*. 2nd Ed. Pergamon Press (1970).
38. BIRKS, J. B. *Theory and Practice of Scintillation Counting*. Pergamon Press, Oxford (1964).
39. CROUTHAMEL, C. E. *Applied Gamma Ray Spectrometry*. Pergamon Press (1960).
40. HEATH, R. L. *Scintillation Spectrometry, Gamma Ray Spectrum Catalogue*. 2nd Ed. USAEC Ref. IDO-16880.
41. MARINELLI, L. D. Radiation dosimetry and protection. *Ann. Rev. Nucl. Sci.* **3**, 249–70 (1953).
42. MOTT, W. E. and Sutton, R. B. *In: Encyclopaedia of Physics* **45**, 86 (S. Flügge, Ed.). Springer, Berlin (1958).
43. QUITTNER, P. *Computer Evaluation of Scintillation and Semiconductor Detector Gamma Ray Spectra*. Hilger Ltd., London (1972).
44. SEIGBAHN, K. (Ed.) *Alpha, Beta and Gamma Ray Spectroscopy*, 501–37. North Holland Publishing Co. (1966).

Liquid Scintillation and Cerenkov Counting

45. BALLANCE, P. E. and Johnson S. A procedure for the Cerenkov counting of high energy γ and β emitting radionuclides in coloured plant extracts. *Planta* **91**, 364 (1970).
46. BIRKS, J. B. *Theory and Practice of Scintillation Counting*. Pergamon Press, Oxford (1964).
47. BUSH, E. T. General applicability of the channels ratio method of measuring liquid scintillation efficiencies. *Anal. Chem.* **15**, 7 (1963).
48. BUSH, E. T. Liquid scintillation counting of doubly-labelled samples—Choice of counting conditions for best precision in two-channel counting. *Anal. Chem.* **36**, 1082 (1964).
49. BUSH, E. T. A double ratio technique as an aid to selection of sample preparation procedures in liquid scintillation counting. *Int. J. Appl. Rad. Isot.* **19**, 447 (1968).
50. CLAUSEN, T. Measurement of ^{32}P activity in a liquid scintillation counter without use of scintillator. *Anal. Biochem.* **22**, 70 (1968).
51. HENDLER, R. W. Procedure for simultaneous assay of two β-emitting isotopes with the liquid scintillation counting technique. *Anal. Biochem.* **7**, 110 (1964) [channels ratio].
52. HERBERG, R. J. Channels ratio method of quench correction in liquid scintillation counting. *Packard Tech. Bull.* **15** (1965).
53. HUTCHINSON, G. W. Cerenkov detection. *Progr. Nucl. Phys.* **8**, 195 (1960).
54. JAFFEE, M. and Ford, L. A. On the nature of quenching. *Int. J. Appl. Rad. Isot.* **21**, 49 (1970).
55. PENG, C. T. Liquid scintillation and Cerenkov counting. (D. I. Coomber, Ed.) *Radiochemical Methods in Analysis*. Plenum Press, New York (1975).

56. MOGHISSI, A. A. and Carter, M. W. Internal standard with identical system properties for determination of liquid scintillation counting efficiency. *Anal. Chem.* **40**, 812 (1968).
57. ROGERS, A. W. and Moran, J. F. Evaluation of quench correction in liquid scintillation counting by internal automatic, external, and channels ratio standardization methods. *Anal. Chem.* **35**, 794 (1963).
58. ROSS, H. H. Measurements of β-emitting isotope using Cerenkov radiation. *Anal. Chem.* **41**, 1260 (1969).
59. SCALES, B. Liquid scintillation counting: the determination of background counts of samples containing quenching substances. *Anal. Biochem.* **5**, 489 (1963).
60. WACHTER, J. de and Fiers, W. External standardization in liquid scintillation counting of homogenous samples labelled with one, two or three isotopes. *Anal. Biochem.* **18**, 351 (1967).

*From Arthur Thomas Company, P.O. Box 779, Philadelphia, PA 19105, USA.
**From Glen Creston Ltd., London, England.
***Christy and Norris Ltd., Chelmsford, Essex, England.

CHAPTER 6

Radioisotopes and Tracer Principles

TRACER PRINCIPLES

THE unique advantage of an isotope is that one can use an extremely small, or trace, amount to follow the behaviour of much greater amounts of the abundant isotope of the same element. From this simple basic conception a number of more elaborate treatments may be developed. In considering the principles of application it is necessary to assume that the tracer isotope will behave in the system similarly to its more abundant counterpart, and that the isotope can be conveniently measured either by the radiation that it emits, or by determining its abundance in the case of "heavy" or stable isotopes. Thus we may distinguish the tracer from the common isotope and obtain an estimate of the amount of tracer present. There are situations where these assumptions do not hold, but in general they are valid and serve as the basis for the present Chapter. Some of the possible difficulties and errors are discussed in the last sections.

The especial advantages of radioisotopes are that it is possible to differentiate between an element or substance that is already present and that which is added to the system; the sensitivity of the method is so high that only small amounts of the test substance need be added to the system and thus the conditions existing normally are not changed; it is possible to visualize or study extremely small parts or units; the adaptability of the method is very great as it is now possible to get almost any chemical or material labelled with a radioisotope. Although the use of stable isotopes as tracers is based on the same principles as radioisotopes, the sensitivity is not nearly so high and the differentiation between the tracer and the element already present in the system not nearly so easy.

Uptake, Movement and Metabolism

Radioisotopes have been used in the study of direct movement of such substances and objects as soil particles, soil water, nutrients, bacteria, viruses and fungi. Such a radiotracer may be used to identify the place, time or amount of movement, deposition, uptake or excretion of the nutrient, metabolite or object under investigation. A radiotracer may also be used to identify enzymes, residues, precursors, metabolites or degradation products and in difficult analyses.

If the movement of an object such as sand or soil particles is to be traced, the radioisotope used may belong to any chemical element. The choice of label will be decided by the ease of attachment to or incorporation into the test object, and by the penetrating power of the radiation, which must be sufficient to overcome either the distance or possible shielding of the object from the detector. Additionally, cost must often be taken into account, and the half-life of the radioisotope should be suitable for the purpose of the investigation.

The inverse principle may also be used. Instead of labelling the object of investigation, a confined amount of radioisotope is placed at the anticipated or potential place of arrival of the moving object. When the object becomes radioactive, it has reached its destination. This principle has, for instance, been used in the study of rooting patterns (see Chapter 12).

Studies of the movement of soil water require that the tracer should not easily become bound to the soil particles, and this is discussed in Chapter 14.

If the test substance to be introduced into the system is an organic material or compound, the radioisotope used can belong to any one of the appropriate elements in the compound. This usually means ^{14}C or ^{3}H, with ^{32}P, ^{35}S, ^{36}Cl or ^{131}I when appropriate. Stable isotopes ^{3}H (deuterium) or ^{15}N may also be used. Such a label may be incorporated in the test substance through metabolic reaction, biological growth or chemical synthesis.

Where the molecule contains more than one atom of a particular element, the compound is called "totally labelled" if all the atoms are labelled. "Uniform labelling" is when all the tagged atoms have the same specific activity. Many biological reactions or systems give rise to totally but non-uniformly labelled compounds if not in equilibrium.

The nature of the labelling of biochemicals is important if adequate and reliable information is to be obtained without the label being lost by exchange, etc. This requires some knowledge of the chemical structure. For example, although amino acids can be obtained with ^{15}N, ^{14}C, ^{3}H and ^{18}O labelling, the nature of the label is important for specific studies. Thus ^{15}N labelled amino acids would be used for determining their contribution to the amine pool, but ^{3}H or ^{18}O labelled material for the study of peptide formation rates. On the other hand, studies involving the fate of the carbon skeleton will require ^{14}C labelled material.

When an element such as a mineral nutrient is to be traced, then the label must be an isotope of the same element. However, in certain cases there may be an exception to this rule, e.g. where ^{86}Rb has been used under certain circumstances as a tracer for potassium for want of a sufficiently long-lived potassium radioisotope.

Radioisotopes have been especially useful for studies of gross rates of movement of biologically important ions and substances, of circulation and rates of removal, and of ion movement where the nett mass transfer is zero i.e. where a recirculation pattern exists. However, it must be remembered when planning an experiment involving "time of arrival", following the introduction of a radioisotope label into a system at a given place and its later detection at another place, that the first sign of activity at the second site will not be the precise moment when the bulk of the labelled material has arrived. This is because the activity will have become diffused during transport.

Despite this, the flow rate can still be estimated by means of the Stewart-Hamilton method:

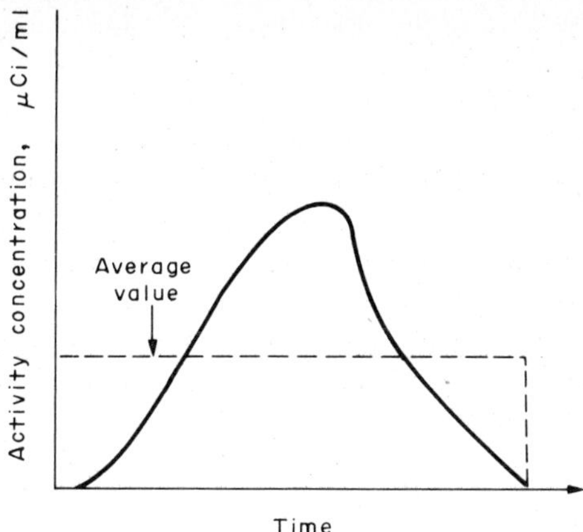

FIG. 6.1 Transport: Tracer concentration with time, following a single injection at another site.

if R = rate of flow, as $\frac{\text{ml}}{\text{time}}$

A^* = amount of radioactive label injected at first site, as μCi

S^* = average activity concentration value, as μCi/ml

a = area under tracer concentration curve, as in Fig. 6.1 = $S^* \times$ time

then $$R = \frac{A^*}{a} = \frac{A^*}{(S^*)(\text{time})} \tag{1}$$

assuming that no tracer is lost between the site of injection and the detection point.

Exceptionally small amounts of element or compound, not measurable by conventional methods, may be determined by means of radioisotope tracers because of the great sensitivity with which the radioactive label can be assayed. Assuming that the specific activity S^* of the test substance has attained the same value at every place of interest in the system under investigation, then the quantity Q of a test substance "A" in any sample is given by

$$Q = \frac{A^* \text{ (c.p.m.)}}{S^* \text{ (c.p.m. per unit of "A")}} \tag{2}$$

where A^* is the count rate of the sample. In order to determine the specific activity it is necessary to be able to take a large enough sample at some point in the overall system, containing sufficient test substance to determine it by chemical or physical methods.

The specific activity is then

$$S^* = \frac{A^*}{A(\mathrm{wt})} \tag{3}$$

where A(wt) is the weight (content) of test substance and A* the count rate. We might typically express S^* as c.p.m./mg"A" or use some other convenient unit (see specific activity, page 4). Note that the determination of absolute activity is almost never required in biological radiotracer work.

Consider a translocation problem. An amount of test substance X is labelled with a known specific activity S^*. An unknown amount of X, say x units, translocates to a site of interest where it inevitably mixes with an unknown amount of unlabelled substance. Nevertheless, we can still determine the amount of x if we measure the activity, x^* c.p.m., at the sampling site,

$$\text{thus } x = \frac{x^*}{S^*} \tag{4}$$

Example

As a practical example we may consider an experiment involving the determination of the uptake of rubidium by portions of excised roots. A 1×10^{-3} M solution of $RbCl_2$, i.e. each 1 ml of solution contains 1 μmole of rubidium, is labelled with ^{86}Rb. If after mixing we take 1 ml of this solution and count it, we can quite simply determine its specific activity as x c.p.m./μmole Rb. (In practice for a 15–60 min experiment we should probably label the solution with about 50 μCi ^{86}Rb/litre of uptake solution, i.e. about 0.05 μCi/ml, but note that it is quite immaterial to know the actual activity of the tracer in order to calculate the specific activity of the test solution). We may then allow 1 g weight samples of excised roots to absorb ions for a given unit time. After the elapse of the appropriate time each root sample is removed from solution and following appropriate procedures (see page 316 for typical details) the sample is counted, giving say y c.p.m. for 1 g roots. Then the amount of rubidium taken up by a root sample

$$= \frac{y \text{ c.p.m.}}{x \text{ c.p.m.}} \text{ μmoles/g wt/unit time} \tag{5}$$

Of course, this example has been simplified both in terms of procedure and solution concentration in order to bring out the essential principle, but such a basic calculation can be developed for a wide range of experimental conditions.

Isotope Dilution Procedures

The principle of *isotope dilution* procedure is that for a given constant amount of radioactivity, the specific activity is inversely proportional to the total amount of test element or substance present in the system. The concept was developed by Hevesy and Hofer[5] in 1934 and is thus as old as the tracer technique itself.

The technique has a number of advantages and applications, but is especially useful for obtaining analytical data when quantitative separations are not possible, or are too difficult or time consuming for the particular circumstances. For example: for such diverse purposes as determining the amount of nutrient in a plant that has been derived from fertilizer; and for recognition and quantitative estimation of intermediates in metabolism studies.

There are basically three forms of isotope dilution procedure: direct isotope dilution, inverse isotope dilution and double isotope dilution. A feature common to all procedures is the basic assumption that after mixing and attainment of equilibrium the system is uniform in respect to the specific activity S^* of the particular test element or molecule.

Direct Isotope Dilution

This method may be used to determine an inactive compound of known chemical identity by means of dilution with a known amount of the active compound. The method may be used for either inorganic or organic materials and has the advantage that only a portion of the compound need be separated; quantitative isolation is not necessary.

The procedure comprises the addition of a known amount of radioisotope labelled element or compound to the system or mixture containing an unknown amount of the same unlabelled substance. A portion of the non-labelled compound is isolated in pure form, but it is not necessary to recover all the compound, provided a known weight of the compound can be separated. The separated portion is then counted and its activity determined; from this can be calculated the specific activity, because the weight is known.

This procedure differs from the practical example given above of an ion uptake experiment using ^{86}Rb, because in that case the added radioactivity was of negligible weight (carrier free) and could be discounted. We now have a situation where the weight of the tracer substance must be taken into account.

If such a system or a mixture contains an unknown weight W of a substance in the sample, and

w^* = known weight of isotopically labelled "spike" added to the sample
S_1^* = known specific activity of labelled "spike"
S_2^* = determined specific activity of sample mixed with labelled "spike"

then, as the total radioactivity is the same before and after, we have

$$(S_1^*)(w^*) = (S_2^*)(w^* + W) \qquad (6)$$

from which is derived

$$W = w^*\left(\frac{S_1^*}{S_2^*} - 1\right) \tag{7}$$

Inverse Isotope Dilution

This is a simple variation of the direct isotope dilution technique whereby it is possible to measure the quantity of isotopically labelled substance in a system by adding a *known amount of unlabelled* substance. The specific activity is determined on portions of the substance separated from the system in pure state, before and after the addition of the unlabelled material.

As with the direct dilution technique, total recovery of test material is not necessary, but an additional advantage of the inverse dilution procedure is that smaller amounts of material can be determined. For this reason a modified procedure can be useful in activation analysis techniques (Chapter 8).

Using similar notation, where

W^* = unknown weight of labelled material
w = known weight of unlabelled material added to the sample
S_1^* = specific activity of an isolated sample before the addition of unlabelled material
S_2^* = specific activity of an isolated sample after the addition and mixing of unlabelled material

then

$$W^* = \frac{(w)(S_1^*)}{S_1^* - S_2^*} \tag{8}$$

The greater the amount of unlabelled material that can be reasonably added, then the more accurate the results. The proper mixing of the unlabelled substance and the purity of the isolated samples are of critical importance.

Double Isotope Dilution Analysis

This procedure was first described by Block and Anker ([2]) and, unlike the previous method, it is not necessary to know the specific activity of the substance being quantified. Although not as precise as the two preceding methods, it does permit quantitative analyses which would not otherwise be possible.

Two different sized aliquots of unlabelled substance, identical in chemical form to that which is being determined, are added to two samples of the mixture, thoroughly mixed, and a portion of each then isolated for the determination of specific activity.

Where,

W^* = unknown weight of labelled material in the sample
w_1 and w_2 = different known weights of unlabelled material added to two samples of the material
S^* = unknown specific activity of labelled material
S_1^* and S_2^* = specific activities of the purified portions taken from each sample to which w_1 and w_2 were added

then,
$$W^* = \frac{w_1 S_1^*}{S^* - S_1^*} \text{ and } S^* = \frac{w_1 S_1^*}{W^*} - S_1^* \quad (9)$$

and secondly,
$$W^* = \frac{w_2 S_2^*}{S^* - S_2^*} \text{ and } S^* = \frac{w_2 S_2^*}{W^*} - S_2^* \quad (10)$$

these are equated,
$$\frac{w_1 S_1^*}{W^*} - S_1^* = \frac{w_2 S_2^*}{W^*} - S_2^* \quad (11)$$

and simplified,
$$W^* = \frac{w_2 S_2^* - w_1 S_1^*}{S_2^* - S_1^*} \quad (12)$$

Relative Contribution of Two Sources

The possibility of using isotope dilution principles to determine the relative contributions of two sources of a substance to a product, such as occurs when a plant takes up both soil-derived and fertilizer-derived nutrient, has given us two practical procedures of wide importance in soil/plant work. Thus the determination of A-value, that is the availability of a given soil nutrient in terms of a standard (Fried and Dean) ([4]), is in effect a direct use of the basic isotope dilution equation (7) given above. The A-value is discussed extensively in Chapter 11, but we may note here that if A is the amount of an available nutrient element in the soil and B is the amount of the same labelled nutrient added to the soil as a standard, then equation (7) can be re-written as

$$A = (B)\, X \left(\frac{S_f^*}{S_p^*} - 1\right) \quad (13)$$

where S_f^* is the specific activity of the nutrient element added to the soil and
S_p^* is the subsequent specific activity of the plant material grown on the soil.

As a further general consequence of the isotope dilution principle, it follows that if an isotopically labelled test substance of known specific activity is added to a system containing some of the same unlabelled substance and the system is subsequently sampled, then the *fraction* in the sample due to the added substance is given by

$$\frac{S_2^*}{S_1^*} \tag{14}$$

where the experimentally determined specific activity of the sample, S_2^*, is divided by the known specific activity, S_1^*, of the added substance.

This can be applied directly to determine the relative contributions of fertilizer or soil-derived nutrient to the plant, even if the total quantity of the nutrient in the soil system is unknown. Thus if we apply labelled fertilizer nutrient, ^{32}P for example, of specific activity S_f^* to the soil, then over a period of time the plant takes up a certain amount of both labelled nutrient and soil nutrient and we may determine its specific activity S_p^* by chemical analysis and radioisotope counting. Then

$$\frac{\text{Specific activity of plant, } S_p^*}{\text{Specific activity of fertilizer nutrient, } S_f^*} \times 100 = \begin{array}{l}\text{percentage of nutrient in}\\ \text{the plant derived from}\\ \text{fertilizer}\end{array} \tag{15}$$

Stated slightly differently, then $\frac{S_p^*}{S_f^*}$ is the fraction of nutrient in the plant derived from the labelled standard nutrient, while $1-\frac{S_p^*}{S_f^*}$ will be the fractions derived from the soil. Such fertilizer studies are considered in greater detail in Chapter 12.

Determination of System Volume

This is another application of isotope dilution principles. The liquid volume of a system may be determined by adding a known volume of known specific activity to the unknown unlabelled volume.

If,

V = unknown volume
v_1 = volume of added isotopically labelled solution
V_m = unknown volume of the mixture
S_1^* = specific activity of added volume
S_2^* = specific activity of the mixture at equilibrium
A^* = total added activity = total activity in mixture

then, $$V_m = \frac{A^*}{S_2^*} \quad \text{where } A^* = v_1 S_1^* \tag{16}$$

and clearly $V = V_m - v_1$

by combining we have,

$$V = \frac{v_1 S_1^*}{S_2^* - v_1} = v_1\left(\frac{S_1^*}{S_2^*} - 1\right) \tag{17}$$

In effect this is the same as equation (7). This procedure may be used to determine various ''volumes'' of biological interest: exchangeable potassium and sodium, chloride ''space'' etc.

Although this discussion has primarily been concerned with radioactive tracers, exactly the same principles and techniques may be employed with stable isotopes, by determining initial and final enrichment as atom % excess.

Tracer Kinetics

This section may be passed over at a first reading.

In biological systems we are concerned with processes of accumulation, removal, turn-over and exchange. The investigation of any of these processes is normally complicated by the fact that often more than one process is proceeding simultaneously.

It is frequently possible to describe such processes mathematically, and these kinetic analyses have long been especially developed in animal physiology studies, where systems and compartments tend to be more clearcut than those met in plant biology ([1,8,10,12]).

The objectives of a kinetic analysis are to identify the processes involved; to determine the transfer rate of the material from one phase to another; and to estimate the quantities in each phase. Only a limited treatment will be attempted here, in order to illustrate some basic generalizations.

It is frequently found that the fractional rate of change of many biochemical and biophysical processes is constant, that is they are exponential. Thus, in a first order reaction, the reaction proceeds at a rate proportional to the quantity of material present, so that if

dQ = change in quantity of substance Q in unit time, dt

k = rate constant for the reaction, i.e., the fractional rate of change of Q with time

then

$$\frac{dQ}{dt} = kQ \tag{18}$$

now, if Q_0 = quantity of material present at zero time

Q = quantity of material present at time t

e = base of the natural logarithm

then integrating between limits gives us

$$\int_{Q_o}^{Q} \frac{dQ}{Q} = -k\int_{o}^{t} dt \tag{19}$$

$$\ln = \frac{Q}{Q_o} - kt \tag{20}$$

$$2.3 \log \frac{Q}{Q_o} = -kt \tag{21}$$

or

$$Q = Q_o e^{-kt} \tag{22}$$

It will be noted that this exponential expression of removal is exactly analogous to the equations for absorbtion of gamma rays (Chapter 14) and radioactive decay. These equations provide the basis on which the treatment of a number of practical experiment situations can be developed.

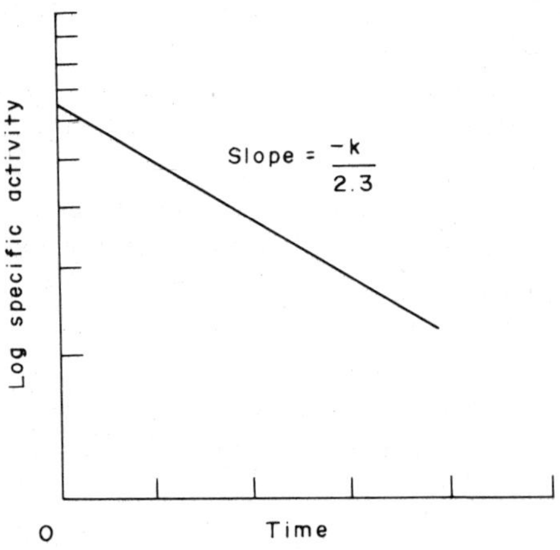

FIG. 6.2 Exponential removal from a single phase; log specific activity versus time

Removal from a Single Phase

If there is a single injection of a radiotracer into an open compartment—we define a compartment as any part of a system having the same specific activity at any specified time—then following equations (21) and (22) we are able to determine simple removal from the "compartment" or phase. After initial mixing of the radio-tracer a semi-logarithmic plot of specific activity versus time would give a line as in Fig. 6.2. As the first order rate constant, k, can be calculated from the slope of the line, then the removal rate of the substance will be kQ, where Q is the quantity in grams of the unlabelled tracee substance. Q is determined by straightforward isotope dilution principle, i.e.

$$Q = \frac{A^*}{S^*},$$ where A^* is the total activity present

in the mixture and S^* is the specific activity.

The half-value time, the constant $t_{1/2}$, is the time required for the removal of half the substance present at any specified time. $t_{1/2}$ may be derived from equation (21).

$$t_{1/2} = \frac{-2.3 \log \frac{1}{2}}{k} = \frac{0.693}{k} \tag{23}$$

Turnover-removal from a Single Phase with Renewal

A state of simple turnover exists where there are equal rates of formation and degradation of the substance under study, resulting in a zero concentration change. This is a familiar situation in the synthesis and utilization of many biological intermediates, and we may wish to determine the turnover rate constant k, or the turnover time, t_t, which is defined as the time necessary to completely renew the substance.

If at zero time a radioisotope label is introduced into a compartment containing the test substance, although the concentration of the test substance will remain the same, the radiotracer will disappear at an exponential rate because it is not being renewed. Therefore, we have to determine the constant rate of renewal (turnover) of the test substance by measuring the disappearance of radioactivity in relation to time.

If dA^* = change in radioactivity of radioisotope A in unit time, dt, this will be given by

$$\frac{-dA^*}{dt} = kA^* \tag{24}$$

and by integration $$A^* = A^*_o e^{-kt} \tag{25}$$

The turnover rate constant k can then be determined by plotting A^*, expressed as nett count rate or activity, as a function of time on semi-log paper. Alternatively, instead of determining total activity A^* it is possible to determine the specific activity S^* on test samples, because the following equation also holds true:

$$S^* = S^*_o e^{-kt} \text{ (e.g. c.p.m. per millimole)} \tag{26}$$

The turnover time, t_t, will then be

$$t_t = \frac{1}{k} \tag{27}$$

Now $t_{1/2}$, the time required for the substance to reach half the zero time value, is, analogous to equation (23):

$$t_{1/2} = \frac{0.693}{k} \tag{28}$$

therefore $t_t = 1.44\, t_{1/2}$ (29)

Product-precursor Relationship: Transfer between Phases

The previous section showed how simple turnover can be estimated by measuring the disappearance of a radiotracer, but it is also possible to determine turnover rate by measuring the appearance of the radioisotope label in the product, providing the time/activity relationships of the immediate precursor are also known.

Thus in biochemistry and physiology, knowledge of the turnover rate and precursors of a metabolite already in a system can be obtained by determining the rate of transfer of an isotopically labelled precursor into one or more products in which the activity is measured. Of course, in many biological systems it is necessary to take into account that simultaneous with the formation of a product there is often a competing process of breakdown and transfer. Therefore, while in a precursor-product system the specific activity of the product substance will in general increase, there will also be simultaneous loss of the isotope label by breakdown.

The importance of precursor product studies resulted in very early development ([12]) of suitable mathematical and graphical treatment, as soon as radioisotopes started to be generally available. Graphic estimation of turnover rate is shown in Fig. 6.3. This is done by estimating the cross-hatched area in the figure and dividing it by S_i^*, the increase in the specific activity of the product from time t_1 to t_2

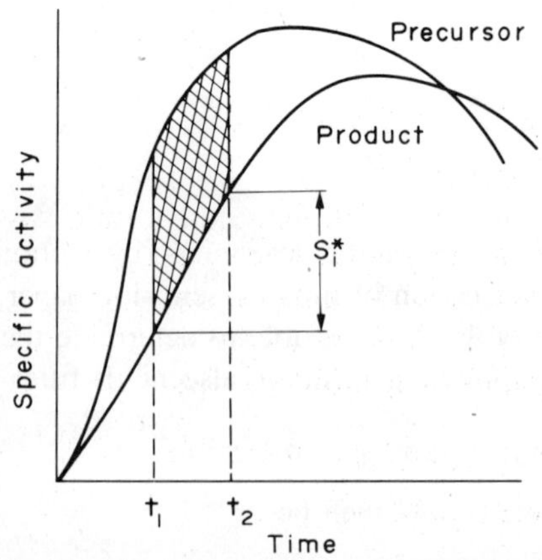

FIG. 6.3 Product and precursor relationships; specific activity versus time ([12]).

$$\text{thus,} \quad t_t = \frac{\text{cross-hatched area}}{S_i^*} \tag{30}$$

This determination makes the assumption that there is only a single precursor; the rate of formation is equal to the rate of disappearance; and the rates are constant.

It will be apparent that if a radioisotope tracer is introduced into a system so that a potential precursor becomes labelled, then the precursor specific activity will exceed the product specific activity until such time as the latter achieves its maximum, when the two specific activities will be the same. Subsequently the product specific activity will exceed that of the precursor (Fig. 6.3). If the system does not conform to this pattern, then the potential precursor studied cannot be the immediate precursor of the product, and another metabolite must be sought.

Simple Accumulation

If a quantity of substance Q is being taken from one compartment or phase and accumulated exponentially by another, then the amount in the second compartment will be $Q_o - Q$. Following equation (22),

$$Q = Q_o e^{-kt}, \text{ so that}$$

$$Q_o - Q = Q_o - Q_o e^{-kt} = Q_o(1 - e^{-kt}) \tag{31}$$

Or if only part of the substance in the first compartment is transferred to the second:

if Q_2 = quantity of substance moved to compartment 2 in time t
Q_{equ} = quantity of substance in compartment 2 at equilibrium

$$\text{then} \quad Q_2 = Q_{equ}(1 - e^{-kt}) \tag{32}$$

or analogous to equation (21)

$$2.\ 3 \log\left(1 - \frac{Q_2}{Q_{equ}}\right) = -kt \tag{33}$$

So for graphical representation, a straight line results from a semi-log plot of $\left(1 - \frac{Q_2}{Q_{equ}}\right)$ versus t. In this case, k, the fractional rate of uptake, is estimated from the slope of the line. The procedure may be used to assess the dependence or independence of uptake mechanisms in relation to experimental variables, because if k is independent then so will be the mechanism.

Exchange

Here we have no nett transfer of test substance, but a steady-state system where the substance is exchanged between two phases within a closed system. We may visualize the ions of the test substance moving randomly from phase 1 to phase 2 and similarly from phase 2 to phase 1.

If a radioisotope tracer is introduced into phase 1 then there will be an increase in specific activity of phase 2 and a decrease in specific activity of phase 1 until equilibrium is achieved. Then the rate of exchange or transport can be determined by following the relative specific activities of the two phases.

Let,

C_1 and C_2 = concentration of test ions in phase 1 and phase 2, respectively
$C_1^*$and C_2^*= concentration of radioisotope labelled test ions in phase 1 and phase 2 respectively, at time t
ρ = rate at which test substance is exchanged

$S_1^* = \frac{C_1^*}{C_1}$ = specific activity of test ions in phase 1, at time t

$S_2^* = \frac{C_2^*}{C_2}$ = specific activity of test ions in phase 2, at time t

and the concentration of labelled test ion in phase 1 at zero time will be

$C_o^* = C_1^* + C_2^*$

Then the rate of movement of the labelled ions from phase 1 will be $-\rho\frac{C_1^*}{C_1}$, while the rate of movement of other ions *into* phase 1 will be $\rho\ \frac{C_2^*}{A_2}$, so that the nett movement of the labelled ions will be

$$\frac{dC_1^*}{dt} = -\rho\frac{C_1^*}{C_1} + \rho\frac{C_2^*}{C_2} \tag{34}$$

or expressed as specific activity

$$\frac{dS_1^*}{dt} = \frac{-\rho t}{C_1}(S_2^* - S_1^*) \tag{35}$$

now $S_o^* = \frac{C_o^*}{C_1}$, so substituting S_2^* in terms of S_1^* and integrating gives us

$$\frac{S_1^*}{S_o^*} = \frac{C_1 + C_2e^{-pt}\,[(1/C_1) + (1/C_2)]}{C_1 + C_2} \tag{36}$$

Then if $t_{1/2}$ is the time required for half the ions exchanged, then the relationship between ρ and $t_{1/2}$ is given by

$$\rho = \frac{0.693\, C_1 C_2}{t_{1/2}(C_1 + C_2)} \tag{37}$$

In practice, a semi-log plot of "specific activity in phase 1 less the equilibrium specific activity" i.e. $(S_1^* - S_{equ}^*)$ versus time gives a straight line, and the slope can then be calculated.

Now $$2.3 \log(S_1^* - S_{equ}^*) = -(\text{slope})t \tag{38}$$

so $$t_{1/2} = \frac{0.693}{(\text{slope})} \tag{39}$$

The exchange rate ρ may then be calculated from equation (37).

RADIOISOTOPES

Choice of Isotope

The characteristics of a number of the principle tracer isotopes used in biological work are given in Table 6.1, including both radio- and stable "heavy" isotopes. The choice of a tracer isotope depends on a number of factors: the actual availability of any tracer isotope; any radioisotope must have a sufficiently long half-life for the experiment contemplated; the commercial (or economic) availability of sufficient of the tracer isotope; the cost of the isotope relative to the experimental budget, the amount required and the importance of the experiment.

The half-life of a radioisotope tracer is often a critical factor in determining whether an experiment can be carried out. As a radioisotope decays to an insignificant level of radioactivity after a period of ten half-lives, a period of six half-lives is normally considered to be the maximum time that can be allowed to elapse before sample counting takes place. A shorter period than this is of course desirable. In practice it is therefore impossible to carry out an experiment which is likely to last say ten days, if the only radioisotope available has for example a half-life of 12 hours (e.g. ^{42}K) nor to carry out an experiment for six months with a radioisotope of $t_{1/2}$ = 14 days (e.g. ^{32}P). It is however quite possible to carry out field experiments with ^{32}P providing that the last harvest is taken at 60–70 days.

Suitable experiments are sometimes practical with short lived isotopes such as ^{11}C ($t_{1/2}$ = 20.5 minutes) providing the experiment can be carried out very near to the production facility, i.e. cyclotron or reactor. Short half-lives can of course be a positive advantage in some cases e.g. in medical or veterinary work, or when it is desired to carry out consecutive experiments on the same system, whether plant or animal.

Isotopes which may be suitable in theory are not always available, either due to preparation in very small quantities e.g. ^{26}Al, or due to relatively infrequent preparation, or preparation to order in the case of isotopes with very short half-life e.g. ^{28}Mg. The use of such isotopes implies a considerable degree of forward-planning of experimental work.

The cost of the isotope may have to be taken into account in the planning of the experiment(s), if the budget is restricted or the amount of isotope required is large. Careful planning may often substantially reduce the amount necessary. In field experiments utilizing ^{32}P for instance it is usual to apply the labelled fertilizer only to small sub-plots, larger sub-plots being taken for yield and other observations. In this way the cost can be kept down to quite a reasonable amount.

Sometimes there are quite big differences in the cost of certain radioisotope preparations, according to specific activity, degree of chemical separation and purity and the chemical form, and these factors must be taken into account for any particular experimental purpose. In general, usually a product of high specific activity is preferred because it can often be regarded as the means of adding just a radioactive tag to a treatment solution, leaving the chemical concentration of the solution unchanged. With low specific activity products it will usually be necessary to take into account changes in the chemical concentration of the test solution, especially if it be dilute. The chemical form of the labelled product may often be important and although a wide range of ready-prepared radioactive tracers are now available, it may be necessary for the user to reconstitute a standard product in the chemical form he requires.

Principal Tracer Isotopes

A very large number of isotopes have been used for tracer experiments at one time or another but in practice not more than about fifty are in common use. A list of these is given in Table 6.1, together with major characteristics.

An extremely large number of ^{14}C and ^{3}H labelled biochemicals are now available, some also with ^{15}N. Many herbicides and pesticides have also been prepared. Increasing complexity of labelling of both radio- and stable isotopes has been a growing feature over recent years. For example the herbicide 2,4,5T has been synthesized with ^{13}C at 10 positions and ^{18}O at three positions (Los Alamos).

Some of the principal suppliers of radioactive tracer products are: The Radiochemical Centre, Amersham, England; Oak Ridge National Laboratory, Union Carbide Nuclear Company, Isotope Sales Department, P.O. Box X, Oak Ridge, Tenn., U.S.A.; Commissariat à l'Energie Atomique, Gif-sur-Yvette (Seine-et-Oise), France; Belgonucléaire, Centre d'Etude de l'Energie Nucléaire, Mol, Belgium; Japan Atomic Energy Research Institute, 1-1 Tamura-cho, Shiba, Minato-ku, Tokyo, Japan; Atomic Energy Research Establishment, Trombay, Bombay, India; New England Nuclear Corporation, 575 Albany Street, Boston 18, Mass., U.S.A.; Bio-Rad Laboratories, 32nd and

TABLE 6.1

Principal Tracer Isotopes for Plant Biology and Soil Studies (Stable isotopes marked with *. Most commonly used isotope is underlined. e, indicates decay by electron capture, see page 18)

Most abundant Isotope	Tracer isotope(s)	Radiation (Mev) Beta	Gamma	Half life	Notes
Aluminium-27	^{26}Al	3.21, 1.16	1.83, 1.12	7.4×10^5 yr	Very restricted availability
Arsenic-75	^{73}As		0.054, 0.14	76 days	
	^{74}As	0.9, 1.36	0.06, 0.64-2.53	17.5 days	
	^{76}As	2.97, 2.41	0.56, 1.21, 0.66	26.8 hr	
Barium-138	^{131}Ba		e, 0.5, 0.122, 0.216	11.6 days	
	^{133}Ba		e, 0.082, 0.36, 0.30, 0.8	7.5 yr	
	^{140}Ba	1.02, 0.48	0.54, 0.16	12.8 days	
Boron-11	^{10}B*			stable	
Bromine-79	^{82}Br	0.44	0.55-1.47	35.7 hr	
Calcium-40	^{45}Ca	0.254		152 days	Weak beta-emitter, count by G-M or gas flow
Carbon-12	^{14}C	0.156		5720 yr	Count by liquid scintillation or by gas flow
	^{11}C	0.96		20.5 min	Cyclotron produced and experiment must be near because of short $t\frac{1}{2}$
	^{13}C*			stable	
Cesium-133	^{137}Cs	0.52, 1.18	(0.662)	30 yr	
Chlorine-35	^{36}Cl	0.71		3×10^5 yr	
	^{38}Cl	4.8, 1.1, 2.8	2.1, 1.6	37.3 min	Not often used as tracer, but arises as neutron activation product of ^{37}Cl
Chromium-52	^{51}Cr		e, 0.32	27.8 days	
Cobalt-59	^{60}Co	0.31	1.33, 1.17	5.27 yr	
	^{58}Co	0.48	0.81, 1.64	71 days	
	^{57}Co		0.122, 0.014, 0.137	267 days	
Copper-63	^{64}Cu	0.57, 0.66		12.9 hr	
Hydrogen-1	^{3}H	0.0181		12.26 yr	Extremely weak beta-emitter. Liquid scintillation counting essential
	^{2}H*			stable	
Iodine-127	^{131}I	0.61, 0.25, 0.81	0.36, 0.08-0.72	8.05 days	
Iron-56	^{55}Fe		e	2.7 yr	
	^{59}Fe	0.46, 0.27, 1.56	1.10, 1.29, 0.19	45 days	
Magnesium-24	^{28}Mg	0.45(2.87)	0.032, 1.35, 0.95	21.3 hr	Usually produced to order
		1.75, 1.59	0.83, 1.01	9.5 min	Arises in neutron activation analysis of ^{26}Mg

Most abundant isotope	Tracer isotope(s)	Radiation (Mev) Beta	Gamma	Half life	Notes
Manganese-55	^{52}Mn	0.6	e, 1.46, 0.94, 0.73, 0.84	5.7 days	
	^{54}Mn		e, 0.84	314 days	
Mercury-203	^{203}Hg		0.3	47 days	
Molybdenum- (96 etc.)	^{99}Mo	1.23, 0.45	0.74, 0.041	66 hr	
Nickel-58	^{63}Ni	0.067		92 yr	Weak beta: concentrate Ni with dimethylglyoxime
Nitrogen-14	^{15}N*			stable	0.366% natural abundance. Determine by mass or optical spectrometry
	^{13}N	1.19		10 min	
Oxygen-16	^{18}O*			stable	Determine by mass spectrometry
Phosphorus-31	^{32}P	171		14.3 days	Hard beta emitter; count by G-M or Cerenkov effect
	^{33}P	0.25		25 days	easily counted,, by G-M or scintillation
Potassium-39	^{42}K	3.55, 2.0	1.52	12.4 hr	
	^{40}K	1.32	1.46	1.3×10^{9}yr	Natural radioisotope; abundance 0.0118%
	^{41}K			stable	6.88% natural abundance
Rubidium-85	^{86}Rb	1.77, 0.7	1.08	18.7 days	Strong beta and gamma. Count by scintillation or G-M
Selenium-80	^{75}Se		e, 0.265, 0.136, 0.280, 0.024-0.58	120 days	
Sodium-23	^{22}Na	0.54	1.28	2.58 yr	Scintillation or G-M counting
	^{24}Na	1.39	2.75, 1.37	15 hr	
Strontium-88	^{89}Sr	1.47		50.4 days	G-M counting
	^{90}Sr	0.54(2.27)		28 yr	Strong beta; count by G-M; long $t_{1/2}$ demands care; allow Yttrium-90 daughter to decay before counting ($t_{1/2}$ = 64.2 hr)
Sulphur-32	^{35}S	0.168		86.7 days	
Zinc-64	^{65}Zn	0.33	e, 1.11	245 days	

Griffin Avenue, Richmond, Calif., U.S.A. (especially ^{14}C and ^{3}H compounds), Junta de Energia Nuclear, Dirección de Quimica y Isotopos, Madrid-3, Spain.

Parent-daughter Nuclides

There are certain radioisotopes that decay to give other nuclide species with different physical e.g. $t_{1/2}$, or chemical properties. The situation most likely to be met in biological work is that of ^{90}Sr ($t_{1/2}$ = 28 yr) and its daughter ^{90}Yr ($t_{1/2}$ = 64.2 hr), as indicated in Table 6.1. This is an example of a parent-daughter relationship in which the daughter nuclide is much shorter-lived than the parent. Both parent and daughter decay simultaneously and after an elapse of time equivalent to 6–10 daughter half-lives an equilibrium, known as *secular equilibrium*, will be reached between the two nuclides at which point the daughter nuclide decays at the same rate as it is produced. In practical terms this means that with tracer work with ^{90}Sr it is usual to delay counting samples for about 3 weeks or so until equilibrium of ^{90}Yr is attained. The strong beta energy of ^{90}Yr is usually then counted, this being directly proportional to the ^{90}Sr content. Growth and decay curves of parent-daughter couples can be treated mathematically ([18]).

Radionuclide Purity

In the early days of radioisotope work the radionuclide purity of the tracer to be used was of some considerable concern. With the increasing sophistication of radioisotope production the likelihood of receiving a sample of a radioisotope from a major commercial source which is not the specified isotope or which is contaminated with another radionuclide is now very small. Nevertheless, users should still take care to check consignments, both for radioisotope and radiochemical purity, if there is any suspicion of anomalous results.

In isotope production, as in activation for analytical purposes (Chapter 8), there may be nuclear reactions other than the principal one desired. As far as possible this is overcome by irradiating target material of the greatest possible purity, and by using the metal or oxide. As the efficiency of production of any radionuclide is dependent on the radiation cross section, the production of a radionuclide contaminant may greatly exceed the proportion of the original impurity. Now classical examples of this are 0.01% calcium phosphate in target $CaCO_3$ resulting in 5% ^{32}P contamination of ^{45}Ca. As ^{45}Ca is such a weak β-emitter with difficult counting characteristics, the presence of ^{32}P would present serious counting problems. Similarly, 1% sodium impurity in a potassium target will give 13% contamination after irradiation ([17]).

For similar reasons the use of unprocessed irradiation units should be avoided whenever possible, which is usually the case nowadays. This is because activities of the target material can give different chemical radioactive species of the main nuclides which would require separation. Thus ^{33}P as a contaminant of ^{32}P would cause errors in counting due to its longer half life. Massive phosphite contamination arises from reactor activation of orthophosphate due to neutron disruption. This makes it impossible to use this form of phosphate for fertilizer studies (Chapter 12) ([20]).

Identification of radionuclides and possible contaminants may be carried out by determination of half-life combined with beta or gamma spectrometry as appropriate. The determination of possible chemical contamination may present more practical difficulties, involving standard semi-micro techniques with appropriate modifications for handling radioactive materials. Ascending or descending paper chromatography may represent the easiest approach, especially for ^{14}C labelled biochemicals.

It should be emphasized that nowadays the need for such checking by most radioisotope users is a rare exception rather than the rule. Nevertheless, there may also be occasions, particularly in metabolic studies, when it is essential to know the extent to which the activity may be distributed between different chemical forms.

POTENTIAL TRACER DIFFICULTIES

Potential Radiation Effects

As was noted in Chapter 3, all radiation is harmful to biological organisms. It is not therefore surprising that when radioisotope tracers first came into use there was some concern that the radiation effect might so alter the normal behaviour or physiological status of the plant or other organism as to make invalid the conclusions drawn from tracer experiments.

It is now established that when radioisotopes are used at the low levels of radioactivity quite sufficient to achieve effective tracing, then there is no measurable effect on the organism or its normal metabolism. The length of the experiment is one factor to be considered and potential radiation effects have to be considered especially carefully in long term experiments with animals, especially as selective accumulation of certain ions takes place.

In the case of plants about 2 μCi ^{32}P/litre of nutrient solution or its equivalent seems to be generally accepted as an approximate maximum level for any but very short experiments, such as ion uptake studies.

Work on the radiation effects of radioisotope tracers in plants is now mostly historical. Blume *et al.* ([14,15]) showed that radiation effects on barley plants, as shown by histological changes was due primarily to the accumulation of ^{32}P in the cells rather than injury to the root by external radiation. The most important factor was the specific activity of the ^{32}P in the solution rather than the actual level of ^{32}P i.e. with high specific activity the plant absorbed more ^{32}P.

Earlier it was reported ([28]) that when barley was grown in nutrient solution containing 10 μCi ^{32}P/litre there was a reduction in the weight of roots, and at even higher levels (50 μCi/l) the content of phosphate in the roots was affected. However, these levels of activity are far greater than needed for effective tracing. It took levels of as much as 100 and 900 μCi/l of ^{32}P and ^{14}C respectively to produce chromosome changes in Tradescantia ([21]).

Various experiments relevant to the use of ^{32}P as a phosphorus tracer for fertilizer experiments established that when used at reasonable specific activities the uptake of phosphorus and general growth were not significantly affected ([19,29,13,15]). One treatment applying as much as 12.5 mCi ^{32}P/g P showed no significant effects ([15]). This is far above the level suggested in Chapter 12 for using labelled superphosphate in field experiments.

In general therefore, radiation effects from radiotracers used with plants can be discounted if intelligent choice is made of the lowest appropriate specific activity. If there should be any doubt as to a possible radiation effect then obviously an experiment should be carried out with either two levels of radioisotope or alternatively with one experiment using only stable isotope.

A particular case of chemical effect of radiation occurs with ^{14}C labelled compounds as they show a greater rate of decomposition during storage than do ^{12}C compounds. This is due to the rupture of chemical bonds by the radiation emitted.

The problem arises mainly with high specific activity products, which may therefore require re-purification. There are three possible solutions: permanent dilution with the stable form of the compound, which may not be desired; add small glass beads to the solution to absorb a major part of the radiant energy; dilute the radioactive substance with an appropriate solvent which may be evaporated to restore the initial activity when required for use.

Isotope Effects

All tracer studies, whether carried out with stable or radioactive isotopes, depend on the assumption that the tracer behaves chemically and physically as does the much larger tracee counterpart. If there is a difference in atomic weight, i.e. in the masses of the respective atoms of isotopes of the same element then they will have a comparable difference in mobility and reactivity. This is particularly relevant in the case of stable isotopes where methods of detection are in fact based on the isotope effect.

Due to its relatively early use as a tracer deuterium, ^{2}H, has been especially studied as a constituent of "heavy" water D_2O. It is found that although primitive plants like algae can be grown in completely deuteriated water, higher plants even when grown with only 50% D_2O show stunted growth and inhibited reproduction. Plants grown in partially deuteriated water exhibit marked selectivity in favour of the normally abundant ^{1}H isotope ([30]). The effects of ^{3}H, tritium, with a greater mass, are even more marked. Thus it was found that the uptake of $^{3}H_2O$ by algae was only 45% of that for $^{1}H_2O$ ([33]).

The extreme isotope effects found with $^{1}H_2$ (hydrogen), $^{2}H_2$ (deuterium) and $^{3}H_2$ (tritium) are of course because the mass difference is a quite sizeable fraction of the atomic weight. In fact $^{1}H_2$ to $^{3}H_2$ which gives the largest isotope effect, shows a difference of 1:3. In large molecules (various tritiated biochemicals for instance) the effect of mass difference is at its least if the tracer is relatively far from the site of reaction of the molecule (secondary effects) as opposed to when it is present at the site of the reaction (primary effect) ([35,36]).

Although such primary and secondary effects can be recognized in chemical reactions this is almost impossible in biological systems. There the corresponding situation is between $^{3}H_2$ in easily exchangeable positions such as O-H or NH, and slow exchange situations such as C-H.

Similar effects are observed for the isotopes of carbon. Very early work ([24,31]), showed that living organisms discriminated against ^{13}C in favour of ^{12}C even though the natural ratio of these stable isotopes is as great as 1:90. This was confirmed with studies ([16]) over several years showing that plants tend to favour the abundant isotope ^{12}C as opposed to the heavier ^{14}C and that preferential selectivity was also shown in comparison with ^{13}C but the effect was not so marked. In general it now appears that plants can utilize $^{12}CO_2$ ([26,27,32]), approximately 17–20% faster than $^{14}CO_2$. Extensive work has shown that it is possible to distinguish between the C-3 and C-4 photosynthetic types of plant by means of differences in the $^{13}C/^{12}C$ ratios, and this is discussed further in Chapter 7.

In contrast, isotope effects between ^{15}N and ^{14}N have generally been considered to be negligible and most tracer studies assume a discrimination factor of unity. However, there now appears to be significant discrimination in nature ([23]). It may prove possible to use differences in $^{15}N/^{14}N$ ratios between plants and organic matter to study nitrogen cycles, pollution sources and N_2 fixation (see Chapter 7).

Although many variations in the behaviour of stable and radioactive isotopes have been summarized ([22]), in practical terms most isotope effects will go unnoticed in short term biological experiments, as the effect is likely to be no more than 5% and probably less. Little work seems to have been done on most radioisotope tracers used in biology, because the mass differences are relatively small. The main effects are likely to be seen with ^{14}C, or ^{3}H and this can be counteracted by having the label on a non-reactive part of the molecule, in the case of organic compounds. For studies with deuterium and tritium it is also possible to correct for the isotope effect by determining the starting and final ratios of $^{1}H_2$, $^{2}H_2$ and $^{3}H_2$.

Isotopic Exchange

Isotopic exchange has two aspects: it can be used for tagging compounds as in the Wilzbach method for tritium labelling of organic molecules; for the unwary it can be a serious source of error in the interpretation of tracer experiments.

Exchange is when atoms of an element interchange with two (or more) chemical forms of the same element. It is a more or less continuous process in biological systems and an isotope can provide a "tag" or marker whereby the extent of exchange can be measured in terms of reaction rates and possible molecular rearrangements. Although in tracer experiments with biological systems equilibrium is ultimately achieved between the tracee element and the tracer isotope, the difficulty arises when the experiment does not go on long enough for attainment of equilibrium.

The now classical work of Overstreet and Broyer ([25]) showed this clearly in an uptake experiment with barley using both stable and potassium-42. With plants grown with high levels of potassium it was found that there was apparent uptake of ^{42}K but no nett uptake of total K. Thus there had been merely exchange of the radioisotope-tagged K for stable K already in the plant. Such situations have to be constantly guarded against in ion uptake and similar experiments with plants, and that is why plants are usually grown in low concentrations of the ion being investigated or in dilute $CaSO_4$ solution.

Similarly, in metabolism studies where a radioisotope tracer is being used to study synthesis of a substance it is important that actual synthesis is proven, rather than just the manifestation of an exchange process. As exchange is not energy dependent but synthesis requires respiratory energy, one approach to obtaining confirmation of synthesis is to conduct a parallel experiment using a respiratory inhibitor such as cyanide.

In the Wilzbach ([34]) method of tritiating compounds in a random manner, tritium gas and the compound are sealed up together in an ampoule and are left to stand for a week or two. During this time exchange takes place and some of the ^{1}H atoms of the compound are replaced by tritium. Bond energy and the configuration of the molecule determines which hydrogen atoms are most easily exchanged and in what proportion.

However, the exchange process which results in tritation of a molecule can also result in loss of the label. When tritium and deuterium atoms are attached to inactive carbon atoms in the molecule they tend to be stable. However, when attached to $-NH_2$, -COOH or -OH groups, or active carbon then the label is rapidly exchanged with unlabelled water. In aqueous media where the label became generally distributed the presence of the label could no longer be taken as representative of the active molecule.

Where tritium is part of a tightly bound organic molecule or attached to sulphur groups such as H_2S, SH^-, cysteine, etc. there is much less risk of random exchange.

The general prevalence of liquid scintillation counting for ^{14}C has removed one potential source of exchange error. Formerly when many ^{14}C labelled samples were prepared for counting as $Ba^{14}CO_3$ there was a risk of loss of activity by exchange with CO_2 from the atmosphere.

REFERENCES FOR FURTHER READING

Tracers

1. BERGNER, P. E. and Lushbaugh, C. C. Eds. *Compartments, Pools and Spaces in Medical Physiology*. USAEC/DTI (1967).
2. BLOCK, K. and Anker, H. S. An extension of isotope dilution method. *Science* **107**, 228 (1948).
3. COMAR, C. L. *Radioisotopes in Biology and Agriculture*. McGraw-Hill, New York (1955).
4. FRIED, M. and Dean, L. A. A concept concerning the measurement of available soil nutrients. *Soil Sci.* **73**, 263–71 (1951).
5. HEVESY, G. von and Hofer, G. *Nature* **134**, 879 (1934).
6. HEVESY, G. *Radioactive Indicators, their Application in Biochemistry, Animal Physiology and Pathology*. Interscience Publishers, New York (1948).
7. KAMEN, M. D. *Radioactive Tracers in Biology*. Academic Press Inc., New York (1951).
8. RESIGNO, A. and Segre, G. *Drug and Tracer Kinetics*. Blaisdell Publishing Co. (1966).
9. SHEPPARD, C. W. *Basic Principles of the Tracer Method*. John Wiley & Sons Inc. (1962).
10. SOLOMON, A. K. The kinetics of biological processes; special problems connected with the use of tracers. *Adv. in Biol. Med. Phys.* **3**, 65–97 (1953).
11. WANG, C. H. and Willis, D. C. *Radiotracer Methodology in Biological Science*. Prentice Hall (1965).
12. ZILVERSMIT, D. B., Entenmann, C. and Fishler, M. C. On the calculation of turnover time and turnover rate from experiments involving the use of labelling agents. *J. Gen. Physiol.* **26**, 325–31 (1943).

Nuclide Effects

13. BOULD, C., Nicholas, D. J. D. and Thomas, W. E. E. Radiation effects in plant nutrition experiments with phosphorus-32. *Nature* **167**, 140 (1951).

14. BLUME, J. M., Hagen, C. E. and Mackie, R. W. Radiation injury to plants grown in nutrient solutions containing phosphorus-32. *Soil Sci.* **70**, 415–26 (1950).
15. BLUME, J. M. Radiation effects on plants grown in soil treated with fertilizer containing ^{32}P. *Soil Sci.* **73**, 299–303 (1952).
16. BUCHANAN, D. C., Nakao, A. and Edwards, G. Carbon isotope effects in biological systems. *Science* **117**, 541–45 (1953).
17. COHN, W. E. Radioactive contaminants in tracers. *Anal. Chem.* **20**, 498–503 (1948).
18. COOK, G. B. and Duncan, J. F. *Modern Radiochemical Practice*. Oxford, Clarendon Press (1952).
19. DION, G., Bedford, C. F., St. Arnaud, R. J. and Spinks, J. W. T. Plant injury from ^{32}P. *Nature* **163**, 906–7 (1949).
20. FRIED, M. and McKenzie, A. J. Contamination in orthophosphate irradiated in a neutron pile. *Science* **111**, 472–93 (1950).
21. GILES, N. H. and Bolomey, R. A. Cytogenetical effects of internal radiation from radioisotopes. *In*: Biological Application of Tracer Elements, Cold Spring Harbour Symposia, *Quant. Biol.* **13**, 104–12 (1948).
22. GUILLOT, P. The radioisotope concentration effect in biology: its consequences in the use of radioactive tracers. *In: Isotope Ratios as Pollutant Sources and Behaviour Indicators*. Proc. Symp. IAEA, Vienna (1975).
23. HOERING, T. *Science* **122**, 1233 (1955).
24. NIER, A. O. and Gulbransen, E. A. Variations in the relative abundance of the carbon isotopes. *J. Amer. Chem. Soc.* **61**, 697–98 (1939).
25. OVERSTREET, R. and Broyer, T. C. The nature of absorbtion of radioactive isotopes by living tissues, as illustrated by experiments with barley plants. *Proc. Nat. Acad. Sci. USA*, **26**, 16–24 (1940).
26. RABINOWITZ, J. L. and Chase, G. D. *et al.* Studies on isotope effects with carbonic anhydrase using ^{14}C sodium bicarbonate. *Atompraxis* **6**, 433 (1960).
27. ROPP, G. A. Effect of isotope substitution on organic reaction rates. *Nucleonics* **10**, 22–27 (1952).
28. RUSSELL, R. Scott and Martin, R. P. Use of radioactive phosphorus in plant nutritional studies. *Nature* **163**, 71–72 (1949).
29. RUSSELL, R. Scott, Adams, S. N. and Martin, R. P. Radiation effects due to phosphorus-32 in fertilizer experiments. *Nature* **164**, 993 (1949).
30. THOMSON, J. E. *Biological Effects of Deuterium*. Pergamon Press, Oxford (1963).
31. UREY, H. C. Oxygen isotopes in nature and the laboratory. *Science* **108**, 489 (1948).
32. WEIGL, J. W. *Relation of Photosynthesis to Respiration*. UCRL-590, 1950.
33. WEINBERGER, D. and Porter, J. W. Incorporation of tritium oxide into growing *Chlorella pyrenoidosa* cells. *Science* **117**, 636–38 (1953).
34. WILZBACH, K. E. Tritium labelling by exposure of organic compounds to tritium gas. *J. Amer. Chem. Soc.* **79**, 1013 (1957).
35. YANKWICH, P. E. Isotope effects in chemical reaction. *Ann. Rev. Nuc. Sci.* **3**, 235–48 (1953).
36. MELANDER, L. *Isotopic Effects on Reaction Rates*. Ronald Press (1960).

CHAPTER 7

Stable Isotopes As Tracers: Mainly The Use Of ^{15}N

INTRODUCTION

THERE are four stable or heavy isotopes of potential interest to researchers in soil and plant studies. These are ^{18}O, ^{2}H, ^{13}C and ^{15}N. Of these ^{2}H (deuterium) and ^{18}O are used for rather specialized applications, but ^{13}C having gone through a period when it lost ground to the newly developed and then cheaper ^{14}C is now enjoying renewed interest following its increased production at a much lower price. ^{15}N is however, the heavy isotope of broad application, but ^{13}C, ^{18}O and ^{34}S are readily available ([1,2,3]).

The radioactive isotopes of nitrogen have too short half-lives to be of any general value, thus for ^{13}N it is 10 minutes, for ^{16}N seven seconds, and for ^{17}N four seconds. Therefore, although there is a theoretical possibility of using ^{13}N for very short uptake experiments it is clear that the stable ^{15}N isotope is the most suitable isotope for long term studies. Comparatively recently the use of ^{14}N-depleted nitrogen compounds has opened up a new approach to studies with nitrogen isotopes.

The stable isotope ^{15}N is used as a tracer in exactly the same way as radioactive isotopes are used except that we identify the tracer atoms by mass differences. However, instead of the specific activity of a sample used in the case of radioisotopes, the term % abundance is used for stable isotopes. The $\%^{15}N$ abundance is the ratio of ^{15}N to ^{15}N + ^{14}N atoms, i.e.

$$\%^{15}\text{N abundance} = \frac{\text{atoms } ^{15}\text{N}}{\text{atoms } ^{15}\text{N} + \text{atoms } ^{14}\text{N}} \times 100$$

Since the natural environment has an ^{15}N abundance of 0.37% (0.3663%), the amount of ^{15}N in a sample is conveniently expressed as $\%^{15}N$ atom excess over the natural abundance of 0.37.

A practical consequence of having to identify ^{15}N by means of mass difference and to define its presence as % abundance is that tracer techniques for stable isotopes are much less sensitive than for radioisotopes, the latter being able to trace at concentration as low as 10^{-11} (1 to 100 billion). However, the newer double collector mass spectrometers have shown great improvements in recent years, and a precision of less than $\pm$ 0.05% has been reported for the Micromass Model 602 C ([84]), which is a considerable advance. But assuming that an older single collector mass spectrometer is capable of determinations with a precision of 0.5 relative % on samples of any ^{15}N abundance down to normal abundance and the normal abundance is 0.37%, then the theoretical limiting excess values which can be detected will be roughly 0.002 atom % excess or somewhat less. In other words an initial dose of nitrogen with 20% atom excess ^{15}N can be diluted only 10,000 times if the ^{15}N is to still remain detectable (not quantitatively) in the samples. In practice the analytical error is about 0.01% atom excess ^{15}N, so for quantitative determinations we cannot work with dilutions as great as this and final samples for analysis should contain 0.15% atom excess ^{15}N, or more, for routine analysis with a single collector.

This greatly decreased sensitivity with increasing dilution inevitably affects the design and practicability of soil-plant experiments with ^{15}N. As mentioned in Chapter 12, 1% atom excess ^{15}N is adequate for fertilizer experiments where the crop takes up a substantial portion of the applied fertilizer. However, for soils experiments, such as studies on turnover processes in the nitrogen cycle; the fractionation and movement of applied nitrogen fertilizer; or studies of gaseous losses, then much higher ^{15}N atom excess material must be used to overcome the enormous dilution effect. Such experiments require material with 30–50% ^{15}N atom excess or even more, and materially influence the cost of the work and the size of the experiment that can be conducted.

DETERMINATION OF ^{15}N ABUNDANCE

The determination of ^{15}N abundance is carried out by either a mass spectrometer or an emission spectrometer. In both instances it is necessary to transform the nitrogen in a sample into nitrogen gas.

In a gas sample labelled with ^{15}N there are $^{14}N^{14}N(^{28}N)$, $^{14}N^{15}N(^{29}N)$ and $^{15}N^{15}N(^{30}N)$ molecules.

$$\text{The \% abundance of } ^{15}N = \frac{^{29}N + 2\,^{30}N}{2(^{28}N + {}^{29}N + {}^{30}N)} \times 100 \tag{1}$$

Mass spectrometer and emission spectrometers have recording systems that give peaks which are a measure of the amount of ^{28}N, ^{29}N and ^{30}N in the gas sample. From equation (1), the $\%^{15}N$ abundance can thus be calculated.

In practice it is not necessary in mass spectrometry to measure the ^{30}N peaks since in equilibrium

$$^{14}N^{14}N + {}^{15}N^{15}N \leftrightarrows 2\,^{14}N^{15}N$$

which has an equilibrium constant, K = 4. ^{30}N can therefore be expressed in terms of ^{28}N and ^{29}N. Moreover, particularly when dealing with low ^{15}N enrichment, the measurement of ^{30}N peaks is not always possible due to interference with other substances of the same mass number.

From the equilibrium reaction it may be seen that

$$^{30}N = \frac{(^{29}N)^2}{4\,^{28}N} \tag{2}$$

Substitution and rearrangement of (2) into (1) gives

$$\%^{15}\text{N abundance} = \frac{100}{2\,\frac{^{28}\text{N}}{^{29}\text{N}} + 1} \tag{3}$$

when $\frac{^{28}\text{N}}{^{29}\text{N}} = R$, the final expression for ^{15}N abundance becomes

$$\%^{15}\text{N abundance} = \frac{100}{2R + 1} \tag{4}$$

The ratio R can be easily derived from mass spectrometric or emission spectrometric measurement. Tables giving the R and corresponding $\frac{100}{2R + 1}$ value facilitate the procedure for determining the $\%^{15}$N abundance.

MEASUREMENT BY MASS SPECTROMETRY

When the nitrogen gas is led into the mass spectrometer, the ^{28}N, ^{29}N and ^{30}N molecules are subjected to an electron beam and obtain a charge. In the electromagnetic field of the spectrometer the charged nitrogen molecules are split up into different beams as a result of their different mass number. At the collector end the discharge of the N molecules takes place on insulated electrodes and the electric pulses are amplified and recorded. By measuring the height of the ^{28}N and ^{29}N peaks corrected for a background reading, the R values are determined and the $\%^{15}$N abundance calculated or read from a table, [4–9].

The Mass Spectrometer

Figure 7.1 shows the block diagram of a double collector mass spectrometer. Ionization of the gas sample is done by using a white-hot tungsten filament as a source of electrons which are then passed across the gas flow by an electric field. Quite low electron energies strip the outer negatively charged electrons from the gas molecules, giving positively charged ions. These ions are then directed through a series of positively charged plates which are divided into segments by slits, which make it possible to focus the ion beam by adjusting the charge on each segment. The first plate is known as the repeller plate and carries the highest charge, and each successive plate is at a lower positive potential, so that the ions are accelerated down a potential gradient. Finally the focussed beam enters the analyzer chamber through an entrance slit.

Within the analyzer chamber a strong magnetic field (about 40,000 gauss) imparts differentially curved paths to ions of different mass. By varying the kinetic energy of the ions and/or the strength of the magnetic field it is possible to focus a beam of ions of the desired mass onto a collector, consisting of an insulated plate connected to an electrometer. A low current is produced proportional to the number of ions collected and this is amplified and recorded. With single collector instruments, by

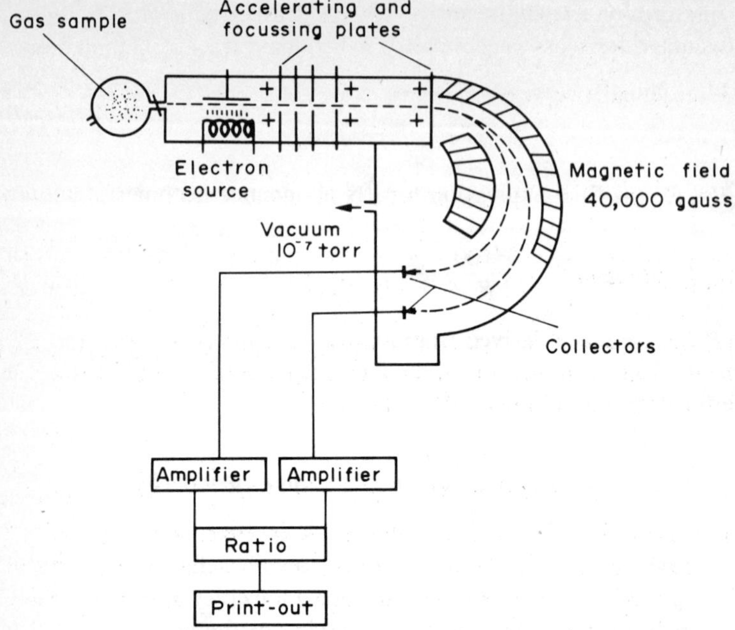

FIG. 7.1 Block diagram of a double collector mass spectrometer.

altering the focus in a regular manner successive ion beams can be caused to fall on the collector, making it possible to record a mass spectrum (Fig. 7.5). Ions of mass 28 are first recorded and then ions of mass 29 etc., the ratio of the two signals then being calculated. With double collector instruments the beam of mass 28 is focussed on one collector and those of mass 29 on the second. The ratio of the number of ions collected on each collector is then directly determined by a potentiometer.

Preparation of Nitrogen Gas from a Sample

Nitrogen gas for isotopic analysis can be prepared from plant samples and soil extracts by several methods.

(i) Classical Kjeldahl digestion—distillation into acid—total nitrogen determination by titration—aliquot taken for transformation into N_2 gas by the *Rittenberg Method*.

(ii) Basic procedure as in (i) above, but aliquot taken for transformation into N_2 gas by the *Dumas Method*.

(iii) Production of N_2 gas and determination of $^{15}N\,^{14}N$ ratio by the *Direct Dumas Method*. This method is suitable if it is desired to know only the $^{15}N^{14}N$ ratio. If total N is required a separate estimation would be necessary, using the Kjeldahl or other method.

The Kjeldahl digestion may follow common practice and use a potassium sulphate digestion mixture, or else by means of a hydrogen peroxide/H_2SO_4 digest. The latter has the advantage for mass spectrometry in reducing trace contamination.

Basic Procedure

(*Digestion mixture*: 50g selenium powder are ground with 250 g copper sulphate, $CuSO_4.5H_2O$, then added to 500 g K_2SO_4 with further mixing). About 4 g mixture are used with 12 ml conc. H_2SO_4 for 0.5 g plant material. Start digestion on low heat, when frothing ceases raise heat gradually as the sample becomes charred. Bring to boil, continue until solution becomes clear, then continue for a further 45 minutes to ensure all organic matter is destroyed. Cool and dilute before the salts crystallize.

Peroxide-Sulphuric Acid Procedure

12 ml conc. H_2SO_4 are added to 0.5 ml plant material and digested until the contents are a homogenous black slurry. Remove the flask from the heater, cool adequately and add 1 ml 30% H_2O_2 drop by drop. Return to heater and continue heating until the liquid becomes black again. Remove, cool, add more peroxide, return to heat and continue this process until the solution remains clear.

H_2O_2/H_2SO_4 Procedure with Technicon Block Digestors

This is a very convenient method for plant material where there are large numbers of samples. Forty samples can be handled simultaneously in the block digestor, incorporating a H_2O_2/H_2SO_4 digestion. Aliquots of the diluted digest may then be taken for total N determination colorimetrically using the Technicon automatic sampling and flow system, another aliquot being taken for $^{15}N^{14}N$ ratio analysis. (*Digestion mixture*: mix 350 ml of 30% H_2O_2, 0.42 g selenium powder and 14 g $Li_2SO_4.H_2O$ in a boiling flask. 420 ml of H_2SO_4 are added carefully with swirling and cooling.) Store at 1°C (Parkinson and Alan method adapted to Technicon system by Jorgensen) ([16,17]). 7 ml of digestion mixture are added to 0.25–0.5 g plant material in 75 ml Technicon digestion tubes. After heating at low temperature until all frothing ceases the temperature is raised to 350°C and heating continued for 45 minutes after the solutions become clear.

Soil samples for total-N or KCl-extracts for exchangeable ammonium may be treated by classical Kjeldahl procedure in similar manner using K_2SO_4 digestion mixture.

Where nitrate is thought to be high in soil or plant samples they should be subjected to a preliminary treatment with salicylic-sulphuric acid mixture (50 g salicylic acid are dissolved in 1 litre of conc. H_2SO_4) (Ashton) ([20]). For 5 g air-dry soil or 0.5 g plant material in a Kjeldahl flask, add 12 ml salicylic/sulphuric mixture, proportionately more for larger soil samples, swirl and allow to stand for 30 minutes. Add 2.5 g sodium thiosulphate and 10–25 ml water. Heat gently until frothing ceases, cool, add K_2SO_4 digestion mixture and proceed as above.

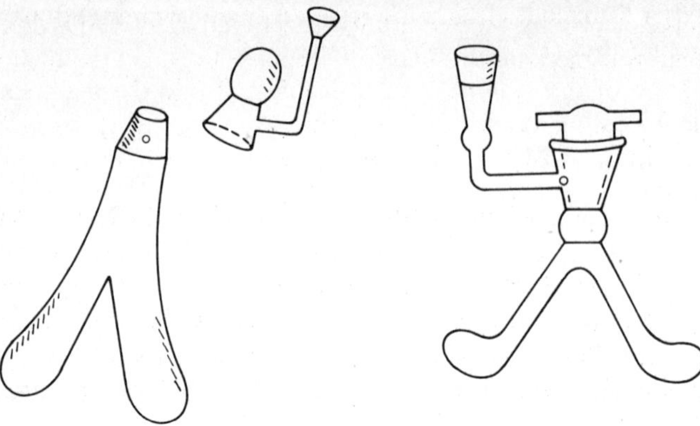

FIG. 7.2 Two patterns of Rittenberg flask with ground glass joints, and either cap or stopper to permit closing off and disconnection from the vacuum line, between N_2 conversion process and mass spectrometry.

After the digests from Kjeldahl procedures are cool, 30 ml of 40% NaOH is added and the NH_3 distilled over into 15 ml of 0.1 N H_2SO_4 and titrated with 0.1 N NaOH using methyl red indicator, to determine total N. The back titrated solution is acidified and an aliquot of this (or from the diluted digest of the Technicon procedure) is evaporated to a small volume of a few ml. An aliquot of this extract containing approximately 1 mg total N is taken for transformation into nitrogen gas, either by Rittenberg or Dumas methos.

(i) Rittenberg Method ([18])

Ammonium compounds are converted into di-nitrogen gas by reaction with alkaline sodium hypobromite in a special reaction vessel. The following reaction takes place:

$$2NH_4CL + 3NaBrO + 2NaOH \longrightarrow N_2 + 5H_2O + 3NaBr + 2NaCl$$

Before letting the N_2 gas enter the spectrometer the Rittenberg flask is frozen with liquid nitrogen to prevent H_2O and traces of other gases (CO, CO_2) from entering the inlet system.

Basic Procedure

A sample aliquot of about 2 ml containing 1 mg of nitrogen is pipetted into one side of a two-arm Rittenberg flask (Fig. 7.2) and 4 ml of NaBrO are pipetted into the other arm. The flask is attached to a vacuum line comprising a rotary vacuum pump plus a water cooled oil diffusion pump (10^{-3} – 10^{-4} Torr) and fitted with a liquid nitrogen moisture cold trap to protect the pump. Low vacuum is applied carefully and when bubbling ceases apply high vacuum and continue for 5–10 minutes after a vacuum of 10^{-3} Torr is indicated. The stop-cock is then closed and the flask is rotated on the ground glass joint in order to mix the sample and reagent. Vigorous formation of N_2 gas takes place, and when the reaction subsides equalize the volume of solution in each arm. After 5 minutes, freeze the reaction vessel with liquid nitrogen to freeze out water. Figure 7.3 shows a typical preparation line.

After evacuation of the system, the N_2 gas is lead into the mass spectrometer via an inlet line incorporating a capillary U-tube moisture trap frozen with liquid nitrogen, so that the gas sample has to pass through it before reaching the ion-source of the mass spectrometer.

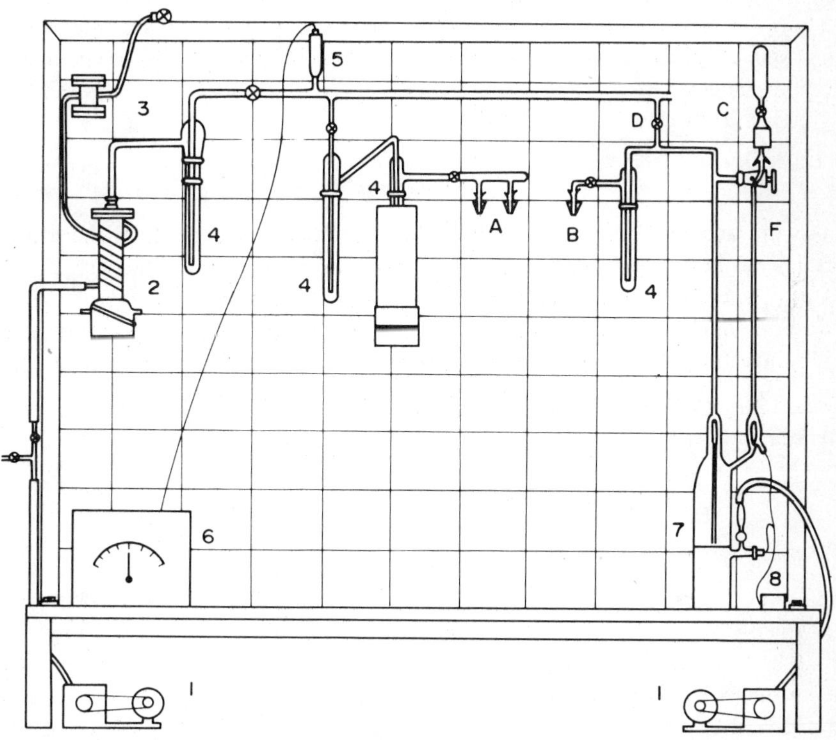

FIG. 7.3 Typical high vacuum line for preparation of N_2 gas by the Rittenberg procedure. (1) mechanical pump (2) diffusion pump (3) safety relay for cooling water (4) liquid nitrogen cooling trap (5, 6) vacuum determination and control (7) Toepler pump for transferring the gas (8) Toepler pump control, (A) connections for Rittenberg tubes (Trivelin *et al.*) [21].

(ii) Dumas Method [14,19]

In this classical method nitrogen is released by means of copper oxide, CuO. CaO is included in the reaction mixture to absorb the H_2O and CO_2 that is also produced, e.g.

$$2NH_4Cl + 4CuO \longrightarrow N_2 + 4H_2O + 2CuCl + 2\ Cu\ (CO_2)$$

Basic Procedure

An aliquot containing 1 mg N as NH_4 is evaporated to dryness in a small glass tube and about 1.5 g CuO + 1.5 g CaO are added. The tube is evacuated on a vacuum line (10^{-4} Torr) and sealed off with a flame. The tube is then heated in a small muffle furnace for about 3 hours at a temperature of 600°C, resulting in transformation of nitrogen into N_2 gas. CuO will oxidize any NH_4 to N_2 whereas the Cu which is liberated during heating of CuO, will reduce any oxidized form of nitrogen (NO, NO_2) to N_2 gas. CaO absorbs H_2O and CO_2 which would interfere in the determination. The tube is broken by means of a crushing device connected to the inlet system of the mass spectrometer and the nitrogen gas is led into the spectrometer for measurement of ^{15}N abundance.

Precautions

Glassware adsorbs ^{15}N, thus resulting in "memory" effects. It is therefore very important that glassware should be thoroughly cleaned between analyses, preferably by steam. The copper oxide used must be heated to 700°C to remove absorbed nitrogen and stored in an evacuated dessicator. Similarly H_2O and CO_2 is absorbed by the CaO and this must be prepared for use by heating to 850–900°C to drive off water from $Ca(OH)_2$ and CO_2 from $CaCo_3$ and stored in a dessicator over silica gel.

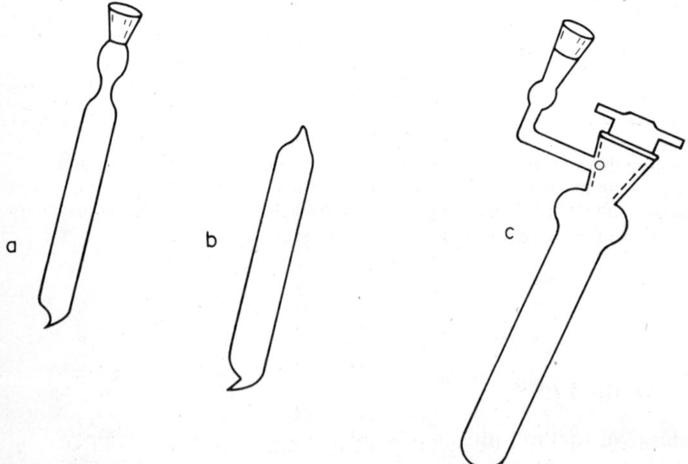

FIG. 7.4 Glassware for Dumas method. (a) combustion tube (b) tube after sealing, note the easily-broken tip (c) expansion tube in which the combustion tube is placed for connection to the inlet system of the mass spectrometer.

Calculation of % ^{15}N Excess

The principles of the calculation of ^{15}N abundance have already been discussed at the beginning of this section. The height of the peaks is measured in mm from the base line to the tip of the peak as shown in Fig. 7.5. It may then be necessary to multiply the determined height by an amplification factor. The same procedure is carried out for the background, which will normally be constant throughout a day's working, and is frequently so low it can be ignored. The background is deducted from the corresponding peak heights. Normally samples are determined in duplicate, so the results will be calculated as the average of two sets of peaks, with a single collector spectrometer.

The R value is calculated as the $\frac{28}{29}$ peak ratio. %^{15}N abundance is then calculated from $\frac{100}{2R + 1}$ or is determined from published tables. As the natural environment has a mean ^{15}N abundance of 0.37%, then to calculate %^{15}N atom excess it is necessary to deduct 0.37 from %^{15}N abundance as determined. Due to isotopic discrimination, the

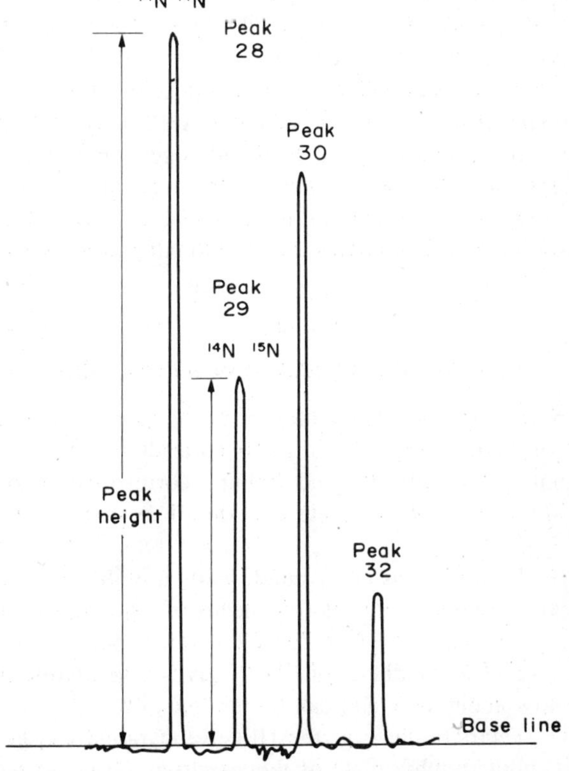

FIG. 7.5 Part of mass spectrometer record chart showing the various N peaks, especially 28 and 29, used for determination. 32 is a contaminant peak due to oxygen.

^{15}N abundance in plant material may not be the same as the atmosphere, therefore in precise work it should be determined on unfertilized control plants. This is not normally necessary for fertilizer experiments.

Comparative Advantages of Rittenberg and Dumas Methods

The Rittenberg procedure although effective involves a long, and rather tedious, conversion process of NH_4 compounds to N_2. Furthermore, although the Rittenberg flask is frozen with liquid air not all traces of CO_2 are frozen out and this can interfere in both mass and optical spectrums.

The Dumas procedure when well organised can be somewhat faster, but in the case of the "Direct" Dumas technique the adsorption of atmospheric nitrogen both on and within the sample will effectively decrease the enrichment. This will be especially serious in the emission spectrometric determination (discussed below) on very small samples. Similarly all traces of CO_2 may not be removed and may interfere. In the direct-combustion spectrometric technique there is an obvious problem in obtaining a representative sample small enough for photospectrometry, especially in samples from whole plants for example. Also in the emission spectrometer procedure, where combustion takes place within the discharge tube, it may be difficult to obtain complete combustion of plant material samples.

In the case of mass spectrometer determinations, as it is usually necessary to determine the amount of total nitrogen present as well as the $^{15}N/^{14}N$ ratio there is a good deal to be said in favour of the Kjeldahl digestion followed by the Dumas procedure on an aliquot. This not only permits the determination of both total N and the $^{15}N/^{14}N$ ratio on the same sample, but it takes advantage of the favourable features of the Dumas procedure while removing some of the disadvantages of both Rittenberg and Dumas ([15]).

MEASUREMENT BY EMISSION SPECTROMETRY

Measurement of the isotopic composition of nitrogen spectroscopically is based on the measurement of the intensity ratio of the bandheads of $^{14}N^{14}N$, $^{14}N^{15}N$ and $^{15}N^{15}N$ molecules, between wavelengths 297 and 299 nm, corresponding to the 2,0-transition of the second positive system. Complete data on all the band systems of nitrogen is available ([22]).

Quartz tubes filled with N_2 gas are excited, using a high frequency (25–100 Mc/s) microwave generator of about 50 watts. By means of a quartz spectrograph the bandheads $^{14}N^{14}N$ at 297.7 nm and $^{14}N^{15}N$ at 298.29 nm are separated for the determination of the ^{15}N abundance. The bandhead of $^{15}N^{15}N$ gives a negligible peak with concentrations of ^{15}N below about 10% and can be ignored.

Normal existing spectrographs e.g. the Hilger apparatus, can be modified for ^{15}N determination, the photographic plate of the spectrograph being replaced by a monochromator and photomultiplier ([23,26]). (Commercial instruments are available, including the Statron NO1-5 and Isonitromat 5200 :Isocommerz GmbH, 1115 Berlin-

Buch, D.D.R., and the Jasco NIA-1 :Japan Spectroscopic Co. Ltd., Tokyo.). Light from the excited gas in the discharge tube is focussed in the monochromator, while provision is made for scanning the selected portion of the spectrum automatically. Light passes from the monochromator to the photomultiplier and the pulses are amplified and recorded on a chart recorder, corresponding to the ^{28}N and ^{29}N peak heights. The calculation of the $\%^{15}N$ abundance is similar to that used in mass spectrometry.

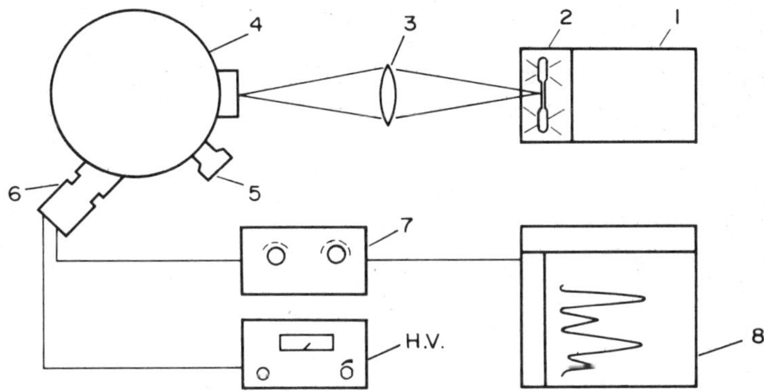

FIG. 7.6 Block diagram of Statron NO1-5 emission spectrometer for ^{15}N. (1) high frequency generator (2) discharge tube (3) condenser lens (4) mirror monochromator dispersing the light into a spectrum (5) drive mechanism which automatically moves the prism plate in the monochromator to permit light of the two band heads to fall successively on the photocathode of the photomultiplier (6), the impulses passing through the amplifier (7) to a chart recorder (8).

Preparation of Samples

In principle the preparation of samples for emission spectrometry is similar to that required for mass spectrometry. However, the total quantity of nitrogen involved is much smaller. In particular when dealing with samples of less than 1 μg N, contamination with nitrogen from air, filter paper, chemicals and nitrogen adsorbed on glass container walls is likely to interfere if the necessary precautions are not taken. Since the quantity of N_2 in which the ^{15}N abundance can be determined is so small, the method can be used for the determination of ^{15}N in fractions separated by paper or thin layer chromatographic techniques. Blanks and ^{15}N labelled standards should be used to check for possible contamination of samples.

Specially designed Perpex glass or quartz tubes about 20 cm long, sealed at one end and with a double constriction (Fig. 7.8) are heated to about 600°C to remove adsorbed nitrogen. The sample in a glass capillary together with CuO and CaO are inserted. The tube is evacuated on a vacuum line, and after evacuation the tubes are sealed and heated in a muffle furnace to liberate N_2 by the Dumas reaction. The tubes are now ready for emission spectrometric determination of ^{15}N abundance.

A preparation apparatus after Faust is now available commercially (Karl Kummer of Thüringen) but the apparatus presents few problems to those with vacuum line

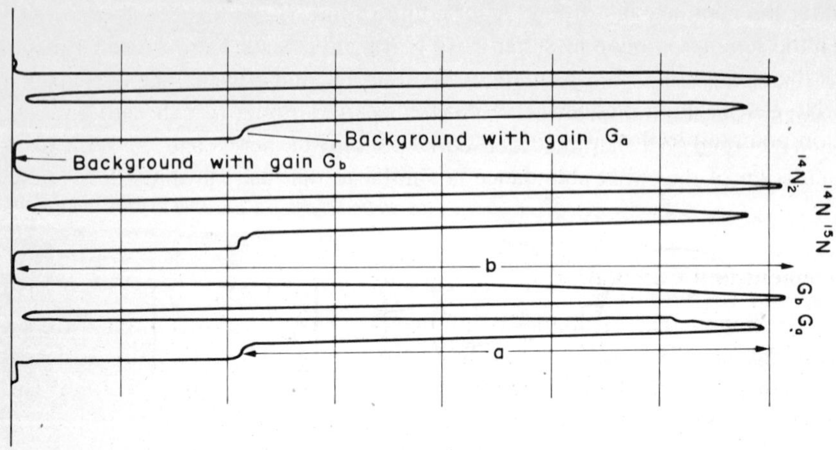

FIG. 7.7 Typical emission spectrum showing the working peaks.

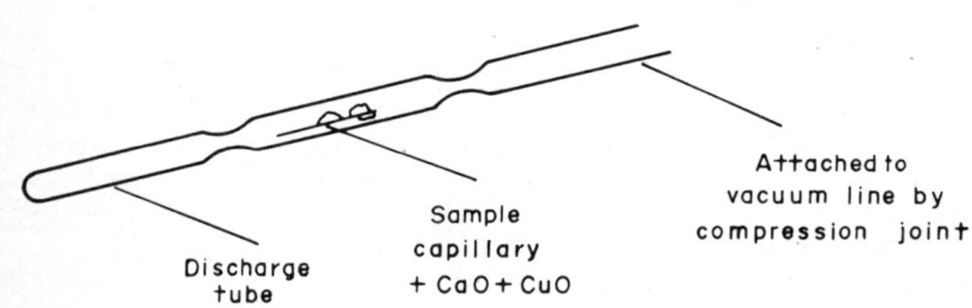

FIG. 7.8 Reaction tube for micro-Dumas N_2 preparation for emission spectrometry.

experience, A vacuum line capable of reducing to 10^{-4} – 10^{-5} Torr is required, preferably by means of a two stage rotary pump and a water cooled oil diffusion pump, as this avoids the danger of contamination from a mercury diffusion pump. A preparation line is supplied with the Jasco instrument.

Basic Procedure

A sample containing 10–15 μg N is taken up into a capillary 1–4 cm long and about 2 mm internal diameter and dried at 50–60°C. The capillary is placed in one portion of the reaction tube (6–7 mm internal diam, 15–25 cm long) and 10 mg of recently degassed CuO wire and 10 mg of CaO, preferably pelleted, is added. Both capillary and discharge tube are roasted out immediately before use. The reaction tube is attached to the vacuum line by a compression joint and after the required vacuum is attained the end of the tube is sealed off at the constriction nearest the line. Heating the tube for 2–3 hours at 550°C in a muffle furnace will ensure complete nitrogen production. The emission tube can then be sealed off at the second constriction and the section containing the capillary discarded. After cooling, the N_2 gas sample is now ready.

When the samples contain less than 1 μg N, the tubes have to be treated with xenon to block the adsorption of N_2 gas to the walls. Since xenon prevents the later excitation of the N_2 gas by high frequency discharge, helium is added. Helium lowers the excitation potential in the presence of xenon for reasons not yet fully understood ([24, 28]).

Measurement and Calculation

It will be found that the intensity of the two bandheads measured differs very greatly, as the natural isotopic composition has a ratio of 130:1, and it is necessary to use a higher gain setting for the $^{14}N^{15}N$ band at 298.29 nm than required for the $^{14}N^{14}N$ band at 297.7 nm. Therefore, the ^{15}N abundance has to be determined from the intensity ratio of the bandheads with a standard curve..

If the intensities of the bandheads—that is the peak heights—are *a* and *b* and the gain setting is *G*, then the ratio

$$R = \frac{bGa}{aGb}$$

can be used to obtain the $\%^{15}N$ abundance from the equation

$$\%^{15}\text{N abundance} = \frac{100}{2R + 1}$$

Reference must then be made to a calibration curve derived from analysis of a number of NH_4 samples with known ^{15}N abundance. This is primarily because of the non-linearity of the amplification.

Application of Emission Spectrometry

The emission spectrometry method has two major advantages. It can determine the ^{15}N abundance of samples with as little as 1 μg N, with the most recent developments pushing this limit down to 0.1–0.2 μg N. Secondly, the method requires less expensive equipment than is required for mass spectrometry. Many laboratories already have spectrographs which are underutilized and can be modified relatively inexpensively.

It is now possible to undertake abundance analysis of small spots taken from chromatograms, to carry out very short term NO_3 and NH_4 ion uptake experiments, etc. thus greatly extending the range of work. Together with established biochemical methods for determining very small amounts of nitrogen it should prove to be an exceptionally powerful tool in tracing pathways and determining rates of assimilation into intermediates, etc.

It should be noted, however, that at the very lowest levels of nitrogen (less than

1 μg), techniques are still experimental. At present, where a mass spectrometer is available and samples contain as much as 1 mg N (e.g. from field experiments), this is still probably the most satisfactory and accurate way for routinely determining a large amount of samples. Apart from its ability to determine very small amounts of nitrogen, the emission spectrometer method is also especially relevant to general laboratories without much experience of vacuum line techniques and without the workshop back-up and instrumentation experience possessed by established laboratories using mass spectrometers. The error term from small samples is also likely to be greater, though the accuracy of the emission spectrometer will of course be improved.

^{15}N ENRICHMENT REQUIREMENTS FOR FIELD EXPERIMENTS

The ^{15}N enrichment used must be considered in relation to ^{15}N analytical limitations. Keeping the ^{15}N enrichment as low as possible is important in order to reduce costs, and in Chapter 12 1.0% ^{15}N excess has been suggested as generally adequate. These requirements were considered in more detail by Fried *et al.* ([39]) based on a series of field experiments.

For analysis by mass spectrometer, the analytical error including sub-sampling is of the order of 0.01% ^{15}N atom excess for a single sample. Now, in practice this means that samples for analysis should contain not less than about 0.15% ^{15}N atom excess. The coefficient of variation from a single sample from a field experiment was rarely below 10% and was more often of the order of 15–25%. In effect then, with a sample containing upwards of 0.15% ^{15}N atom excess the contribution to the coefficient of variation from the analytical error (0.01%) is comparatively small. With these parameters it is possible to estimate the amount of ^{15}N required for any particular field experiment, based on the following estimates.

a. The total N taken up by the crop = anticipated yield × anticipated N content, e.g. normally 100–200 kg N/ha.
b. The % fertilizer N taken up by the crop. This will usually be 30–50% and is not likely to be less than 10%.

In Fig. 7.9 the quantity of ^{15}N to be applied has been plotted against the total N taken up by the crop, for varying per cent fertilizer utilization, and assuming a field variation of 10% for a single sample. For example, if 80 kg N/ha are to be applied in an experiment where the total N uptake is likely to be of the order of 100 kg/ha and the expected utilization of *N* fertilizer, 30%, then from the graph 0.33 kg/ha of ^{15}N is required. Therefore, the enrichment required for a rate of application could be as low as 0.41% ^{15}N atom excess $(\frac{0.33}{80} \times 100)$. If, however, one of the treatments likely to result in only 10% utilization of the applied N fertilizer, then the % ^{15}N atom excess required will be greater, 1.0% ^{15}N atom excess or higher, as can be determined in the same manner.

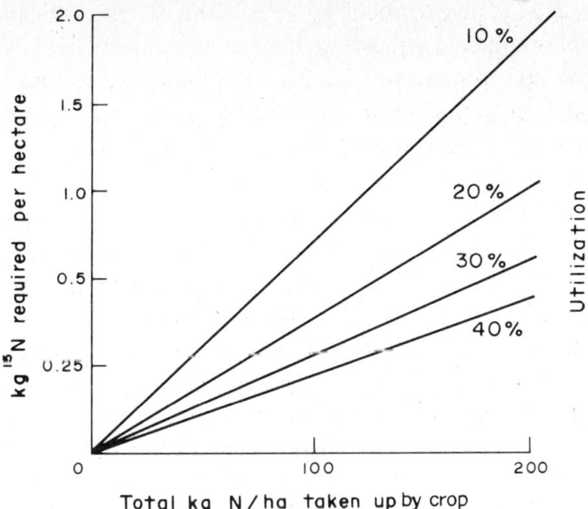

FIG. 7.9 The amount of N-15 required for field experiments as a function of the predicted total nitrogen uptake by the crop and % utilization of the applied fertilizer (Fried *et al.*) ([39]).

USE OF DEPLETED ^{15}N FERTILIZERS

It has only comparatively recently been realized that it was possible to use nitrogen *depleted* in ^{15}N as a tracer, in other words the use of isotopic ^{14}N. The implications are quite considerable for the field use of ^{15}N depleted materials. As they are the by-product of the production of much smaller amounts of ^{15}N enriched materials they are much cheaper and likely to be available in larger amounts, thus making much larger experiments possible ([29,30,31,32]).

Broadbent and Carlton ([100]) found spatial variability of ^{15}N in unfertilized soils to be sufficiently low that calculations of uptake of ^{15}N-depleted fertilizers are not seriously affected. Moreover, they were able to detect ^{15}N depleted fertilizer in soil solutions at depths down to 3 meters. However, as one would expect, ^{15}N-depleted fertilizer lost its isotopic identity when incorporated into the organic pool.

It will be recalled that on usual isotope dilution principles, in studies with ^{15}N enriched fertilizers the per cent plant nitrogen derived from the fertilizer is calculated on the basis:

$$\%\text{ plant } N \text{ derived from fertilizer} = \frac{\%\ ^{15}\text{N excess in sample}}{\%\ ^{15}\text{N excess in fertilizer}} \times 100 \qquad (5)$$

In the case of ^{15}N depleted materials the calculation requires modifying so that:
% plant N derived from the fertilizer =

$$\frac{(N_u - N_t)}{N_u - \dfrac{N_f}{n}} \times 100 \qquad (6)$$

where N_u = atom% ^{15}N in unfertilized plants
N_t = atom% ^{15}N in fertilized plants
N_f = atom% ^{15}N in the fertilizer (for example 0.006%)
n = the plant discrimination factor between ^{14}N and ^{15}N. If it is assumed that there is no discrimination between ^{14}N and ^{15}N, then $N = 1$.

There are some practical restrictions on the use of ^{15}N depleted material, primarily related to the sensitivity of determination, and the N-dilution factor of any particular experiment. Thus, analysis has to be done by mass spectrometry as the emission spectrometer is not sufficiently sensitive at present. Double collector mass spectrometers are desirable though not essential. The type of experiment that can be undertaken is restricted to those that can be completed in one growing season, such as fertilizer uptake studies in annual crops, loss of fertilizer nitrogen, etc. Studies involving residual soil nitrogen are not practicable with depleted material, due to the high dilution factor.

PLANT-SOIL STUDIES WITH ^{15}N AS A TRACER

^{15}N has now been so widely used as a tracer in soil-plant studies that there is hardly any major area where this isotope has not been applied. Only a few representative references can be given here. In field experiments the factors influencing the efficient utilization of applied N, including placement, time of application and fertilizer source have been studied, e.g. for rice and maize and many other crops ([38,39,50,55]). Such factors have a direct connection with N transformation and possible N losses and in this area detailed studies have been connected with:

— soil conditions and placement which encourage direct volatilization of NH_3-nitrogen ([33,37]);
— NO_3-nitrogen losses by leaching, denitrification, and other chemical or biological processes ([35]);
— the relative amounts of soil and fertilizer nitrogen in the pool of available nitrogen ([46]);
— symbiotic nitrogen fixation by legumes in the field ([85,86]);
— fixation of NH_4-nitrogen on clays;
— the degree of biological activity and immobilization ([38,39,56]);
— transformation of N ([44,47,54]);
— the available nitrogen status of soils ('A'-value) ([40,52]);
— recovery of fertilizer-*N* ([38,40]);
— balance sheet and residual effects ([43]);
— efficiency of *N*-sources ([51,53]).

Detailed reviews of the fate of nitrogen applied to soils have been given by Allison ([34]) and Legg ([47]) and many of the references apply to ^{15}N work. A review of Hauck and Bremner ([41]) is devoted entirely to tracers in soil and fertilizer nitrogen research.

Studies involving very small amounts of N and short uptake times (5–120 min) have been carried out comparing the uptake of NO_3 and NH_4 by excised roots ([40]), and by seedlings ([42]). We may expect that the development of the emission spectrometer technique will encourage many biochemical and physiological studies of N-uptake, utilization, turnover and protein formation in plants.

STUDIES ON N_2-FIXATION

Since the isotope ^{15}N became generally available there has been widespread although comparatively limited use of it in N-fixation experiments, both with legumes and non-legume N_2-fixing plants. References ([57]) and ([58]) provide examples and techniques. Direct evidence of associative N-fixation in tropical grasses ([59]) and sugar cane ([60]) has also been proven comparatively recently by the use of ^{15}N.

The use of ^{15}N labelled atmospheres to determine actual N_2-fixation must now be regarded as the primary standard against which other methods are judged. The principle is simple: the whole plant or at least the root is exposed to ^{15}N enriched atmosphere in a chamber for an appropriate period of time, followed by determination of the ^{15}N concentration in the plant. Either the whole plant or its separate parts may be analyzed, then the per cent total nitrogen in the plant sample fixed during the experimental period is given by:

$$N_{fix} = \frac{\text{sample atom } \%^{15}\text{N excess}}{\text{gas atom } \%^{15}\text{N excess}} \times 100 \tag{7}$$

Knowing the total amount of N present in the sample enables calculation of the actual amount fixed during the experimental period.

The former limitations of cost and the availability of a mass spectrometer are no longer real problems in the use of nitrogen-15 in N_2-fixation studies. The present disadvantages of using $^{15}N_2$ remain the problem of devising experimental arrangements for a large number of plants, and the difficulty of using the method in field situations.

Due to the dilution effect and the relatively short time of most N_2-fixation experiments it is necessary to use at least 30–50% ^{15}N atom excess material, and up to 90% for very short exposure times. $^{15}N_2$ is now conveniently purchased in a cylinder or may be converted from ammonium salts by sodium hydroxide solution, passing over hot copper oxide to complete oxidation, removing contaminating gases by passing through a liquid air cold trap.

N_2-fixation can also be determined through ^{15}N-labelled fertilizer experiments, making use of an extension of equation (5). This enables us to measure the proportion of total nitrogen in the plant or crop that has been derived from the applied fertilizer and how much from the soil. In the case of N_2-fixing plants, although a perfectly valid determination of fertilizer-derived *N* is obtained, the value obtained for "soil-*N*" is in fact the sum of soil pool-*N* and nitrogen derived from N_2-fixation. The problem therefore is to separate these two components. Fried and Mellado ([98]) proposed a method to do this, it being essentially a refinement of an earlier procedure based on A-values ([85]). The validity of the concept has been confirmed ([99]).

Equation (5) above holds good when there are two sources of nutrient, ^{15}N-labelled fertilizer and unlabelled soil-*N*. With a dinitrogen fixing crop a third source, unlabelled atmospheric nitrogen, must now be taken into account. In fact this cannot be done unless a non-fixing test crop is grown simultaneously with the N_2-fixing crop. In an experiment then, ^{15}N labelled fertilizer of the same % atom excess ^{15}N will be applied to the nitrogen fixing crop and to the non-fixing test crop, then following Fried and Mellado's notation:

where nitrogen sources are, S = Soil-*N*
F = Fertilizer-*N*
A = Atmosphere, fixed-*N*

and

s_o, f_o = proportion of nitrogen in the plant, from soil and fertilizer respectively, when only *F* is applied and there is no N_2-fixation (i.e. with non-fixing test crop)

s_a, f_a = proportion of nitrogen in the plant, from soil and fertilizer respectively, when plant is incorporating *A* (i.e. dinitrogen fixing crop)

a = proportion of *N* in the plant derived from the atmosphere

then,

$$s_o + f_o = 1$$

$$s_a + f_a + a = 1$$

$$\frac{s_o}{s_a} = \frac{f_o}{f_a} = \frac{s_o + f_o}{s_a + f_a} = \frac{1}{1 - a} \qquad (8)$$

$$\text{and, } a = 1 - \frac{f_a}{f_o} = 1 - \left[\frac{\%\text{ atom excess }^{15}\text{N in fixing crop}}{\%\text{ atom excess }^{15}\text{N in test (non-fixing) crop}}\right] \qquad (9)$$

The requirement for a non-fixing test crop establishes limitations almost identical to the use of non-effective isolines as in soybeans, but the isotope technique is not affected by the fact that use of non-effective isolines may underestimate nitrogen fixation due to greater root growth, and that methods of determining N_2-fixation "by difference" work best when plants are grown on low-*N* soils. The method could be especially useful for screening cultivars for N_2-fixing capacity, or for screening the effectiveness of legume-*Rhizobium* strain combinations.

STUDIES INVOLVING STABLE ISOTOPE RATIOS

Stable isotope ratios are providing in a number of instances a type of tracer which enable us to draw conclusions as to the effect of environmental and other factors on carbon, nitrogen and oxygen cycles, movement of these elements, and the processes involved in the synthesis and exchange of compounds formed by the elements in nature.

Molecular species formed by different isotopes have special physical and physiochemical properties, and in fact due to various selection and discrimination processes the natural isotopic composition of these elements shows small variations.

Considerable evidence now exists that there is long term discrimination by plants between ^{12}C and ^{13}C isotopes, leading to variable concentrations of these isotopes. In general, plants are depleted in ^{13}C compared with the proportion of ^{13}C to ^{12}C in the atmosphere, so clearly they preferentially assimilate the lighter of the two isotopes.

As variations in concentration are in fact rather small a more sensitive measure must be used rather than just % abundance of the isotope; this is the δ value, where

$$\delta\ ^{13}C\%_o = \left[\frac{^{13}C/^{12}C \text{ sample}}{^{13}C/^{12}C \text{ standard}} - 1\right] \times 1000 \tag{10}$$

Plant species show further discrimination according to the photosynthetic system. Thus C_3 (Calvin cycle) species, in which the primary carboxylase is ribulose-1,5-diphosphate carboxylase, discriminate to a greater extent than C_4 (Hatch and Slack pathway) [65,66]) species, where the primary carboxylase is phosphoenolpyruvate carboxylase ([62,63,64]). Measurements *in vitro* have further confirmed that RuDPCase discriminates to a greater extent than PEPCase ([61]). .

Plants which have Crassulacean acid metabolism (CAM) show variable patterns of discrimination, sometimes as much as C_3 plants or no more than C_4 plants. This is apparently due to environmental factors probably, because a feature of the CAM C-assimilation pattern is to behave like C_3 plants during the day and at night have a capacity for CO_2 fixation involving PEPCase in a C_4-like mechanism. Environmental factors such as water stress and day/night temperatures influence the relative importance of dark and light CO_2 fixation, and this in turn affects the $^{13}C/^{12}C$ ratio ([64,68]).

C_4 plants have δ ^{13}C values around -11 to $-14^0/_{00}$ while C_3 plants have δ ^{13}C values of -25 to $-30^0/_{00}$. CAM plants are around δ $^{13}C = -17^0/_{00}$.

In general most temperate crop plants are of the C_3 type, but maize and many tropical grasses are of the C_4 pattern. There are grounds for thinking that the C_3 carbon fixation process is more primitive than the C_4 process and that C_4 plants tend to predominate under arid conditions. Reference 68 provides a good review of various aspects of C_3 and C_4 plant types in relation to ecology and δ ^{13}C values.

It is apparent that with this variation a substantial field of investigation has been opened up for studies in plant ecology, adaptation, phylogeny, comparative biochemistry, palaeoecology of peat deposits, the study of soil organic matter and the CO_2 cycle in forests and soils, etc.

Although variations in the stable isotope ratio of nitrogen in natural environments has long been recognized ([72,73]) and quite a lot of work carried out, relatively few general conclusions have emerged. This is probably because the variations are much smaller, harder to define and more difficult to interpret. The nitrogen of the atmosphere has a constant isotopic content over variations in location and altitude ([75]), but there are isotopic variations in the nitrogen of nitrate and ammonium ions from rain water ([74]).

In an exactly similar manner to carbon isotope ratios, $\delta^{15}N^0/_{00}$ is given by

$$\delta^{15}N^0/_{00} = \left[\frac{^{15}N/^{14}N \text{ sample}}{^{15}N/^{14}N \text{ standard}} - 1\right] \times 1000 \quad (11)$$

or alternatively $\delta^{15}N^0/_{00}$ values have been calculated on the basis of

$$\delta^{15}N^0/_{00} = \frac{(\text{at }\%\ ^{15}N)\text{ sample} - (\text{at }\%^{15}N)\text{ atmosphere}}{(\text{at }\%^{15}N)\text{ atmosphere}} \times 1000 \quad (12)$$

The latter expression gives slightly higher values.

It has been found that differentiation occurs during nitrogen fixation by legumes ([70]) so that legumes not fixing nitrogen (or non-legumes) have $\delta^{15}N$ values similar to the soil, but legumes which are actively fixing nitrogen have $^{14}N/^{15}N$ levels similar to nitrogen of the atmosphere. The application of isotope dilution principles to natural nitrogen isotope abundance determinations may provide an estimate of plant-N_2 fixation under some circumstances. Considering the various soil microbiological reactions which affect the fractionation process of N isotopes in soils, it has now been generally established that the nett effect is to give a slight increase in the ^{15}N abundance of soil nitrogen (see reference 41 for a summary) compared with atmospheric nitrogen. Therefore, plant nitrogen coming from the soil should have a greater abundance of ^{15}N than plant nitrogen directly fixed from the atmosphere. Then if N_{fix} is the nitrogen fraction in the plant derived from N_2-fixation, following isotope dilution principles,

$$N_{fix} = 1 - \left[\frac{\delta^{15}N^0/_{00} \text{ in plant leaves} - \delta^{15}N^0/_{00} \text{ atmosphere}}{\underset{\text{(test plant)}}{\delta^{15}N^0/_{00} \text{ available soil N}} - \delta^{15}N^0/_{00} \text{ atmosphere}}\right] \quad (13)$$

This method has been used to estimate associative N_2-fixation in sugarcane ([87]). The method clearly cannot be used in all circumstances as it has two prerequisites: that there is a high initial $\delta^{15}N$ value to make it possible to derive a value essentially obtained by difference, and secondly that there is no variation in $\delta^{15}N$ value down the profile of the active rooting zone (variation in deep profile beyond the rooting zone is not relevant). Relatively small differences in $\delta^{15}N$ values that are encountered within the rooting zone should not theoretically affect the overall result provided very adequate soil sampling is undertaken. Similarly, the extensive plant root system is an extremely effective sampling and averaging device. A deficiency of the method is that only a small variation in $\delta^0/_{00}\ ^{15}N$ corresponds to a comparatively large difference in N_2 fixation.

The method adopted by Legg and Sloger ([86]) essentially followed the same reasoning, but in order to ensure an adequate variation between soil ^{15}N and atmosphere ^{15}N they added ^{15}N labelled fertilizer to the soil sometime previously and waited until the newly incorporated ^{15}N was in organic form before carrying out the experiment.

Amarger *et al.*, ([95]) working with *Lupinus luteus, var. sulfa*, grown in sand culture with nutrient solution, established the fractionation factor, α, and enrichment coefficient, ε, for nitrate uptake by the plant and N_2 fixed by the plant, which were: $\alpha = 0.9989$, $\varepsilon = -1.1$ and $\alpha = 0.9991$, $\varepsilon = -0.9$ respectively for each process. It was possible to calculate the rate of nitrogen fixation, and the method proved to be more accurate than methods based on total nitrogen or acetylene reduction. Later work with inoculated and non-inoculated soybeans in soil suggested that the method can work in soil if a non-fixing control plant is present (personal communication).

At present these methods are experimental, and due in some circumstances to low soil $\delta\ ^{15}N^{0}/_{00}$ values or to extreme variability of values in field situations, it is clear that they will not be universally applicable. Nevertheless, for some special problems such as determining associative N-fixation in sugarcane, which has a long growing season, and is a very large plant for experimentation both from the point of view of dilution of ^{15}N in tracer experiments, and general handling and culture, it probably has a valuable role to play. Probably $\delta^{15}N$ values of available soil nitrogen should be determined on the nitrogen taken up by a non-fixing test crop. Lettuce has been used for this purpose, but growing the non-fixing test crop alongside the N_2-fixing crop would be better when possible (96).

In the soil, denitrification tends to produce nitrogen gas depleted in ^{15}N with the residual nitrate slightly enriched, while nitrification leads to nitrite and nitrate depleted in ^{15}N with the residual ammonium enriched in ^{15}N ([70,79]). However, much work is now in progress attempting to explain and make use of these values in better understanding of the nitrogen cycle within an ecosystem, involving nitrogen fixation in the soil, and nitrogen losses by denitrification, volatilization and leaching. Such work may assist in identifying N-fertilizer excess and the contribution of this to the nitrate content of surface water, NO_3 being a serious pollutant in certain circumstances. It is suggested that this can be done by means of the differences in the natural isotopic composition of soil nitrogen and fertilizer-N ([69,76,88]). While this may be true in particular circumstances other work strongly suggests that this is seldom likely to be a valid approach ([89,90]).

Current work suggests ([69,71]) that plant nitrogen derived from microbiological fixation in the soil has a lower $^{15}N/^{14}N$ ratio than when derived via the soil nitrogen pool. In contrast, losses by denitrification within soil profiles (as opposed to leaching) give higher $^{15}N^{14}N$ ratios. This offers the possibility of being able to estimate N fixation/loss ratios by measuring $\delta^{15}N$ values down representative profiles in conjunction with plant data. Nevertheless, the value to environmental studies of measuring variation in natural N isotopic ratios has by no means been universally agreed ([77,78,84]), and much further work is required before adequate interpretation is possible.

The use of $^{18}O/^{16}O$ ratios, as reflecting the fractionation of water into $H_2{}^{18}O$ and $H_2{}^{16}O$ by means of evaporation and condensation, is proving a useful tool in hydrological studies ([80,81]).

The study of the D/H i.e. $^{2}H_2/^{1}H_2$ ratios in the cellulose of plants, particularly of trees, shows promise that the δD values of plants can be used as indicators of climatic change ([82,83]).

SUPPLIERS OF STABLE ISOTOPES

Stable isotopes, especially ^{15}N, can be obtained from a number of sources: Bureau de Stable Isotopes, Saclay, France (^{15}N); Isocommerz GMBh of the Institute for Stable Isotopes, 1115 Berlin Linderberger Weg 70, East Germany, D.D.R.; BOC Ltd. (Prochem) Deer Park Road, London SW19 3UF, England; Junta de Energia Nuclear, Dirección de Quimica y Isotopos, Madrid 3, Spain; Monsanto Research Corporation, Stable Isotope Sales, Mound Laboratory, P.O. Box 32, Miamisburg, Ohio 4532, U.S.A.; products of Los Alamos Scientific Laboratory are sold through Monsanto, Mound Laboratory.

NITROGEN-13

The major radioactive isotope of nitrogen ^{13}N ($t_{1/2}$ = 10.05 min; β 1.24 MeV) has been used in a limited number of experiments ([91,93,94]). There is continuing interest in ^{13}N ([91]), despite the problem of its production and its short half-life, because of the much greater sensitivity of determination of the radioactive isotope as opposed to the stable ^{15}N.

^{13}N may be prepared by a number of different routes. One reaction is through the bombardment of carbon by deuterons in a cyclotron:

$$\frac{{}^{2}_{1}D + {}^{12}_{6}C \longrightarrow {}^{13}_{7}N + {}^{1}_{0}n}{\text{with} \quad {}^{13}_{7}N \longrightarrow e^{+} + {}^{13}_{6}C}$$

Alternatively ^{14}N can be used as a target for bombardment by protons, in the reaction $^{14}N(p;\ p,n)^{13}N$. In order to overcome the low target density of gaseous or liquid nitrogen, a solid target of high nitrogen content, such as melamine ($C_3H_3N_6$), is used. Theoretically ^{13}N can also be produced by the reaction $^{14}N(n,2n)^{13}N$ commonly used in activation analysis of nitrogen, but the low target density and the poor cross section would require a reactor with a high flux. Whatever method is adopted the $^{13}N_2$ has to be produced continuously during the experiment due to the short half-life.

REFERENCES FOR FURTHER READING

Stable Isotopes and Mass Spectrometry

1. Hammond, A. L. Stable Isotopes: expanded supplies may lead to new uses. *Science* **176**, 1315 (1972).
2. Edmunds, A. O. and Lockhart, I. M. Separation of stable isotopes and the preparation of labelled compounds with special reference to ^{13}C, ^{15}N and ^{18}O. *In: Isotope Ratios as Pollutant Sources and Behaviour Indicators*. Proc. Symp. IAEA, Vienna, 279–93 (1973).
3. Maturyoff, N. A., Cowan, G. A., Ott, D. G. and McInteer, B. B. Problems and Potentialities of Stable Isotopes as Tracers for Studying Pollutant Behaviour under Field Conditions. *In: Isotope Ratios as Pollutant Sources and Behaviour Indicators*, 305–25, IAEA, Vienna (1975).
4. Nier, A. P. C. The mass spectrometer and its application to isotope abundance measurements in tracer isotope experiments. *In:* Wilson *et al.*, *Preparation and Measurement of Isotopic Tracers*. Symposium, Edwards, Ann Arbor, Michigan (1948).

5. ROBERTSON, A. J. B. *Mass Spectrometry*, Methuen's & Co. Ltd. London (1954).
6. DUCKWORTH, H. E. *Mass Spectrometry*. U.P. (1958).
7. FIEDLER, R. and Proksch, G. *Anal. Chim. Acta* **78**, 1–62 (1975).
8. ROBINSON, C. F. Mass Spectrometry. *In: Physical Methods in Chemical Analysis*, vol. I, W. G. Berl, Ed. Academic Press, New York, London, pp. 463–545 (1960).
9. WHITE, F. A. *Mass Spectrometry in Science and Technology*. Wiley, New York, London, 352 pp. (1968).

Analysis of ^{15}N

10. GLASCOCK, R. F. *Isotopic Gas Analysis for Chemists*. Academic Press, New York (1954).
11. SMITH, J. H., Legg, J. O. and Carter, J. N. *Equipment and Procedures for ^{15}N Analysis of Soil and Plant Material with the Mass Spectrometer* (1962).
12. PROKSCH, G. Routine analysis of ^{15}N in plant material by mass spectrometry. *Plant and Soil*, **XXXI**, 380–84 (1969).
13. BREMNER, J. M. Inorganic forms of nitrogen and isotope ratio analysis of nitrogen in ^{15}N investigations. *In*: C. A. Black, *Methods of Soil Analysis*, Part II (1969).
14. PROKSCH, G. Application of mass and emission spectrometry for $^{14}N/^{15}N$ ratio determinations in biological material (Dumas Method). *In: Isotopes and Radiation in Soil-Plant Relationship Including Forestry*. Symposium, IAEA, Vienna, 217–25 (1972).
15. FIEDLER, R. and Proksch, G. The determination of nitrogen-15 by emission and mass spectrometry in biochemical analysis: a review. *Analytica Chim. Acta* **78**, 1 (1975).
16. PARKINSON, J. A. and Allen, S. E. A wet oxidation procedure suitable for the determination of nitrogen and mineral nutrients in biological material. *Commun. Soil Science and Plant Analysis*, **6**(1), 1–11 (1975).
17. JORGENSEN, S. S. *Guia Analítico*. Metodologia utilizada para análises químicas de rotina. pp. 24. Centro de Energia Nuclear na Agricultura, Piracicaba, S.P., Brasil (1977).
18. RITTENBERG, D. The preparation of gas samples for mass spectrographic isotope analysis. *In*: Wilson *et a. Preparation and Measurement of Isotopic Tracers*. Symposium. J. W. Edwards, Ann Arbor. Michigan (1948).
19. FAUST, H. Isotopenpraxis **3**, 102 (1967) (Dumas method).
20. ASHTON, F. L. [Salicylic Method for Nitrate], J. Agric. Sci. **26**, 238-48 (1936).
21. TRIVELIN, P. C. O., Salati, E. and Matsui, E. Preparo de amostras para análise de ^{15}N por espectrometria de massa. *Boletim Didático*, Centro de Energia Nuclear an Agricultura. Piracicaba, Brasil, pp. 41 (1973).

Emission Spectrometer Analysis

22. PEANE, R. W. B. and Gaydon, A. G. *The Identification of Molecular Spectra*. 3rd Ed. Chapman & Hall, London (1963).
23. LEICHNAM, J. P. *et al. Int. J. Appl. Rad. Isot.* **19**, 235–47 (1968).
24. GOLEB, J. A. and Middleboe, V. *Analytica Chim. Acta* **43**, 229–34 (1968).
25. MIDDLEBOE, V. High resolution optical nitrogen-15 analysis. *Appl. Spectrometry*, **28**, 274 (1974).
26. SCOTT, T. A. and Humpherson, H. Determination of ^{15}N by electrodeless discharge. *Laboratory Practice*, 703–15 (1974).
27. LEICHNAM, J. P., Middleboe, V. and G. Proksch. *Anal. Chim. Acta* **40**, 487 (1968).
28. FIEDLER, R. and Proksch, G. Emission spectrometry for routine analysis of nitrogen-15 in agriculture. *Plant and Soil* **36**, 371–78 (1972).

Studies with ^{15}N Depleted Material

29. STARR, J. L., Broadbent, F. E. and Stout, P. R. A comparison of ^{15}N-depleted and ^{15}N-enriched fertilizer as tracers. *Soil Sci. Soc. Amer. Proc.* **38**, 266 (1974).
30. EDWARDS, A. P. and Hauck, R. B. 1968. ^{15}N-depleted vs ^{15}N-enriched ammonium sulphate as tracers in N-uptake studies. *Agronomy Abstracts* (1968) 101.

31. CARLTON, A. B. and Hafez, A. A. R. *Calif. Agric.* **27**, 10–13 (1973).
32. HAUCK, R. D. and Kilmer, V. J. *Proc. 2nd Int. Conf. Stable Isotopes Chem. Biol. Med.* (1976).

Representative Studies Using ^{15}N

33. ALEKSIS, Z., Broeshart, H. and Middleboe, V. Shallow depth placement of $(NH_4)_2SO_4$ in submerged rice soils as related to gaseous losses of fertilizer nitrogen and fertilizer efficiency. *Plant and Soil*, **29**, 338 (1968).
34. ALLISON, F. E. The fate of nitrogen applied to soils. *Adv. Agron.* **18**, 219–58 (1966).
35. ANDREEVA, E. A. and Scheglova, G. M. Use of ^{15}N and nitrification inhibitors in soil and fertilizer uptake studies. Trans. 9th Int. Congr. Soil Sci. II Sydney (1968).
37. BARTHOLOMEW, W. V. ^{15}N in research on the availability and crop use of nitrogen. *Nitrogen-15 in Soil-Plant Studies* (Proc. Panel Sofia, 1969) 1, IAEA, Vienna (1971).
38. CARTER, H. N., Bennett, O. L. and Pearson, R. W. Recovery of fertilizer nitrogen under field conditions using ^{15}N. *Soil Sci Amer. Proc.* **31**, 50 (1967).
39. FRIED, M., Broeshart, H., Cho, C. M. and Caldwell, A. C. Field experiments with N-15 on efficiency of fertilizer utilization for maize. *Geoderma* (1968).
40. FRIED, M., Zsoldos, F., Vose, P. B. and Shatokhin, J. L. Characterizing the NO_3 and NH_4 uptak process of rice roots by use of ^{15}N labelled NH_4NO_3. *Physiol. Plantarum* **18**, 313–20 (1965).
41. HAUCK, R. D. and Bremner, J. M. Use of tracers for soil nitrogen research. *Advances in Agronomy* **28**, 219–66 (1976).
42. JACKSON, W. A., Flesher, D. and Hageman, R. H. Nitrate uptake by dark-grown corn seedlings. *Plant Physiol.* **51**, 120 (1973).
43. JANSSON, S. L. Balance sheet and residual effects of fertilizer nitrogen in a six-year study with ^{15}N. *Soil Sci.* **95**, 31 (1963).
44. KETCHESON, J. W. and Jakovljevic, M. Transformation of NO_3 and NH_4 in Soils. *In: Isotopes and Tradiation in Soil-Organic Matter Studies.* IAEA, Vienna (1968).
45. KORENKOV, D. A. Utilization of ^{15}N in agrochemical research. Proc. Soviet-British Symp. on *Agrochemical Research and Fertilizer Application.* Moscow, May 4–7 (1970).
46. LEGG, J. O. and Stanford, G. Utilization of soil and fertilizer N by oats in relation to the available N status of soils. *Soil Sci. Soc. Amer. Proc.* **31**, 215–19 (1967).
47. LEGG, J. O. *Fate of Fertilizer Nitrogen Applied to Soils. In*: Technical Report IAEA-120, Vienna (1970), p. 146–155 (1969).
48. MERZARI, A. H. and Broeshart, H. Utilization by rice of nitrogen from ammonium fertilizers as affected by fertilizer placement and microbiological activity. *In: Isotope Studies on the Nitrogen Chain*, IAEA, Vienna (1968).
49. MYERS, R. J. K. and Paul, E. A. Plant uptake and immobilization of ^{15}N-labelled ammonium nitrate in a field experiment with wheat. Nitrogen-15 in Soil-Plant Studies (Proc. Panel, Sofia 1969) 55, IAEA, Vienna (1971).
50. NISHIGAKI, S. Use of N-15 as a tracer in fertilizer efficiency studies in Japan. *In*: Technical Report IAEA-120, Vienna (1970), p. 161–69 (1969).
51. RENNIE, R. J. and Rennie, D. A. Standard isotope vs nitrogen balance for assessing the efficiency of nitrogen sources for barley. *Can. J. Soil Sci.* **53**, 73 (1973).
52. RENNIE, D. A. The significance of the 'A'-value concept in field fertilizer studies. *In*: Technical Report IAEA-120 (1969).
53. SAPOZHNIKOV, N. A. The application of ^{15}N and ^{32}P in studies on fertilizer conversion in the soil and mineral nutrition of plants. Application of Isotopes and Atomic Radiation in Agriculture, *Atomizdt*, 135, Moscow (1971).
54. STAN, J. L., Broadbent, F. E. and Nielson, D. R. Nitrogen transformation during continuous leaching. *Soil Sci. Amer. Proc.* **38**, 283 (1974).
55. VOSE, P. B. Review of the coordinated research programme on the application of isotopes to rice fertilization studies, 1962–1968. *In*: Technical Report IAEA-120, Vienna (1970) p. 6–14 (1969).
56. BROADBENT, F. E. Nitrogen immobilization in relation to N-containing fractions of soil organic matter. *In: Isotopes and Radiation in Soil Organic Matter Studies*, IAEA, Vienna, 131 (1968).
57. BURRIS, R. H. *Methods of Enzymology* (A. San Pietro, Ed.), vol. 24, 415–31. Acaemic Press, New York (1972).
58. BURRIS, R. H. *In: The Biology of Nitrogen Fixation* (A. Quispel, Ed.) 9–33, North-Holland Amsterdam (1974).

59. DE-POLLI, H., Matsui, E., Döbereiner, J. and Salati, E. Confirmation of nitrogen fixation in two tropical grasses by $^{15}N_2$ incorporation. Soil Biol. Biochem., in press (1977).
60. RUSCHEL, A. P., Henis, Y. and Salati, E. Nitrogen-15 tracing of N-fixation with soil-grown sugar can seedlings. *Soil Biol. Biochem.* **7**, 181–82 (1975).

Stable Isotope Ratios

61. WHELAN, T., Sackett, W. M. and Benedict, C. R. *Plant Physiol.* **51**, 1051 (1973).
62. BENDER, M. M. *Photochemistry* **10**, 1239 (1971).
63. SMITH, B. N. and Epstein, S. *Plant Physiol.* **43**, 380 (1971).
64. TROUGHTON, J. H. *In: Photosynthesis and Photorespiration*, M. D. Hatch, C. B. Osmond & R. O. Slatyer (Eds.) Wiley Interscience, pp. 124–29 (1971).
65. HATCH, M. D. and Slack, C. R. *Biochem. J.* **101**, 103 (1966).
66. HATCH, M. D. and Slack, C. R. Photosynthetic CO_2 fixation pathways. *Ann. Rev. Pl. Physio.* **21**, 141–62 (1970).
67. HAUCK, R. D. Nitrogen tracers in nitrogen cycle studies. Past use and future needs. *J. Environ. Qual.* **2**, 317–27 (1973).
68. Carnegie Institution, Ann. Rept. Director Dept. of Plant Biology, *Carnegie Institution Year Book 1973/74*, pp. 768–846 (1974).
69. KOHL, D. H., Shearer, G. B. and Commoner, B. Fertilizer nitrogen: contribution to nitrate in surface water in a corn belt watershed. *Science* **174**, 1331 (1971).
70. DELWICHE, C. C. and Steyn, P. L. *Environ. Sci., Tech.* **4**, 929 (1970).
71. RENNIE, D. A. and Paul, E. A. Nitrogen isotope ratios in surface and sub-surface soil horizons. *In: Proc. Symp. Isotope Ratios as Pollutant Sources and Behaviour Indicators.* IAEA, Vienna (1974).
72. HOERING, T. *Science* **122**, 1233 (1955).
73. HOERING, T. *Proc. 2nd Congr. on Nuclear Processes in Geologic Settings* **39** (1955).
74. HOERING, T. *Geochim. Cosmochim. Acta* **12**, 97 (1957).
75. DOLE, G. *et al.*, *Geochim. Cosmochim. Acta* **6** 65 (1954).
76. SHEARER, G., Kohl, D. H. and Commoner, B. Use of variations in the natural abundance of ^{15}N to study sources, transformations and movement of nitrogen in a plant-soil-water system. *In: Origin and Fate of Chemical Residues in Food, Agriculture and Fisheries.* IAEA, Vienna, (1975).
77. HAUCK, R. D. *et al.* Use of variations in natural isotope abundance for environmental studies: a questionable approach. *Science* **177**, 453–54 (1972).
78. EDWARDS, A. P. Isotopic tracer techniques for identification of sources of nitrate pollution. *J. Environ. Qual.* **2**, 382–87 (1973).
79. WELLMAN, R. P., Cook, F. D. and Krouse, H. R. *Science* **161**, 269–70.
80. MATSUI, E., Salati, E., Friedman, I. and Brinkman, W. L. F. Isotopic Hydrology in Amazonia. 2. Relative discharges of the Negro and Solimões Rivers through ^{18}O concentrations. *Water Resources Research* **12**, 781–85 (1976).
81. STOLF, R., Leal, J. M., Fritz, P. and Salati, E. Water and Salt Balance of an Artificial Reservoir in the Semi-Arid Region of the Northeast of Brazil Using the Natural Variations of ^{18}O and D. Proc. of U.N. Conference on Water, Mar del Plata, Argentina, 1977 (in press).
82. EPSTEIN, S., Yapp, C. J. and Hall, J. H. The determination of the D/H ratio of non-exchangeable hydrogen in cellulose extracted from aquatic and land plants. *Earth and Planetary Science Letters* **30**, 241–51 (1976).
83. EPSTEIN, S. and Yapp, C. J. Climatic implications of the D/H ratio of hydrogen in C-H groups in three cellulose. *Earth and Planetary Science Letters* **30**, 252–61 (1976).
84. FREYER, H. D. and Ally, A. I. M. *In: Isotope Ratios as Pollutant Sources and Behaviour Indicators*, 21–23, IAEA, Vienna (1975).
85. BROESHART, H. *Netherlands J. Agr. Sci.* **22**, 245–54 (1974).
86. LEGG, J. O. and Sloger, C. A tracer method for determining symbiotic nitrogen fixation in field studies. Proc. 2nd Int. Conf. Stable Isotopes, Oak Brook, Illinois, E. R. Klein and P. D. Klein, (Eds.), 661–66 (1975).
87. RUSCHEL, A. P. and Vose, P. B. Present situation concerning studies on associative nitrogen-fixation in sugarcane, *Saccharum officinarum* L., CENA, Boletim Científico BC-045, July 1977, Piracicaba, S.P., Brasil (1977).
88. KOHL, D. H., Shearer, G. B. and Commoner, B. Variation of ^{15}N in corn and soil following application of fertilizer nitrogen. *Soil Sci. Soc. Amer. Proc.* **37**, 888–92 (1973).

89. MEINTZ, V. W., Boone, L. V. and Kurtz, L. T. Natural ^{15}N abundance in soil, leaves and grain as influenced by long term additions of fertilizer N at several rates. *J. Environm. Qual.* **4**, 486–90 (1975).
90. BLACK, A. S. and Waring, S. A. The natural abundance of ^{15}N in the soil-water system of a small catchment area. *Aust. J. Soil Res.* **15**, 51–7 (1977).

Nitrogen-13

91. AKKERMANS, A. D. L., van Straten, J. and Roelofsen, W. Nitrogenase activity of nodule homgenates of *Alnus glutinosa*: a comparison with the *Rhizobium*-pea system. *Abstract L36* II Int. Symp. N_2 fixation, Salamanca, Spain, September (1976).
92. CAMPBELL, N. E. R., Dular, R., Lees, H. and Standing, K. G. The production of $^{13}N_2$ by 50-MeV protons for use in biological nitrogen fixation. *Can. J. Microbiol.* **13**, 587–99 (1967).
93. NICHOLAS, D. J. O., Silvester, D. J. and Fowler, J. F. Use of radioactive nitrogen in studying nitrogen fixation in bacterial cells and their extracts. *Nature*, **189**, 634–35 (1961).
94. RUBEN, S., Hassid, W. Z. and Kamen, M. D. Radioactive nitrogen in the study of N_2 fixation by non-leguminous plants. *Science*, **91**, 578–79 (1940).

Additional

95. AMARGER, Noëlle, Mariotti, A. and Mariotti, Francoise. Essai d'estimation du taux d'azote fixé symbiotiquement chez le Lupin par le traçage isotopique naturel (^{15}N). C. R. Acad. Sc. Paris, **284**, 2179–82 (1977).
96. VOSE, P. B., Alaides P. Ruschel and Salati, E. Determination of N_2-fixation especially in relation to the employment of nitrogen-15 and of natural isotope variation. Proc. II Latin American Botanical Congr. Brasilia (1978).
97. IAEA. *Stable Isotopes in the Life Sciences*. Proc. Tech. Comm. Meeting, Leipzig 1977, pp. 456, IAEA, Vienna (1977).
98. FRIED, M. and Mellado, L. (1977) *In: Int. Symp. on Limitations and Potentials of Biological Nitrogen Fixation in the Tropics*. Basic Life Sciences, Vol. 10. Plenum Press, New York (1978).
99. RUSCHEL, A. P. *et al.*, Unpublished (1978).
100. BROADBENT, F. E. and Carlton, A. B. Field trials with isotopically labelled nitrogen fertilizer. *In: Nitrogen in the Environment*, Vol. I, D. R. Nielsen & J. G. MacDonald (Eds.), 1–41, Academic Press, New York (1978).

CHAPTER 8

Activation Analysis for Biological Samples

INTRODUCTION

ACTIVATION analysis is a method of quantitative analysis in which a sample is exposed to a flux of ionizing particles, usually slow neutrons in a reactor or fast neutrons from a neutron generator. The stable isotope of the element being determined is thereby converted into a radioactive isotope which can be assayed by the usual counting methods. Comparison with standards containing a known amount of the stable element enables a quantitative determination to be made.

One great advantage of slow neutron activation analysis is its potential ability to determine minute amounts of element at levels which are far below the power of other analytical methods. Thus samples which are both small and contain only trace amounts of the element can be satisfactorily analyzed. For example, the limit of detection for manganese, copper and arsenic is of the order of 10^{-10}g or less. In forensic medicine, activation analysis has made it possible to analyze a single hair, while in biology and agriculture it has made it practicable to analyze a single root, or 1–2ml of actual soil solution. The method is also useful for elements which cannot be conveniently determined by chemical means.

After quite a lot of comparatively early work the interest in the application of slow neutron activation analysis to agricultural and biological samples has declined, partly because reactors have been available to only relatively few workers, partly because determinations have involved tedious separations or "spectrum stripping" counting procedures, and especially because the development of the atomic absorption spectrometer has made many micro-determinations routine which were previously rather difficult.

More recently with the development of 14 MeV high output tubes, it has come to be appreciated that certain activation techniques involving activation by fast neutrons from a neutron generator, can lead to substantial savings in analysis time and can also be virtually non-destructive of the sample. The determination of nitrogen by the $(n, 2n)$ reaction is an example ([12,16,25]). We are likely to see a considerable extension of these methods, although a much larger sample is required than for slow neutron activation.

The use of "indicator" activation analysis is also an area which will probably grow when the technique becomes more widely appreciated. In this method inactive tracers, boron for example, are used as markers which after sampling are detected by neutron activation analysis. The method has been applied e.g. to tracing the movement of air, plant pollen and fungal spores, and the sorption and movement of ions in soil ([4,14,18,26]). The advantage of the indicator activation method is that there can be no risk of radiation effects on biological organisms, and it can also be used for experiments in public areas, or in experiments involving an inhalation risk, where radioactive tracers might not be acceptable.

General Principles

It will be appreciated that when a sample is irradiated every inducible element in the sample is activated, not merely the particular one that is being analyzed. This is an advantage in one respect, if analysis of more than one element is required, however, identification and quantitative estimation of one or two out of a number of radioactive products presents problems. In the case of slow neutron activation techniques this is often overcome by chemical separation of the nuclide(s) sought from other interfering radioactive products, although pulse height gamma spectrometry (page 106) is used where possible. Other procedures for counting one radionuclide in the presence of another may also be used if practicable, but due to the complex nature of most agricultural and biological samples some chemical separation is normally required with slow neutron activation.

Theoretically, activation analysis can be carried out by means of any radiation with sufficient energy to induce radioactive species when the sample is exposed to a flux of the bombarding particles. The nature of a number of induced nuclear reactions has been discussed in Chapter 2 and that should be read in conjunction with this section. In practice, slow neutrons have hitherto been most frequently used in activation analysis. In this case the stable element is converted to a radionuclide by the (n, γ) process, e.g. $^{63}CU(n,\gamma)^{64}Cu$, $^{55}Mn(n,\gamma)^{56}Mn$, and $^{59}Co(n,\gamma)^{60}Co$. Slow neutron activation cannot be used for elements lighter than sodium, as the half-lives of the products are too short, while for elements like calcium, sulphur, and iron the neutron cross sections are too low.

Fast neutron activations are of increasing relevance and importance. They include such reactions as $^{14}N(n,2n)^{13}N$, $^{16}O(n,p)^{16}N$, $^{37}Cl(n,p)^{37}S$, $^{19}F(n,p)^{19}O$, $^{63}Cu(N,2n)^{62}Cu$, $^{28}Si(n,p)^{28}Al$ and $^{11}B(n,\alpha)^{8}Li$. Activations such as these are specifically considered in the second part of this chapter. Amongst other particles that can be used, deuterons can activate nitrogen and sulphur, while proton activation of carbon, oxygen, boron and fluorine is possible. Relatively few workers have access to this type of facility, however.

Consider the case of a sample being analyzed for a certain element, which is exposed to a uniformly high flux of slow neutrons in a reactor. There are two opposed reactions then taking place: the formation of radioisotope by activation, which is determined both by its cross section (page 21) and the neutron flux; and its concurrent disintegration or decay as expressed by its half-life or its decay constant. This is shown graphically in Fig. 8.1. It is found that the activation curve is the mirror image of the decay curve for the same nuclide. If the radioisotope produced by irradiation did not decay then the amount present would increase linearly, as indicated by the straight dotted line.

Due however to the simultaneous decay it accumulates in fact at a decreasing rate until theoretical equilibrium or saturation activity is achieved at infinite time. At this point the radionuclide would decay as fast as it was produced.

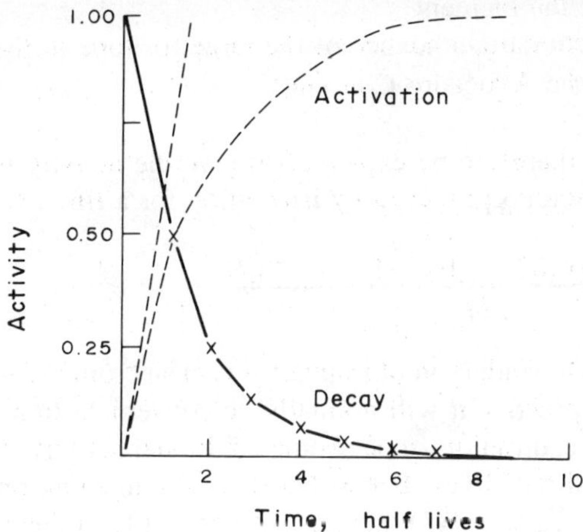

FIG. 8.1 Neutron activation curve.

It is therefore necessary to activate a sample long enough to give a measurable amount of product radionuclide, taking into account the simultaneous decay. The activity after an infinite time of irradiation, A_∞, will be:

$$A_x = N. I. \sigma$$

where N = the number of target nuclei able to form the radionuclide
I = the neutron flux, as particles/cm²/sec
σ = activation cross section, in barns/nucleus

However, we have seen that the saturation activity $A\infty$ cannot be achieved with a finite time of irradiation because of the concurrent decay of the product radionuclide, so after an interval of time, t, the activity produced, A_t, is given by:

$$A_t = A\,(1 - e^{-\lambda t}) \qquad (2)$$

or

$$A_t = N. I. (1 - e^{-\lambda t}) \qquad (3)$$

where λ = the decay constant for the product radionuclide
e = 2.71828

but N may be derived from

$$N = \frac{m}{M} \cdot E \cdot (6.02 \times 10^{23}) \tag{4}$$

where M = atomic weight of the element
m = mass of the element
E = the fractional abundance of the target isotope in the natural element
and 6.02×10^{23} is the Avogadro Constant.

Equation (3) may therefore be expanded to give the activity in disintegrations per second of the radioisotope produced by irradiation for a time, t, as follows:

$$A_t = \frac{m \cdot I \cdot \sigma (6.02 \times 10^{23})(1 - e^{-\lambda t})\, E}{M} \tag{5}$$

In terms of practical irradiation of samples it is clear from theoretical considerations and Fig. 8.1 that in practice it will normally be convenient to activate for about one half-life of the produced radioisotope. Some additional activity will be achieved after irradiation for several half-lives, but with strictly diminishing return.

It is apparent from Equation (5) that it is possible, at least theoretically, to calculate the amount, m, of the element being determined by substitution of known and experimentally determined values. However, as mentioned at the beginning of the chapter, normal everyday procedure is to use a standard comparison sample containing a known amount of the element being analyzed. This is irradiated in the same position in the reactor as the unknown sample, and the two samples are subsequently manipulated and counted in identical manner. The amount of the element in the unknown can then be easily calculated, as the radioactivity of the samples is directly related to the amount of the element present, the amount in the standard being known.

The production of a radionuclide for activation analysis is influenced by the reactor neutron flux, the radiation cross section of the element being irradiated and the half-life of its product. Now, as the majority of reactors have a neutron flux of about 10^{12}–10^{14} n/cm^2 it is apparent that the practicality of activation analysis really depends on the cross section of the element and the product half-life, together with the abundance (E in equation 5) of the target isotope in the naturally occurring element. If the half-life is too short there is inadequate time for subsequent manipulation, separation and counting before the activity becomes too low, while if it is too long then the cost of irradiation and the analytical delay become prohibitive. In general we need a half-life of between a few minutes and several days. It is obviously more practicable to irradiate for several half-lives in cases where the half-life of the isotope is comparatively short. Of the elements for which activation analysis is practical, phosphorus-31, has one of the longest half-lives of 14 days. It should be noted that the higher the abundance, E, then the greater is the sensitivity for activation.

TABLE 8.1
Nuclear activation data for the (n,γ) reaction of some biologically important elements (slow neutrons) ([20,28,37])

Element	Abundance (%)	Cross section (barns: 1 barn = $10^{-24}cm^2$)	Isotope produced	Half-life
Arsenic-75	100.00	4.2	^{76}As	26.8 hr
Calcium-44	2.06	0.7	^{45}Ca	152.0 day
Chlorine-37	37.3	0.5	^{38}Cl	37.3 min
Cobalt-59	100.0	37.0	^{60}Co	5.27 yr
Copper-63	69.06	4.5	^{64}Cu	12.8 hr
Iodine-127	100.00	5.5	^{128}I	25.0 min
Iron-58	0.31	0.9	^{59}Fe	45.0 day
Magnesium-26	11.3	0.026	^{27}Mg	9.45 min
Manganese-55	100.00	13.0	^{56}Mn	2.59 hr
Molybdenum-98	23.78	0.15	^{99}Mo	2.85 day
Nickel-64	1.08	1.6	^{65}Ni	2.6 hr
Phosphorus-31	100.0	0.19	^{32}P	14.3 day
Potassium-41	6.91	1.1	^{42}K	12.5 hr
Sodium-23	100.0	0.51	^{24}Na	15.0 hr
Strontium-86	9.86	1.3	^{87}Sr	2.8 hr
Sulphur-34	4.21	0.2	^{35}S	87.1 day
Zinc-68	18.57	1.1	^{69}Zn	57.0 min

Table 8.1 gives the relevant nuclear characteristics of some biologically important elements, the cross sections being for slow neutrons. Manganese, with high abundance, high cross section and having a radionuclide with an adequate half-life is typical of an element very suited to slow neutron activation analysis. On the other hand iron, with low abundance, a low cross section, and having an isotope with a half-life of 45 days clearly presents problems.

Neutron sources other than nuclear reactors may also be used to activate samples. Neutron generators have already been referred to. They consist of small particle accelerators producing high energy deuterons which are accelerated by high voltage. The deuteron stream is directed at a zirconium hydride-^3H target, i.e. Zr^3H_4. Neutrons are then produced by the reaction

$$^3H(d, n)^4He$$

Other small neutron sources are based on the alpha-ray reaction (α, n) on beryllium. This reaction has been discussed in Chapter 2. In practice, alpha sources used include radium-226, plutonium-239, polonium-210, actinium-227, americium-241, thorium-228 and antimony-124. Of these nuclides, the half-lives of ^{124}Sb ($t_{1/2}$ = 60 days) and ^{210}Po ($t_{1/2}$ = 140 days) are really too short for utility. In this respect ^{228}Th is somewhat better with $t_{1/2}$ = 1.9 years, but many users will prefer the longer half-life offered by the radium and plutonium sources. They are of course more expensive. The americium/beryllium source is largely free of γ contamination. In these sources the finely prepared beryllium is intimately mixed with the alpha-emitting radionuclide. The source is held in a shielded container with an access port for inserting samples.

Nevertheless, the very low neutron flux of such sources restricts their use to elements of high cross section and where great sensitivity is not required. For example, beryllium/radionuclide sources give a neutron flux of about only $10^3 - 2.5 \times 10^5$ n/cm^2/sec and these can never be of wide application, though they offer the possibility of determinations by prompt (n, γ)technique.

The use of Californium-252 sources for activation analysis is rapidly growing, although still comparatively experimental ([41]). ^{252}Cf is a spontaneous fission neutron emitting radioisotope, and gives a high neutron flux of about 10^7n/cm^2/sec per miligram. It has an important advantage over Ra-Be sources in that the accompanying gamma flux is less by a factor of about 700.

The comparatively recent introduction of 14 MeV neutron tubes with a fast neutron flux of $3 \times 10^{10} - 10^{11}$n/cm^2/sec has totally altered the possibilities of activation analysis for many laboratories, hitherto without access to a reactor. However, this flux compares quite favourably with the 10^{12}–10^{14}n/cm^2/sec or more of reactors and the 5×10^8–10^{12}n/cm^2/sec of the cyclotron or Van den Graaf type generator ([31]).

Further discussion on the general principles and practice of activation analysis can be found in references 5, 24, and 27.

FACTORS AFFECTING PRACTICAL ANALYSIS BY SLOW NEUTRON ACTIVATION

The basic limiting factors such as the cross section of the isotope being activated, its abundance in the natural element, and the half-life of the product radionuclide have already been considered. This section considers some additional factors which must be taken into account in practical analysis. It primarily relates to slow neutron activation, but many of the factors also have relevance to fast neutron activation and analysis.

Irradiation and Cooling Time

As different elements have different neutron cross sections and different half-lives it follows that the optimum time of irradiation will vary according to the element which is being determined. However, when more than one element is to be determined in the same sample then more compromise in irradiation time becomes necessary. Where two elements have widely different half-lives and therefore irradiation requirements, e.g. manganese-56 and phosphorus-32, it may be more practicable to sub-divide the sample and irradiate the sub-samples for different times.

Biological samples are normally very complex and there are always problems of other elements present becoming activated, and yielding products with similar half-lives or radiation characteristics. Additionally there are usually a number of very short-lived radionuclide products, of no analytical interest, also formed during irradiation. In order to partly overcome these problems it is customary to allow a "cooling" period after irradiation to allow for the decay of interfering nuclides of short half-life.

In general, the time of irradiation and cooling must be chosen so that there is optimum activation of the desired isotope(s). Thus short irradiation times and a short cooling period will give an enhancement of short-lived components, while allowing a long cooling time will tend to relatively enhance the nuclides with long half-lives. For example, ^{32}P, ^{35}S and ^{45}Ca are isotopes with comparatively long half-lives and consequently it is possible to allow many other potentially interfering nuclides to decay to a negligible amount.

In the case of reactor activation the nature of the neutron flux must also be taken into account. Although with irradiations in the thermal column of the reactor the majority of the neutrons may be of slow or thermal type, an appreciable proportion of the flux can be of fast neutrons. This can give rise to a source of error as described in the next section, and it is necessary to know what proportion of the flux is made up of fast neutrons. The nearer to the centre of the reactor core the higher is the proportion of fast neutrons.

It is usually not necessary for the individual analyst to determine the neutron flux when working with a reactor, as normally reactor operators have adequate knowledge of the neutron flux under different operating conditions and at various positions in the reactor. However, when using a small laboratory source, especially one based on a relatively short-lived isotope, it may be necessary to calculate the neutron flux. This can be done using indium foil. A piece of indium foil of known weight is irradiated for about two half-lives (^{115}In, 95.8% abundance $\longrightarrow ^{116}$In, $t_{1/2}$ = 54 min) the precise time of irradiation being noted. The activity of the irradiated foil is determined by means of comparing it with a standard beta reference source. Due to the short half-life of ^{116}In the precise times when the counts are taken must be carefully noted, so that the actual activity of the foil at the end of the irradiation period can be calculated with correction for decay. The flux can then be calculated by substituting known values in equation 5.

Interfering Reactions

Fast neutrons can give rise to transmutations other than of the (n,γ) type characteristic of slow neutrons, as has been discussed in Chapter 2. A certain proportion of fast neutrons will be present in reactor irradiations and therefore these potentially interfering reactions must be taken into account. It will be appreciated that some of the fast neutron reactions which interfere in slow neutron determinations are precisely the same reactions which make determinations by fast neutrons possible.

An example of this is the $^{31}P(n,\gamma)^{32}P$ determination of phosphorus, where there is interference from

$$^{32}S(n,\ p)^{32}P \text{ and } ^{35}Cl(n,\ \alpha)^{32}P$$

these two reactions both giving ^{32}P as a product. It will be noted incidentally that the reaction $^{32}S(n,\ p)^{32}P$ offers a means of determining sulphur, which is not usually possible by the $(n,\ \gamma)$ process as the abundance of ^{34}S (4.21%) is too low. In this case

$$^{32}S(n,\ p)^{32}P \text{ and } ^{35}Cl(n,\ \alpha)^{32}P$$

$$^{31}P(n,\ \gamma)^{32}P \text{ and } ^{35}Cl(n,\ \alpha)^{32}P$$

Where attempt is made to determine sulphur by the $(n,\ \gamma)$ reaction there is interference from chlorine by means of

$$^{35}Cl(n,\ p)^{35}S$$

Neither phosphorus nor sulphur interfere in the determination of chlorine, because both phosphorus and sulphur form pure beta-emitters after a long irradiation period, while ^{38}Cl is a gamma-emitter which is formed during a short activation, much too short a period to activate ^{32}P and ^{34}S.

One means of overcoming problems of interference due to reactions other than (n, γ), i.e. reactions due to fast neutrons as opposed to slow neutrons, is by means of a double irradiation technique (Bowen and Gibbons). (5) This technique consists of irradiating the samples and standards at two separate positions in the reactor, these positions having different proportions of fast to slow neutrons. Thus one set of samples will be placed in the core of the reactor while the second set is irradiated towards the edge. The values for the activities obtained from irradiation at the two positions can then be substituted in simultaneous equations for the determination of the unknowns. This method has been used for the simultaneous determination of *P* and *S*, as described in the next section.

There are a number of potentially interfering reactions in the determination of some of the cations, but possibly not so troublesome as in the case of phosphorus and sulphur just considered. Thus in the determination of manganese by the $^{55}Mn(n, \gamma)^{56}Mn$ reaction there can be

$$^{56}Fe(n, p)^{56}Mn \text{ and } ^{59}Co(n, \alpha)^{56}Mn$$

In practice, the determination of manganese is one of the least troublesome techniques.

Copper determination by $^{63}Cu(n, \gamma)^{64}Cu$ can have interference from $^{64}Zn(n, p)^{64}Cu$. Determination of potassium by $^{41}K(n,O\ \gamma)^{42}K$ can theoretically be affected by $^{42}Ca(n, p)^{42}K$ and $^{45}Sc(n, \alpha)^{42}K$, but such interference is usually negligible as few biological samples contain ^{45}Sc in any amount and the abundance of ^{42}Ca is only 0.64%. Much the same considerations apply to the determination of calcium by $^{44}Ca(n, \gamma)^{45}Ca$ where there can also be the reactions $^{45}Sc(n, p)^{45}Ca$ and $^{48}Ti(n, \alpha)^{45}Ca$. Here again neither biological nor soil samples contain sufficient ^{45}Sc or ^{48}Ti to be of any significance. The major problem of calcium is its soft beta-ray counting characteristics.

Methods of Analysis

It will already have been realized that there is considerable variation in the manner that an activation analysis can be carried out, depending on the sample, the determinations required, and the instrumentation that is available. There is no "hard and fast" correct way, and the examples that follow have been chosen to illustrate the different approaches possible.

Preparation of Samples and Standards

A major advantage of activation analysis is that the sample is usually taken without any preliminary handling except for the necessary sub-sampling. For this reason possible contamination is reduced to a minimum. Care should be taken to avoid contamination in any sampling or grinding procedure and samples should be handled in the manner which has long been customary for spectrographic analysis.

Samples are normally placed in sealed polyethylene or quartz vials. The size of sample used for homogenous plant and soil materials varies between about 50–250 mg, a sample of 100 mg being an appropriate amount for first trials, especially if it is intended to use a separation procedure to analyze for a number of elements. Fourcy *et al.*, ([14]) irradiated 50 mg of plant ash for their systematic radiochemical separation scheme.

Standard solutions are prepared from spectrographically pure nitrate or chloride salts of cations. For the anions, a potassium salt may be used for phosphate, and magnesium for sulphate and chloride. The short half-life of ^{27}Mg ($t_{1/2}$ = 9.45 min) makes it quite suitable for a chloride standard as it decays during the cooling period. Moreover, the ^{27}Mg photopeak does not interfere with the 1.62 MeV photopeak of ^{38}Cl. Standards may be irradiated as solution or "taken up" in the vial on a small piece of ashless filter paper.

The concentration of the standard should be calculated to approximate the level anticipated in the samples. As a trial, 5–20 μg can be taken for macro-elements and 1–5 μg for microelements, though some e.g. cobalt, would require even less. Where the sample is to be analyzed for several elements then a composite "cocktail" standard containing the desired elements may be made up. However, if there is any doubt about interference, either through secondary reactions or the coincidence of photopeaks at counting, then standards of single elements should be irradiated as well, at least when procedures are being developed.

Irradiation

The principles concerning irradiation and cooling time have already been considered. In practice this means that, assuming a flux of approximately 1×10^{13} n/cm^2, appropriate times are likely to be within the following ranges:

1. *Isotopes with favourable cross-sections and/or products with very short half-life:*
   e.g. Manganese, Chlorine, Copper, Zinc: 5–30 min irradiation time;
   0.5–2.5 hr cooling
2. *Isotopes with short half-life products:*
   e.g. Potassium, Sodium: 12–24 hr irradiation time; about 4 hr cooling
3. *Isotopes with long half-life products:*
   e.g. Calcium, Phosphorus, Sulphur (by ^{32}S (n,p) ^{32}P reaction):
   3–5 days irradiation time;
   4–24 hr cooling
4. *Isotopes with unfavourable cross-section or abundance, or with long half-life products:*
   e.g. Cobalt, Iron:
   5–14 days irradiation time;
   3 days cooling

The above list is clearly not exhaustive, but it will serve as a guide when setting up a procedure. There is nothing hard and fast about either the irradiation or cooling times, and it is easy to find in the literature examples quite different from those above. Again, they are a guide. It will be obvious too that in any scheme for the analysis of a series of elements from a single irradiated sample it is always necessary to compromise on irradiation and cooling times. That is, between those ideal for short half-life products and those desirable for isotopes with unfavourable cross sections or products with long half-lives.

Quantitative Determination

It is possible to quantitatively determine a number of elements without radiochemical separation of the sample ([8]), but the majority of investigators have found it desirable to have at least a group separation when handling complex biological samples. To a certain extent the degree of separation necessary depends on the sophistication of the counting equipment that is available. At the one extreme a laboratory with only elementary G-M or basic-type scintillation counters available will have to separate each nuclide before counting. This often involves the addition of an aliquot of stable carrier of the desired element, with a succession of classical wet-chemistry steps leading up to the precipitation and separation of the nuclide to be counted. This is a tedious process at best and probably only justified when activation analysis is clearly the best means of carrying out the determination, e.g. because of very small sample size.

The approach that the majority of procedures have favoured is the separation of the element or groups of elements by means of exchange resins. Some workers have assayed the nuclide on the resin, while others have removed the element(s) by selective elution before assay. The ion exchange separation scheme of Samsahl ([33,34,35]) has been used in a number of procedures, and may be automated ([42]).

Nowadays most laboratories undertaking much activation analysis are usually equipped for gamma spectrometry, with a multi-channel instrument and computation facilities. This equipment enables the ready identification of nuclides, direct evaluation of interfering nuclide activities, and quantitative measurement of the desired nuclide, despite other activities. Such sophisticated equipment makes possible the most effective use of the activation technique. Nevertheless, methods requiring only simple equipment should not be overlooked, especially be laboratories that need to employ activation techniques only infrequently.

The assy of ^{45}Ca and ^{32}P obviously demands a beta counting technique, but for the remainder of the nuclides likely to be of interest, gamma counting techniques are usually used for preference. With basic equipment, and assuming complete separation of the nuclide of interest, then the unknown weight of element (W_u) in the sample is given by:

$$W_u = \frac{C_u}{C_s} \times W_s \tag{6}$$

Where W_s = weight of element in the standard
C_u = count rate of unknown sample
C_s = count rate of standard

When a gamma spectrometer is available more elaborate treatment is possible. The γ-spectrum, being characteristic for each nuclide, can be initially used to identify the nuclide sought, then the size (area) of the appropriate photopeak can be measured. Typically, a photopeak area such as Fig. 8.2 for ^{56}Mn will be obtained. The area of this photopeak can be calculated empirically by the equation

$$A = 1.07 \times h \times b$$

where h = height of the photopeak
and b = width of the photopeak at half height

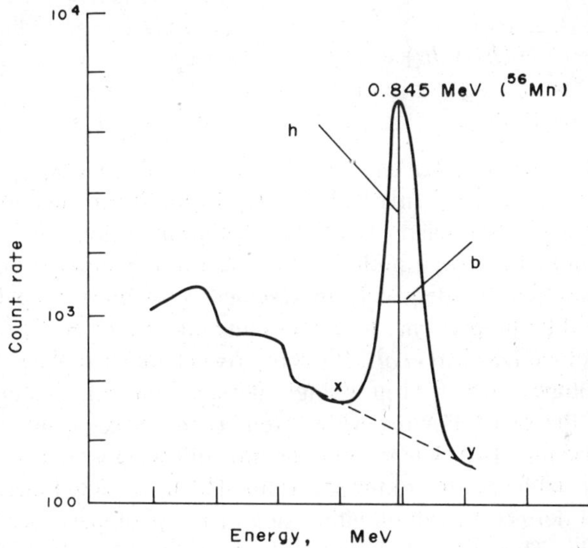

FIG. 8.2 γ-ray spectrum for quantitative determination of ^{56}Mn at the 0.845 MeV photopeak.

The calculation of the unknown weight of the element (W_u) in the sample then follows the pattern of equation (6):

$$W_u = \frac{A_u}{A_s} \times W_s \tag{7}$$

where, W_s = weight of element in the standard
A_u = area of photopeak (unknown sample)
A_s = area of photopeak (standard)

Measurements such as this are satisfactory when either there is only one nuclide present, or when there is more than one, the nuclides have widely different energies or advantage can be taken of differences in half-life, i.e. by allowing short half-life nuclides to decay before counting a long half-life product. However, if the energies are not widely separated there is in effect a multicomponent γ-ray spectrum in which the abundance of each component nuclide is masked by the contribution of the others. For such cases the technique of "spectrum stripping" was developed by Covell ([9]), and has been applied to analysis of plant tissue samples by Dieckert *et al.*, ([11]) and Fourcy *et al.*, ([14]). A multi-channel spectrometer is necessary.

The original communication of Covell ([9]) and also that of Dieckert *et al.* ([11]) should be consulted for a full appreciation of the technique, but briefly, the method is based on the concept that a multi-component spectrum equates with the algebraic sum of its component spectra. If therefore a photopeak region is taken to be the sum of a linear and a non-linear step function, then the area under the non-linear function will be given by:

$$C = \sum_{i=1}^{n} a_i - \frac{n(b_1 + b_n)}{2} \tag{8}$$

where a_i = area of the steps (channels) of the chosen photopeak region
b_1 and b_n = areas of the first and last step of the linear function
n = number of steps (channels) in the photopeak region

On the assumption that the contribution to the spectrum from pulses other than from the nuclide desired can be adequately represented by a linear step function in the region, then C will be proportional to the absolute count rate of the nuclide.

In using this method Dieckert *et al.*, ([11]) chose five channels in the photopeak region so that if the photopeak occurred in a single defined channel, this channel plus the two in front and the two following were taken. If the peak occupied two channels then either the preceding two channels and the one following were taken or vice versa, depending on a visual inspection of the spectrum. The linear step function was chosen by seeking a well defined trough on either side of the photopeak of the spectrum of a standard sample. These trough channels were used to establish the linear step function, the mid-points of the steps of the linear function corresponding to points on the line. Figure 8.3 from Dieckert *et al.*, ([11]) illustrates this, where area B, the linear step function in the photopeak region is represented by the second term of equation 8.

The spectrum stripping technique is in effect a sophisticated development of what can be done visually on a recorder scan of the λ-spectrum. Thus the dotted line intercept *xy* in Fig. 8.2 equates with a linear step function. For example, Nilubol and Kafkafi ([29]) found that manganese in their (separated) samples was proportional to the area above such an intercept.

In practice, even assuming favourable distribution of the photopeaks, it has seldom been possible for γ-ray spectrometry alone to distinguish quantitatively between more than five or six elements at the most. This has usually made it necessary for at least a partial radiochemical separation of the irradiated sample. A number of such schemes have been successfully applied to biological material.

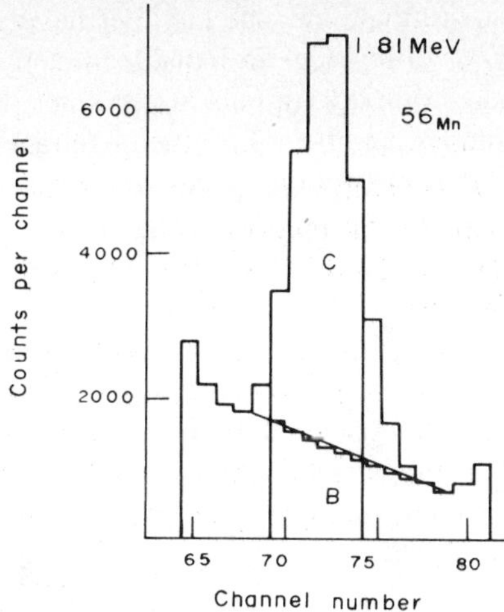

FIG. 8.3 From Dieckert *et al.*, ([11]) γ-ray spectrum to show the selection of a photopeak region and a linear step function for spectrum stripping. The count for a given channel is represented by the area under each step.

The development of computer programmes for comparative evaluation by least squares analysis of gamma spectra from samples and prepared standards of known content has greatly enhanced the possibility of purely instrumental determinations ([22,23,30,36]). This is especially significant for fast neutron activation, when the half lives of the radioelements to be counted are frequently very short, and these methods have been applied to plant material ([6,36]).

Analytical Schemes

This section examines one or two analytical schemes that have been successfully used, plus a number of procedures for single elements.

(i) *Phosphorus, Sulphur and Chlorine*

The problems of interference in the determination of phosphorus and sulphur have already been discussed. There is also the inherent difficulty of determining *S* by the $^{34}S(n, \gamma)^{35}S$ reaction. The method described here due to Cho and Axmann ([7]) determines sulphur by the $^{32}S(n, p)^{32}P$ reaction and makes use of the double irradiation technique to overcome the effect of interfering reactions on *P* assay. The rather complete separation technique, based on the work of Samsahl (*loc.cit.*) makes it practicable to use basic counting equipment:

1. *P, S* and *Cl* standards plus sample vials are irradiated at two places in the reactor, one near the core and the other towards the edge, giving different ratios of fast to slow neutrons. Irradiation time is about 3 days assuming a flux of 10^{13} n/cm^2.
2. P is separated by placing a 1 ml aliquot of sample on a Dowex 2 × 10 NO_3

resin column of 1 cm diam × 7.5 cm. Cations are eluted with 25 ml water, the eluate being discarded. Phosphate is then eluted from the column using 20 ml of 0.1 N HNO_3 at a rate of 2 ml/min. ^{32}P may then be counted by liquid G-M counting after bringing all samples to constant volume; by Cerenkov counting; or alternatively after precipitating and filtering of the *P* as ammonium phosphomolybdate following the addition of 5 mg of *P* as carrier. Samples of unknowns and standards are treated similarly.

3. The *S* standards should be assayed for ^{32}P in exactly the same manner as the *P* standards (recall: ^{32}S (n,p) ^{32}P).

4. In the case of *Cl*, the ^{38}Cl activity of both samples and standards is determined by counting the resin after transferring to counting vials. The photopeak area of 1.6 MeV is then used for γ-counting. Additionally, the ^{32}P activity of the *standards* (recall: ^{35}Cl (n, α) ^{32}P) must *also* be assayed for β-counting.

5. The calculation is somewhat complex, involving simultaneous equations as follows:

$$Cu_1 - Col_1 \times \frac{Acl_1}{A\dot{c}l_1} = \frac{Cp_1}{W_p} x_p + \frac{Cs_1}{W_s} x_s \qquad (9a)$$

$$Cu_2 - Col_2 \times \frac{Acl_2}{A\dot{c}l_2} = \frac{Cp_2}{W_p} x_p + \frac{Cs_2}{W_s} x_s \qquad (9b)$$

where,

Cu_1 and Cu_2 are the ^{32}P activities of the same amount of *unknown sample* irradiated at reactor positions 1 and 2.
Cp_1 and Cp_2 are the ^{32}P activities of the *standard P samples* at positions 1 and 2.
Cs_1 and Cs_2 are the ^{32}P activities of the *standard S samples* at positions 1 and 2.
Col_1 and Col_2 are the ^{32}P activities of the *standard Cl samples* at positions 1 and 2.
Acl_1 and Acl_2 are the ^{38}Cl activities of the *unknown samples* at positions 1 and 2.
Acl_1 and Acl_2 are the ^{38}Cl activities of the *standard Cl samples* at irradiation positions 1 and 2.
W_p and W_s are the weights of the phosphorus and sulphur standards.
x_p and x_s are the unknown weights of phosphorus and sulphur in the samples.

Thus the amounts of sulphur and phosphorus being calculated from the simultaneous equations, the chlorine content can be determined directly from the ^{38}Cl of standard and sample.

(ii) *Manganese, Potassium and Sodium (plus Chlorine)*

This scheme, also from Cho and Axmann ([7]) is of particular interest as its separation procedure makes it possible to use simple counting techniques.

1. Vials of sample and a "cocktail" standard solution containing Mn, K and Na

as chloride or nitrate salts are irradiated for 30 min at 10^{13} n/cm^6, with a cooling time of 20 min.

2. 1 ml aliquots of the samples and standards are taken and placed on Dowex 2-X10 Cl^- resin columns, of 2 cm internal diam with 10 ml wet resin. If in solid form then the samples must first be ashed and made up to known volume. The resin is washed with 15 ml of water and the effluent directly passed onto a Dowex 2-X10 OH^- column of the same dimensions.

3. If it is desired to determine chlorine the resin from the first column is transferred to a counting vial and the ^{38}Cl photopeak area of 1.6 MeV determined by γ-spectrometer.

4. The Dowex 2-X10 OH^- resin is then washed with 20 ml H_2O at a rate of 2 ml/min, the effluent being collected. The resin may then be transferred to a counting vial and manganese determined by γ-spectrometry using the 0.845 MeV photopeak area of ^{56}Mn.

5. To the effluent is added 3 mg of carrier-K and 3 ml of 3% sodium dipicrylate solution, and then warmed. Precipitation is carried out by cooling the solution very slowly from about 50°C to 20°C, to minimize the co-precipitation of Na with K. The K-dipicrylate ppt is filtered off using a filter assembly that will accommodate a 2.5 cm diam filter paper disc, the ppt being washed with saturated K-dipicrylate solution. The paper plus ppt is then placed in a G-M counter for determination of ^{42}K activity.

6. The filtrate from the previous stage is passed directly through a Dowex 50-X12 H^- resin column, the resin transferred to a counting vial and ^{24}Na determined by γ-spectrometer using the 1.37 MeV photopeak area.

7. Calculations are all straightforward and follow equation 6.

(iii) *Systematic multiple cation separation*

Fourcy *et al.*, ([14]) developed a systematic multiple cation separation procedure based on principles suggested by the work of Helwig *et al.*, ([19]) Auboin ([1]) and Auboin and Laverlochere ([2]). In this scheme, iron, cobalt, copper and zinc are removed by passage in 8N HCl through an anion exchange resin; the eluate is diluted to 0.1 N HCl, the other cations being fixed on a cation exchange resin and systematically eluted with successively increasing concentrations of HCl.

The general scheme is shown in Fig. 8.4. Briefly, the procedure involves the irradiation of 50 mg of plant ash for about 4 days at a high neutron flux, together with appropriate "combined" standards. The samples are allowed to cool for as much as 3 days to reduce the high activity due to ^{24}Na and ^{42}K. The ash is then taken up in 8N HCl and passed through Dowex 1 anion exchange resin to remove heavy metals, the other cations and phosphorus passing through in the eluate. Co, Fe and Zn are then systematically eluted from the Dowex 1 resin using decreasing concentrations of HCl. In the meantime the initial eluate is diluted to 0.1 N HCl and the cations fixed by passing through a Dowex 50W-12 cation exchange column, with the phosphorus being discharged in the eluate. Systematic elution of the cations from the column by increasing concentrations of HCl is carried out by means of a constant

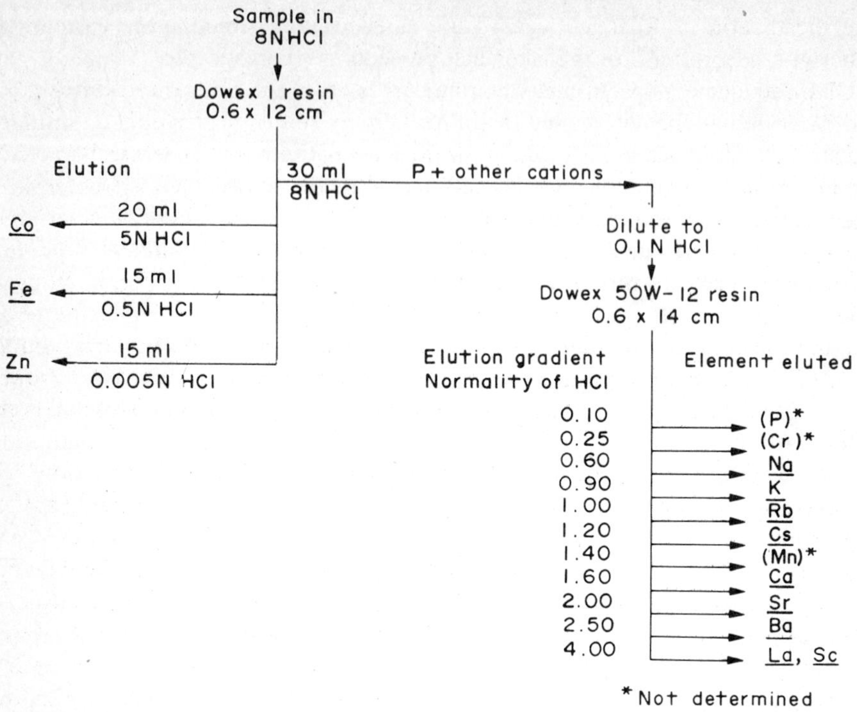

FIG. 8.4. Systematic Multiple Cation Separation (Fourcy *et al.*) ([14])

volume pump and a fraction collector, the tubes from the fraction collector passing in front of a gamma counter coupled to a ratemeter with chart recorder. This gives a continuous record of activity *vis-à-vis* the elution curve, which permits calculation of the amount of element present in comparison with the standards. ^{45}Ca must of course be counted separately by β-counting procedure.

Gaudry *et al.*, ([15]) developed a procedure for determining twenty-four elements in liver, following an intense 24 hr irradiation by a neutron flux of 2.5×10^{14}n cm$^{-2}S^{-1}$. Nitric acid digestion was followed by ion exchange chromatography, a separation apparatus handling four samples simultaneously in a shielded cell.

(iv) *Methods for Single Elements*

Although the greatest benefits of activation analysis are probably obtained when a series of elements are determined, there will be cases when an investigator is only interested in one element, and is attracted to activation analysis by its effectiveness for extremely small samples. This section briefly considers the general approach to a number of biologically important elements. Where a wet chemistry separation procedure is mentioned it is taken for granted that about 5 mg of stable element is added after irradiation in order to "carry" the labelled atoms through the various stages. Otherwise adsorption losses make such procedures impracticable.

Given adequate γ-spectrometer equipment, *manganese, potassium, copper, sodium* and *chlorine* may be determined instrumentally by spectrum stripping or similar techniques, providing that use is made of the different half-lives and energy characteristics. Thus ^{56}Mn ($t_{1/2}$ = 2.59 hr) would be determined first on the 0.845 photopeak, when after further decay to reduce interference ^{42}K ($t_{1/2}$ = 12.5 hr) might be determined on the 1.53 MeV photopeak, and/or ^{64}Cu ($t_{1/2}$ = 12.8 hr) at the 0.51 MeV photopeak. There is normally no problem in determining ^{24}Na ($t_{1/2}$ = 15 hr) as a photopeak can be chosen well apart from ^{42}K, while other potential activities are allowed to decay. Sodium can seriously interfere with ^{42}K determination and the potassium/sodium ratio must exceed 100 for a purely instrumental determination to be possible (Zmijewska and Minczewski) ([40]). If the ratio is less than 100 then separation on an ion exchange column followed by systematic elution is necessary. Sulphate and phosphate and other cations do not normally interfere.

Determination of ^{38}Cl($t_{1/2}$ = 37.3 min) depends on the level of associated K and Na. Within normal limits potassium will not interfere in the determination of ^{38}Cl, especially if determination is carried out on the 2.15 MeV photopeak. Serious interference will occur if there is sodium present, although Zmijewska and Minczewski (*loc. cit.*) found that determination was possible provided the sodium/chlorine ratio did not exceed one. When sodium is present it is not practical to use the 2.15 MeV photopeak because of coincidence with the 2.24 MeV ^{24}Na escape peak, and consequently the 1.62 MeV photopeak is used. This in turn implies that the level of K is very low, because of interference from the ^{42}K 1.53 MeV photopeak.

In biology and agriculture there are relatively few conditions in which chlorine occurs in significant amounts without high potassium or sodium, so in practice it will often be necessary to separate chlorine from sodium. This may be done e.g. by passing the liquid sample through a Dowex 2-X10 Cl^- resin column when the ^{38}Cl is retained on the resin, or alternatively by passing through an Amberlite IR-120 (H) column when the K and Na are removed and chlorine can be collected in the effluent.

If circumstances make it desirable to separate out manganese and copper this may be done either with exchange resins or chemically. Thus Cho and Axmann ([7]) took up ^{56}Mn on Dowex 2-X10 OH^- resin, and counted the resin, as K and Na passed through in the effluent. Alternatively Mn can be precipitated as MnO_2, as was the method of Nilubol and Kafkafi ([29]).

In the case of copper, the element may be separated by an ion exchange procedure developed by Samuelsson ([35a,35b]) for Cu in milk. In this method the ashed sample plus carrier Cu is taken up in 1N HCl containing 0.3% H_2O_2 and passed through an anion exchange resin in chloride form, which is then washed with further 1N HCl. The eluate is evaporated to dryness, dissolved in 1N HCl, a drop or two of H_2O_2 added and adjusted to pH 3.5 with 0.05N NaOH. This solution is then passed through Dowex 2-X10 resin in citrate form, the resin washed with water and 0.01N NaOH, and then eluted with 3N HCl. The eluate is made alkaline, tetraethylene-pentamine (TEPA: blue colour with copper in alkaline solution) added and diluted to volume. Activity is determined on an aliquot of the solution, while the colour can be determined spectrophotometrically at 635 nμ. Carrying out a colorimetric determination simultaneously (recall that Cu carrier is added) enables the chemical yield to be checked and the possibility of correction for any losses during the procedure.

Despite poor cross section and low abundance a useful method was developed for selenium, $^{74}Se(n, \gamma)^{75}Se$, in plant material, with separation based on nitric/perchloric digestion followed by extraction from 4M HBr solution using benzene containing 1% phenol ([44]).

II. ANALYSIS BY FAST NEUTRON ACTIVATION

The possibilities of fast neutron activation analysis were appreciated for a long time, but as a practical method suitable for routine use it required the development of new generation sealed long-life neutron tubes ([31]). Previous to this the neutron fluxes available restricted analyses, there were problems of rapid target decay, interrupted analytical time, and for routine use a potential health hazard involved in the essential changing of the tritium targets.

The attraction of fast neutron activation lies in the possibility of quick non-destructive analysis, and substantial data exists on the sensitivity and suitability of elements for this procedure ([3,17]). Reactions for some of the commonest elements are given in Table 8.2.

TABLE 8.2
Nuclear activation data for fast neutron induced reactions of some biologically important elements ([3,17]).

Element	Reactions	E_γMeV	Half-life
Aluminium	$^{27}Al(n,\alpha)^{24}Na$	1.37–2.75	15 hr
	$^{27}Al(n,p)^{27}Mg$	0.843–1.015	9.45 min
Boron	$^{11}B(n,p)^{11}Be$	2.12–4.16	13.6 sec
Calcium	$^{44}Ca(n,p)^{44}K$	1.15–2.1	2.5 min
Chlorine	$^{37}Cl(n,p)^{37}S$	3.09	5.04 min
Cobalt	$^{59}Co(n,\alpha)^{56}Mn$	0.085–2.13	2.58 hr
Copper	$^{63}Cu(n,2n)^{62}Cu$	0.66–2.24	9.8 min
Fluorine	$^{19}F(n,p)^{19}O$	0.200–1.366	29 sec
Iron	$^{56}Fe(n,p)^{56}Mn$	0.085–2.13	2.58 hr
Nitrogen	$^{14}N(n,2n)^{13}N$	0.511	10 min
Oxygen	$^{16}O(n,p)^{16}N$	6.13–7.13	7.3 sec
Phosphorus	$^{31}P(n,2n)^{30}P$	0.511–2.24	2.5 min
	$^{31}P(n,\alpha)^{28}Al$	1.78	2.27 min
Potassium	$^{39}K(n,2n)^{38}K$	0.511–2.16	7.7 min
Lead	$^{208}Pb(n,p)^{208}Tl$	0.582–2.62	3.1 min
Magnesium	$^{24}Mg(n,p)^{24}Na$	1.37–2.75	15 hr
Mercury	$^{200}Hg(n,p)^{200}Au$	0.367–1.36	48 min
Silica	$^{28}Si(n,p)^{28}Al$	1.78	2.3 min

Considerable effort has been put into the determination of nitrogen by the $^{14}N(n,2n)^{13}N$ reaction, one of the possible applications being for use in plant breeding for improved protein ([12,16,25,38,39]). Nitrogen in dry seeds could be determined non-destructively by keeping the neutron dose low enough to obviate the induction of mutants, the seeds then being sown in the normal way.

The samples are bombarded with fast neutrons and the induced ^{13}N decays by the emission of a positron with a 10 min half-life which annihilates with an electron in the sample with the emission of 0.51 MeV gamma rays. The irradiation and times required are about 2–10 minutes each, one procedure being to irradiate a standard reference sample simultaneously followed by sequential counting with the unknown sample to determine N-content by comparison. Interference comes principally from phosphorus, silicon, and potassium. In the case of phosphorus and silicon this can be minimized by allowing the sample to decay for 3–4 times the period of irradiation, while for potassium which is small, a correction factor is applied. Additionally, interference can come from the hydrocarbon content of the sample though a secondary reaction in which recoil protons react with ^{13}C to give ^{13}N. This is overcome by subtracting a blank.

In general, increasing the size of the plant material sample increases the resolution of the method and typical samples have been normally larger than that used for the average Kjeldahl N-determination. This would be a disadvantage for plant breeders but welcomed by millers and animal feed compounders because it provides better sampling of the bulk.

Developments in fast neutron activation are now such that most elements can be determined given suitable equipment e.g. counting nuclides with half-lives of only a few seconds requires a fast pneumatic sample transfer system and computer facilities. Relevant elements include Ag, Al, B, Ca, Cd, Co, Cr, Cu, F, Fe, Hg, N, O, P, Pb, Mg, Si, Va, Zn. The comparatively large number of trace elements which can be determined by activation analysis has attracted interest from the point of view of monitoring environmental contamination from heavy metals.

Very many elements typically found in biological material can be activated by a flux of 14 MeV, by (n,p), (n,α) and $(n,2n)$ reactions. Indeed a single nuclide can be formed from several stable elements which have neighbouring atomic numbers. The major problem is therefore both to eliminate or allow for the interference, and to breakdown the spectrum in order to determine the desired energy peak(s), and the contribution of the element(s) under examination.

One method of doing this is by spectrum stripping and reference has already been made to the basic principle of this procedure. The short half-lives of many of the nuclides rule out the possibility of slow hand methods and a computer is essential. For any counting system the spectra of a considerable number of gamma rays are determined and stored in the memory of a multi-channel analyzer or computer. The spectrum of a sample is then stripped by fitting the highest energy gamma ray at the photopeak with a known response function and taking this away from the overall total spectrum. The next lowest energy photopeak is then fitted and the spectrum of the gamma ray responsible for that photopeak is also subtracted. Proceeding successively from higher to lower energy peaks the process is continued. As the computer has all the response functions stored in its memory it is programmed to proceed by trial and error to fit all the peaks simultaneously with the appropriate response functions, until it finds the best combination ([22,23,30,32]).

Using a computer and the method of least squares to reduce the complex spectra, the Grenoble group ([6,36]) have developed procedures for the analysis of N, P, K, Mg and Ca in plant material on a virtual routine basis.

Sometimes a major interference can be handled quite simply by determining what is in effect a "blank". An example of this is the determination of oxygen in material which also contains fluorine. The normal reaction for determining oxygen is $^{16}O(n,p)^{16}N$, but the simultaneous fluorine reaction $^{19}F(n,\alpha)^{16}N$ also gives the same nuclide ^{16}N. One possible answer to this is to determine total fluorine by means of the $^{19}F(n,p)^{19}O$ reaction and to subtract this from the total ^{16}N activity.

The establishment of standards and relating sample counts to them is a problem in fast neutron activation analysis, especially with analyses involving nuclides with short half-lives. Comparison of sample with standards must also take into account possible variations in neutron flux. One answer to this is the monitor foil technique principally developed by Philips. With this method the foil activity is made to correspond with that of the standard or sample by irradiating them together simultaneously, and after irradiation they are separated, and the separate activities measured. Special "rabbits" enable monitor foils to be placed either side of the standard or sample for simultaneous irradiation.

As the ratio of standard to foil activity is known to correspond to a certain element content, and as the activity per unit mass is the same from one foil to another, then it follows that the ratio of sample to foil activity can give the amount of the element in the sample. The use of foils makes it possible to take account of small variations in generator output, but the principal advantage is that only a single standard is necessary to calibrate a wide range of element content. This is because if the sample activity is too low for good statistics then increased irradiation can be given, and as the ratio of activities will remain the same, the same standard can be used. In this manner the foil activity matches the sample activity instead of the standard matching the sample.

Fast neutron generators can also be used for slow neutron activation by surrounding the target with a moderator such as paraffin wax, thus slowing down the fast neutrons. The flux is usually not more than 10^8–10^9n/cm^2/sec.

In summary, it will be realized that although activation analysis is basically simple, there are quite a number of difficulties that must be taken into account and overcome in practical analysis. The sensitivity of the method and the small size of sample required are the major positive features of slow neutron activation techniques with reactors. The advantages of fast neutron activation procedures are speed of analysis, usually minutes or even seconds, and its non-destructive nature and lack of sample preparation.

The investigator's choice of activation analysis for any particular problem will clearly depend on the element(s) he is interested in; the size of sample available to him; the cost of analysis; the availability and reliability of other procedures; the speed of analysis required and the possibility of using other physical methods of analysis such as atomic absorption spectroscopy or X-ray fluorescence spectrography.

REFERENCES FOR FURTHER READING

1. AUBOIN, G. *Radiochim. Acta* I, 117–23 (1963).
2. AUBOIN, G. and Laverlochere, J. *Rept. C.E.A.* 2359, C.E.A., Grenoble, France (1963).
3. AUDE, G. and Laverlochere, J. *Spectres gamma de radio-éléments formés par irradiation sous neutrons de 14 MeV*. Presses Universitaires de France, Paris (1963).
4. BORS, J. Untersuchungen über die Ausbreitung von Pilzsporen im Gewächshaus mit Hilfe einer Indikator—Aktivierungsmethode. *Gartenbauwissenschaft* **32**, 513 (1967).
5. BOWEN, H. J. M. and Gibbons, D. *Radioactivation Analysis*. Clarendon Press, Oxford (1963).
6. BREYNAT, G., Fourcy, A. and Garrec, J. P. Dosage rapide du magnesium et de l'aluminium. *In: Nuclear Activation Techniques in the Life Sciences*. Proc. Symp. Amsterdam, 81, IAEA, Vienna (1967).
7. CHO, C. M. and Axmann, H. *Tech. Rept. Series* **48**, 125–30, IAEA, Vienna (1965).
8. COOPER, R. D., Linekin, D. M. and Brownell, G. L. Activation analysis of biological tissue without chemical separation. *In: Nuclear Activation Techniques in the Life Sciences*. Proc. Symp. Amsterdam, IAEA, Vienna (1967).
9. COVELL, D. F. *Analyt. Chem.* **31**, 1785–90 (1959).
10. CROUTHAMEL, C. E. *Applied Gamma Ray Spectrometry*. Pergamon Press, Oxford (1960).
11. DIECKERT, J. W., Derrick, K. S., Davis, R. C. and Singh, J. *In: Isotopes in Plant Nutrition and Physiology*. Proc. Symp. Vienna, 1966 81–92, IAEA, Vienna (1967).
12. DOTY, W. H., Wood, D. E. and Schneider, E. L. Neutron activation analysis of nitrogen in feedstuffs. *Journal AOAC* **52**, 205 (1969).
13. FENDRIK, I. and Glubrecht, H. Investigations of the propagation of plant pollen by an indicator activation method. *In: Nuclear Activation Techniques in the Life Sciences*. Proc. Symp. Amsterdam, 1967, 325 IAEA, Vienna (1967).
14. FOURCY, A., Fer, A., Barbe, R. and Neugerger, M. *In: Isotopes in Plant Nutrition and Physiology*. Proc. Symp. Vienna, 1966 57–67, IAEA, Vienna (1967).
15. GAUDRY, A., Maziere, B. and Comar, D. Multi-element analysis of biological samples after intense neutron irradiation and fast chemical separation. *J. Radioanal. Chem.* **29**, 77–87 (1976).
16. GEORGI, B., Kaul, A. K. and Christoffers, D. Use of (n,2n) activation analysis for the determination of the nitrogen content in seed grains. *Genetica* **6**, 79 (1974).
17. GILLESPIE, A. S. and Hill, W. W. Sensitivities for activation analysis with 14-MeV neutrons. *Nucleonics*. November (1961).
18. GOTTSCHALK, J. Markierung von Luft under Benutzung der Reaktion $^{10}B(n,\alpha)^{7}Li$. *Atomkernenergie* **14**, 47 (1969).
19. Helwig, H. L., Ashirawa, J. K. and Smith, E. R. UCRL Rept. 2655 (1954).
20. HUGHES, D. J. and Schwartz, R. B. *Neutron Cross Sections*, BNL-325, 2nd Ed. (and supplements) (1958).
21. IAEA Report on the Contract between IAEA and the International Rice Research Institute, Parts I and II, FAO/IAEA Division, IAEA, Vienna, Mimeo. (1965).
22. JUNOD, E. Programmes FORTRAN-SAR d'analyse par activation et de spectrométrie gamma MOCASSIN. Note CEA-CENG-DR/SAR G./66.6 (1966).
23. JUNOD, E. Éléments de programme en spectrométrie gamma et analyse par activation. Rapport CEA-R-3955 (1970).
24. KOCH, R. C. *Activation Analysis Handbook*, Academic Press, New York (1960).
25. KOSTA, L., Ravnik, V. and Dumanoić, J. Determination of nitrogen in plant seeds by fast neutron activation analysis. *In: New Approaches to Plant Breeding*, Proc. Panel Röstang, 161 IAEA, Vienna (1969).
26. KÜHN, W. K. G. Determination by indicator activation analysis of the sorption conditions, the movement and the de-enrichment factor of anions at a site on the Danube. *In: Isotope and Radiation Techniques in Soil Physics and Irrigation Studies*, Proc. Symp. Istanbul, 271, IAEA, Vienna (1967).
27. LYON, W. S. (Ed.) *Guide to activation analysis*. Van Nostrand Co. New York (1964).
28. MEINKE, W. W. and Maddock, R. S. Neutron activation cross-section graphs. *Anal. Chem.* **29**, 1171 (1957).
29. NILUBOL, A. and Kafkafi, U. *In: Isotopes and Radiation in Soil-Plant Nutrition Studies*. Proc. Symp. Ankara, 1965 63–69, IAEA, Vienna (1965).
30. QUITTNER, P. *Computer Evaluation of Scintillation and Semiconductor Detector Gamma Ray Spectra*. Adam Hilger Ltd. London (1972).

31. REIFENSCHWEILER, O. A high output sealed-off neutron tube with high reliability and long life. *Proc. Int. Conf. Modern Trends in Activation Analysis*, **2**, 905–910, Washington D.C. (1968).
32. SALMON, L. Analysis of gamma-ray scintillation spectra by the method of least squares. *Nuclear Instrum: Meth.* **14**, 193 (1961).
33. SAMSAHL, K. Rept. AE-54, AB Atomenergi, Studevik, Nyköping, Sweden, (1961a), Rept. AE-56, *ibid.* (1961b).
34. SAMSAHL, K., Brune, D. and Wester, P. O. Simultaneous determination of 30 trace elements etc. Rept. AE-124, AB Atomenergi, Studevik, Nyköping, Sweden (1963).
35. SAMUELSSON, E. G. *Milchwissenschaft* **21**, 31 (1966a), and: Proceedings of 1966 Vienna Seminar *on Radioisotopes in Dairy Science and Technology*, 221–33, IAEA, Vienna (1966b).
36. SRAPENYANTS, R. A., Kovtoun, Y. L., Vernin, E., Aude, G. and Axelrad, C. Methode et installation de dosage automatique par activation neutronique de N, P, K, Ca dans les vegetaux. *In: Nuclear Activation Techniques in the Life Sciences*. Proc. Symp. Bled 373, IAEA, Vienna (1972).
37. STROMINGER, D., Hollander, J. M. and Seaborg, G. T. *Revs. Mod. Phys.* **30**, 585 (1958).
38. WOOD, B. E., Jesen, P. L. and Jones, R. E. [Kaman Nuclear] Industrial application of fast neutron activation analysis for protein content of food products. 52nd Annual Meeting Amer. Assoc. Cereal Chemists, Los Angeles (1967).
39. WOOD, D. E. Protein determinated by neutron activation analysis. *Isot. and Rad. Techn.* 9(3) (1972).
40. SMIJEWSKA, W. and Minczewski, J. *Tech. Rept.* Series no. **48**, 110–22, IAEA, Vienna (1965).
41. LUTZ, G. J. Activation analysis with a Californium-252 source. *In: Some Physical Dosimetry and Biomedical Aspects of Californium-252*. Proc. Seminar Karlsruhe, 14–18 April 1975. IAEA, Vienna (1976).
42. SAMSAHL, K. *Analyst* **93**, 101 (1968).
43. SOETE, D. De, Gijbels, R. and Hoste, J. *Neutron Activation Analysis*, pp. 836, Wiley (1972).
44. COOK, K. A. and Graham, E. R. A neutron activation method for determining submicrogram selenium in forage grass. *Soil Sci. Soc. Amer. J.* **42**, 57–60 (1978).

CHAPTER 9

X-Ray Fluorescence Spectrography For Plants And Soils

INTRODUCTION

THE use of X-ray fluorescence spectrography is the preferred method of analysis for a number of biologically important elements e.g. zinc (22) and strontium (21). Nevertheless, the development of this technique has been overshadowed by simultaneous development of atomic absorption spectroscopy. The latter has the advantage of demanding much less expensive equipment, and possibly requiring a less-knowledgeable operator. Consequently, the full development of X-ray spectrography for agricultural and biological analyses has not been achieved, although the X-ray technique has found widespread use in the analysis of rocks and ores, and in industrial analyses where near-instantaneous determinations of several elements are desired, as in process control and steel-making.

The basic advantage of X-ray spectrography over the well-established emission spectrographic techniques is that it is basically non-destructive. The material is presented to the X-ray beam as a finely ground powder, and after irradiation remains unchanged and available for any chemical tests that may be required. Further advantages are that little preparation other than grinding the material is necessary and there is no risk of contamination arising from extraction reagents, buffers, electrodes, glassware, etc., as the emission techniques. Additionally, concentration of the element is not usually required, and there is no need for great precision in taking aliquots for analysis. It seems likely, too, that when an X-ray spectrograph has been set up for any analysis it can be operated by less skilled assistance than is required for the emission instruments.

A disadvantage of "conventional" wave length dispersive X-ray methods has been the requirement for a larger sample size with a greater concentration of elements, than that needed for emission spectrography. Later developments have now made very small samples practicable, but not yet routine. Meanwhile, "energy dispersive" methods are gaining ground.

GENERAL PRINCIPLES

The principles by which X-ray fluorescence spectrography might be used for analysis have been known for a long time, and derive from the work of Barkla who found that characteristic X-rays are produced when cathode rays of high velocity are directed at a target of the particular element, and also Moseley's discovery that the wavelengths for a particular type of X-ray vary regularly from one element to the next with increasing atomic number. The use of the technique in chemical analysis was discussed by Eddy *et al.*, ([9]) and by Von Hevesy ([11]), while papers by Birks *et al.*, ([3]) and Kemp and Andermann ([14]) for example, gave impetus to the practical development of the technique. General texts are references 15, 24 and 25.

Now, as every element emits its own characteristic secondary X-ray spectrum when bombarded by primary high energy X-rays, it is clear that with a complex sample it will be necessary in practical analysis for any element to "separate" and identify its specific characteristic radiation before it will be possible to determine the amount of that particular element present. The basic method utilises a technique familiar to those conversant with X-ray crystallography, in which the various component *wavelengths are dispersed* by means of a moveable analyser crystal. Counting methods utilize scintillation spectrometry and pulse height analysis. In X-ray *energy dispersive* analysis, use of semi-conductor detectors and a multi-channel analyzer enables analysis for many elements simultaneously. Energy dispersive systems are discussed in the last section.

The sample of material to be analyzed is irradiated with a beam of high energy X-rays (some modern instrumentation also uses electron beam excitation) and the fluorescent X-rays excited in the material fall upon a curved crystal of known grating constant, frequently lithium fluoride. They are then reflected into the detector, which may be either a proportional or scintillation counter. The wavelength of the fluorescent X-rays is defined by the angular setting of the crystal and detector, according to the Bragg equation:

$$n\lambda = 2d \sin \Theta$$

where λ is the wavelength of the X-rays, d is the interatomic spacing (grating space) between successive planes of atoms in the crystal, Θ is the glancing angle of the secondary X-ray beam striking the crystal, and n is an integer, 1, 2, 3, etc. As the element being analyzed is known from the operational wavelength, as fixed by the goniometer angle and the use of a pure reference sample as check, its concentration can be determined by the relative intensity of the reflected X-rays. Recording may be carried out by a chart recorder when a continuous scan is made for general composition, or by conventional scalers and print-out when counts are made at fixed angles.

Radioisotope X-ray fluorescence spectrometry has also been developed, but the low intensity of the X-radiation originally made it too insensitive for the relatively low levels of elements encountered in biological materials. Such instruments have hitherto been used for determinations in ores and mineral processing where comparatively high concentrations are found. Current development are now greatly broadening the scope of application.

Use is made of the K and L radiation spectra, and their intensity varies according to the place of any particular element in the Periodic Table. X-ray spectra and absorption energy data may be found in standard reference texts e.g. *Handbook of*

Chemistry and Physics. Thus, the most sensitive group of elements suitable for X-ray spectrography lie between scandium (at. no. 21) and barium (56). Within this range the radiation of the K spectra can penetrate air at atmospheric pressure, and the middle elements of this series, e.g. zinc (30), strontium (38) and molybdenum (42) are ideal for the X-ray fluorescence technique. It is apparent that the biologically important element calcium (20) lies outside these limits, and to analyze for calcium it is necessary to introduce a vacuum or helium path, as the fluorescence intensity of elements 20 and below is reduced beneath practical values by passage through air. The lowest elements it is practical to analyze are magnesium (12), or down to oxygen (8) in suitable samples, as below this number in the Periodic Table the *K* radiation is too soft for detection, even using helium or vacuum paths. Above barium (56) the *K* spectra have too high energies for measurement and the *L* series spectra must be used with some loss of sensitivity.

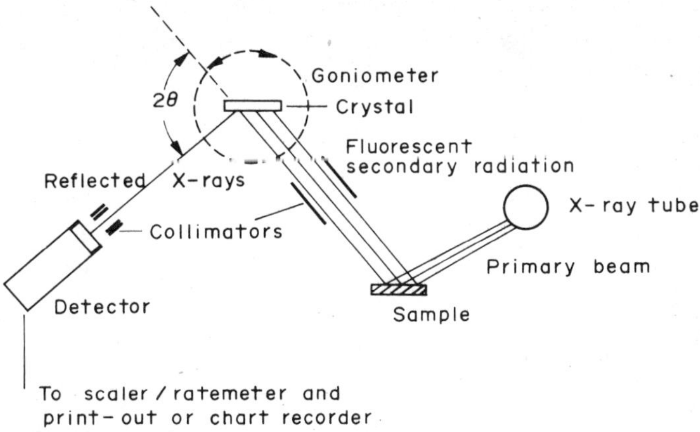

FIG. 9.1 Block diagram of the general layout of a wavelength-dispersive X-ray fluorescence analyzer.

INSTRUMENTATION

Equipment now varies enormously in its degree of complexity but the basic equipment consists of a source of high energy X-rays generated by electron bombardment of a metallic cathode, often of tungsten. Such X-ray tubes are usually operated at a potential of 50–70 kV and with 50–80 mA current, and a highly stabilized generator is essential.

The sample is held at a constant angle to the X-ray tube so that the primary X-rays from the tube are reflected by the sample to the analyzer crystal as secondary fluorescent X-rays. Analyzer crystals are frequently lithium chloride but topaz may be used for elements of high atomic number and Eddt is especially used for elements of low atomic number with X-rays of long wavelength. The analyzer crystal angle may be varied in relation to the fluorescent X-ray beam by turning the goniometer either manually or automatically. It should be noted however that the geometry of the system is arranged so that the angle and distance between the analyzer crystal and the sample always remains the same as the angle and distance between the analyzer crystal and

FIG. 9.2 An advanced X-ray fluorescence spectrograph featuring electron beam excitation, multiple crystals, vacuum and helium systems, pulse height analysis and programmable for 10–20 elements (Hilger).

the detector. This is achieved by synchronous movement of crystal and detector. Often provision is made for automatically scanning an arc, and as the goniometer is very slowly rotated the pulses from the counter are received by a ratemeter and chart recorder. Such an arrangement permits rapid qualitative scanning of a sample for its major constituent elements.

Additionally, instruments may be programmed to count sequentially at a number of pre-determined angles, for the precise analysis of a number of elements. Those instruments designed to analyze elements covering the major part of the Periodic Table may be equipped with a number of crystals, the appropriate crystal being selected automatically.

Following diffraction by the analyzer crystal the now-dispersed X-rays are intercepted by a detector. A scintillation or proportional counter is desirable, both from the point of view of counting efficiency, the ability to accept high count rates, the proportional nature of the amplification and also to permit counting in the appropriate pulse-height channel, rather than counting in integral mode. Pulse height analysis is essential for the analysis of low atomic number elements.

When it is desired to analyze for calcium (20) or elements of lower atomic number it is essential that the apparatus be equipped with either helium flow or vacuum

FIG. 9.3 Ortec energy dispersive system with radioisotope source excitation, in conjunction with a semiconductor detector and a multi-channel analyzer.

facilities, the latter being most common. Provision may be made to introduce new samples by means of an air-lock to avoid loss of counting time in achieving a working vacuum. Automatic sample changing and a means of sample rotation to reduce differences due to lack of sample homogeneity are also now available.

Well known instruments include those of Phillips, General Electric and Rank-Hilger. Elaborate instruments are able to analyze for a number of elements simultaneously and may be equipped with either internal or external standards. Such instruments may be programmed for the different analyzes required by means of plug-in circuit boards. Varied data recording and computer processing systems are available, the more elaborate ones converting "counts" direct into element concentration for print-out.

LIMITATIONS IN PRACTICAL ANALYSIS

Sample self-absorption was discussed earlier in the Chapter on sample preparation and the counting of radioactive samples (page 95). Similar considerations apply to X-ray analysis of plant and other samples. There is absorption both of the primary exciting X-rays and also of the fluorescent secondary X-rays that are emitted to the analyzer crystal. Absorption of the primary radiation reduces the penetration of the beam and therefore limits the number of atoms excited; while the secondary radiation emitted by the particular element being analyzed tends to be absorbed by elements which may have a higher coefficient of absorption for that wavelength.

It is scarcely practicable to have a sufficiently small sample of plant material i.e. an infinitely thin sample, in which all the atoms of the element being determined are excited and in which secondary radiation is not absorbed. Practical analysis is therefore carried out on infinitely thick samples, that is samples which are infinitely large in relation to the depth of penetration of the exciting X-rays. In the same way that in counting infinitely thick radioactive samples the observed activity, corrected for background, is proportional to the specific activity of the sample, so in X-ray analysis the use of infinitely large samples determines the proportionate amount of the element contained in the sample. Thus relative comparisons of total element content are obtained directly, and if the absolute content of any one standard sample is known then the absolute content of other samples of the same broad composition can be calculated.

Matrix effects are also of considerable importance. Different matrices, that is samples of different general composition, absorb radiation to a varying extent. Differential absorption of the primary radiation and the selective absorption of secondary fluorescent radiation have already been mentioned. Additionally, there is variation in the level of scatter or background radiation dependent on the general composition of the sample. It should be noted however that analyses are not influenced by the chemical form of the element being analyzed.

If samples are of the same general composition then the background will be constant. Other components of the scatter radiation are the characteristic radiation of the cathode element of the X-ray tube, and low-intensity continuous X-radiation. This background component remains constant, regardless of sample composition, but is influenced by variations in voltage or current on the X-ray tube. Such variations cause fluctuations in the intensity of the fluorescent radiation and, therefore, precisely defined operating voltage and current is essential.

METHODS OF ANALYSIS

In practice, quantitative analysis may be carried out by using standards of known composition to obtain a working curve, with which the unknown specimen may be compared. It is possible to count on the peak of an element, that is to measure the fluorescent line intensity at the correct wavelength for the element, as fixed by the angle at which the goniometer is set to receive the radiation from the sample. The count is recorded as the number of counts in a preset time, and knowing the number of counts given by a standard sample it is possible to convert this count to percentage composition. This simplified procedure is often not desirable in practice, as there is usually "other element" interference which influences the amount of scatter radiation and the size of the background count.

As the whole basis of quantitative analysis is that the intensity of the fluorescent radiation is a constant function of element concentration, methods adopted must be such that interference due to matrix effects is compensated for, either by addition of an internal standard element or by other means. In the analysis of crushed rock it has indeed been found possible to count directly, by a method known as "borax dilution" ([11,12]). It has been found that the massive addition of borax to a sample greatly reduces the matrix effect. In this method one part of sample is added to nine parts of borax, the mixture being fused and then finely ground. Such a method could be used for analysis of the "total" element present in soil, though this data is not often of value to the soil scientist or agronomist.

The use of an internal standard element was one of the earliest techniques adopted ([5,1]), but in the case of plant analysis, some investigators have preferred to use the scatter radiation as the internal standard ([4,21,22]). The basic theory is that scatter radiation of specific wavelength may be chosen, such that it is absorbed in almost identical manner to the fluorescent radiation of the element being analyzed. Thus the line/scatter ratio F/S is used as the basic unit of calculation. The line/scatter ratio may be more completely defined as the ratio of the observed intensity of a given fluorescent line, F, to the intensity of the scatter radiation, S, in a region free from lines. The F/S ratio at the given wavelength is constant per unit of element, assuming that the only interference is that deriving from absorption of the element's fluorescent radiation.

In this way it is possible to take account of the fluctuations in the background count that may occur with slight changes in the mains voltage, other chemical constituents of the sample, etc. In scanning a sample to select a known fluorescent line or peak for counting, it is desirable that the peak chosen should be fairly sharp, so that the angles selected for counting the scatter radiation intensity may be spaced reasonably close to the angle of the known fluorescent line.

The plant material is presented so that 1–2 sq cm of surface is exposed to the X-ray beam. The sample may simply be finely ground and contained in a sample holder so as to present an infinite thickness to the X-ray beam, or else may be compressed in the form of discs. The formation of compressed discs of the material has the advantage that it eliminates any possible variability due to surface irregularity, fineness of grinding or compaction. Further, it is possible to support the disc at the edges, thus lessening the possibility of background interference due to the backing of the sample holder. The main disadvantage of discs is the trouble of making them; weighing out the ground sample to obtain uniformity and compressing the material. When ground plant material is used without discing the choice of sample holder is important, as the material from which it is made must not contain any of the element under investigation. Brandt and Lazar ([4]) used a holder with a backing of 0.00025 in thick Mylar plastic film while stainless steel planchets were used by Vose and Koontz ([21]) in analysis for strontium. Bakelite may also be used. Either equal weights or equal volumes of plant material may be taken. In practice, equal volumes are sufficient to achieve uniformity, but fineness of grinding and the adoption of a standard technique of filling the sample holder and smoothing the surface is desirable.

The preparation of standard curves may be approached in a number of different ways: (1) by the chemical analysis of a sample or preferably a number of standards to obtain known values of the element under investigation, (2) by adding known amounts of the element to an organic matrix of, say, starch or mannitol, (3) by adding known amounts of the element to plant material of the type under investigation, either of known or unknown composition.

The first method ([4]) is satisfactory when it is possible to analyze for the required element by chemical means, but falls down when the element is difficult or impossible to determine chemically. The second technique is superficially attractive but is impractical, due to the difficulty of achieving a homogenous sample when mixing an infinitely small aliquot into a relatively great bulk. A further difficulty in using standards of both these types is that no adequate account can be taken of "other element" effects or the general composition of the sample. The third method, of adding known aliquots of the element to plant material of the same type as that being fluoresced, enables a standard curve to be plotted by extrapolation, without the need for chemical analysis of the unknowns. This method was used ([21]) in analysis for strontium where chemical analysis is virtually impossible, and a variation of the method ([22]) for zinc, where chemical analysis is unsatisfactory. Matrix effects are fully taken into account.

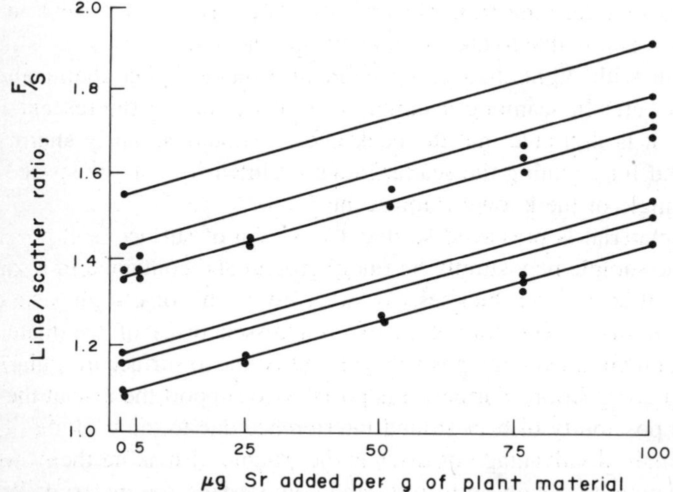

FIG. 9.4 The ratio of the strontium $K\alpha$ (first order) radiation intensity to the scatter radiation intensity as a function of added strontium to samples of plant material of six species, from 21.

The technique adopted by Vose and Koontz ([21]) was thoroughly to moisten the dry finely ground plant material with acetone and add the required strontium aliquot from a pipette, followed by complete mixing and stirring. The paste was then placed on a steam bath until the acetone had been driven off, followed by drying in the oven at 70°C. Mixing was carried out at every stage and the dry material was crushed to powder with a pestle. Separate sub-samples of standards produced by this method showed good reproducibility. When the line/scatter ratio, $\frac{F}{S}$, of a number of standards based on the same plant material was plotted a straight line resulted, which could be extrapolated to obtain the strontium content of the original plant material. This is shown in Fig. 9.4, and it can be seen that standards based on different plant species do not give super-imposed lines, but parallel lines. It is thus possible to calculate a constant factor to enable the determination of the strontium content of any plant material. In the present case the mean increase in *F/S* ratio per 100 μg of added strontium for various plant materials was 0.383 *F/S* units,

therefore, 1 μg *Sr* = 0.00383 *F/S* units (1)
and 1 μeq *Sr* = 0.168 *F/S* units (2)
thus for any plant material,

$$\frac{F/S - 1}{0.168} = \mu\text{eq } Sr \text{ per g} \quad (3)$$

The reason for the standard curves in Fig. 9.4 being parallel for different plant species instead of superimposed is of course primarily because of the differing amounts of the element in the samples, but matrix effect is also responsible.

The procedure for any analysis therefore, is to establish by means of the "element addition" standards a general expression relating element content to F/S units as in equation (3). It is then possible to determine the F/S ratio for the "unknown" plant samples, and from this calculate the amount of element present. For any element the intensity of the fluorescent radiation, F, is found by counting at the appropriate wavelength as determined by the proper angle setting, 2Θ, of the diffraction crystal. The intensity of the scatter radiation, S, is then also determined at an angle setting in close proximity, but in a region free from lines. The F/S ratio can then be calculated.

For K_α radiation and a lithium fluoride crystal, the analyzer is typically set to diffract the radiation at the angle, 2Θ, of 113° for Ca; of 57° for Fe; of 63° for Mn; and of 41.8° for Zn. With a topaz crystal, at the angle, 2Θ, of 37° for strontium. With an Eddt crystal, at the angle 2Θ, of 117° for K; and of 89° for P.

The procedure adopted by Whittig *et al.*, ([22]) differed in detail in the means of establishing the relationship between *F/S* ratio and units of element. Thus in this method the sample is analyzed before and after the addition of a known quantity of element standard, e.g. 20 μg of added zinc per gram of plant material. The addition of the standard is accomplished as in the previous method, by adding the required amount in solution using acetone as a spreader and taking about 10 g of plant material. Then the difference between the *F/S* ratios determined on the two samples established the response to the added standard element per unit weight of sample. To be able to calculate the level of the element in the unknown sample it is necessary also to determine the *F/S* ratio at zero element concentration. This is carried out by the analysis of pure reference materials of supposedly similar basic matrix composition as plant samples, and cellulose, starch and mannitol have been used. It was found that these three materials had the same *F/S* ratio of 0.91.

With the *F/S* ratio of the sample determined before and after the addition of the standard, the initial concentration of element in the sample can then be calculated by the formula:

$$X = \frac{Y(F/S)\mu - 0.91}{(F/S)y - (F/S)\mu}$$

where, X = unknown element concentration in the sample (μg/g)
Y = zinc added (μg/g)
F = radiation intensity (c.p.s.) on the "peak"
S = scatter radiation intensity
μ = "unknown" sample
y = "unknown" sample + Y
0.91 = F/S ratio at zero zinc concentration

Similarly to the previous method described, when the constant relationship of F/S ratio to added standard has been established for a number of samples, it is then possible to use this value in subsequent analyses without using the method of addition for every sample analyzed.

DEVELOPMENTS IN X-RAY SPECTROGRAPHY

X-ray spectrographic analysis of biological samples did not develop initially as fast as might have been anticipated, allowing for the attractions of easy sample preparation and the capability of automatic multiple element analysis. This was only in part a consequence of the expense of the equipment but more particularly the rapid development of atomic absorption spectrometry which, especially in its non-flame (graphite furnace) mode, has been able to determine down to much lower element concentrations. Tölg ([19]) gave comparative limits of detection as follows:

X-ray fluorescence spectrography	10^{-9} g
emission spectrography	10^{-10} g
flameless atomic absorbtion spectrometry	10^{-13} g
neutron activation analysis	10^{-14} g

In routine practice, within the constraints of the Periodic Table, i.e. between atomic numbers 8–96, most elements can currently be determined down to a concentration of a few p.p.m., with elements detected at lower levels in qualitative "scans". However, some of the more recent techniques being developed can undoubtedly show X-ray methods in a more favourable light than this.

The development of charged-particle induced X-ray fluorescence (CPXE) has greatly improved the determination of the lighter elements ([7]). This method uses bombardment by energetic particles to excite the atoms of the sample being analyzed. For low atomic number elements the development of routine commercial instrumentation which uses an electron beam to bombard the samples results in an increased X-ray yield of several magnitudes, giving much greater precision and reduced counting times ([16]). Other fast ion beams such as protons (PIXE = proton induced X-ray emission) or alpha particles from accelerators of 1–5 MeV, such as the Van de Graaff type, are also being used ([8,20]). These latter methods are still largely in the hands of physicists and have yet to make the jump into routine use.

This account has been almost entirely concerned with wavelength dispersive spectrometers but much current work is now being put into energy dispersive spectrometers. This has been brought about by the development of much improved commercially available semiconductor detectors. These high efficiency, high resolution Si (Li) and Ge (Li) detectors having energy resolution of 180 eV within a range of 1–10 keV can effectively separate individual X-ray lines. In contrast then to wavelength dispersive systems, an energy dispersive spectrometer makes simultaneous multi-element analysis possible by means of a multichannel pulse height analyzer. The X-ray spectra obtained are analogous to those obtained in gamma spectrometry and similar methods of fitting the spectra by computer ([13]) can be employed, with reduction of the data for specific elements.

One particularly interesting development has combined energy dispersive spectrometry with excitation of the sample by radioisotopes such as 20 mCi ^{55}Fe, 20 mCi ^{109}Cd and 30 or 100 mCi ^{238}Pu (10). A helium chamber is used for elements of low atomic number. Airborne particulate samples were collected on filter paper, plant (moss) samples were counted as ash on filter paper or as compressed pellets, while aqueous samples were concentrated by ion exchange or precipitation techniques. Determination of calcium, iron, copper, lead, zinc, strontium, sulphur and chlorine were amongst the elements analyzed. The detection and estimation limits were comparable to other trace element analytical techniques, but the procedure has the advantage of being able to determine several elements simultaneously with relatively simple sample preparation.

Other work has concentrated on estimating very small amounts of elements in very small volumes. Rather basic work has used energy dispersive spectrometry coupled with X-ray total reflection analysis ([23]) and reflected polarized X-rays ([2]). Results have yielded with 120 sec counting times minimum detectable concentrations of Cr (0.8 p.p.m.), Mn (0.8 p.p.m.), As (1.1 p.p.m.), Br (1.8 p.p.m.) and Sr (4.5 p.p.m.) in samples as small as 5 μl, and increased counting time brought detectable Mn concentration down to 1 ng/5 μl (0.2 p.p.m.) ([23]). Minimum detectable quantities for Co (1.06 ng), Fe (1.1 ng), Mn (1.3 ng) and Cr (1.3 ng) were determined in 5 μl liquid samples which were deposited on quartz-glass substrate and dried, the area covered being 0.5 cm^2 ([2]).

Proton induced X-ray emission analysis coupled with energy dispersive spectrometry has enabled determinations of S, Cl, K, Ca, Fe, Br and Pb to be determined with sensitivities between 10–100 ng in samples of road-side aerosol/dust samples as small as 2 × 5 mm in area ([8]). In a period of 4 hours as many as 85 separate multiple element analyses could be carried out.

Clearly the potential for X-ray fluorescence has changed from being mainly suited to multiple element detection and survey work with relatively large samples (with specific determination of such elements as strontium, zinc, copper, nickel, iron, cobalt and manganese), to being now capable of trace element determinations at the nanogram level in very small samples. How far these developing techniques will make a significant contribution to agricultural, environmental and general biological analysis it is impossible to say at present.

For the moment the different types of X-ray fluorescence technique should not be seen in competition, but as complementary to each other and fulfilling different functions. Much of the practical development of energy dispersive systems has been done in relation to environmental pollution problems and, as convenient to this work, most of the samples have been infinitely thin thus largely avoiding matrice and particle size effects which are especially critical for this ([18]) technique. Whether this limitation can be overcome and the preparation and analysis of more general samples can be made competitive in time and precision with other methods is an open question.

REFERENCES FOR FURTHER READING

1. Adler, I. and Axelrod, J. M. *Norelco Reporter* **3**, 65–67 (1956).
2. Aiginger, H., Wobrauschek, P. and Brauner, C. Energy-dispersive fluorescence analysis using Bragg-reflected polarized X-rays. *In: Proc. Symp. Measurement, Detection and Control of Environmental Pollutants*, 197–212, IAEA, Vienna (1976).
3. Birks, L. S., Brooks, E. J. and Friedman, H. *Anal. Chem.*, **25**, 692 (1954).
4. Brandt, C. S. and Lazar, V. A. *Agric. and Food Chem.*, **6**, 306 (1958).
5. Campbell, W. J. and Carl, H. F. *Anal. Chem.*, **26**, 800–5 (1954).
6. Claisse, F. Dept. Mines, Province of Quebec, Canada, **327** (1956).
7. Deconninck, G., Demortier, G. and Bodart, F. Application of X-ray production by charged particles to elemental analysis. *Atomic Energy Review* **13**, 367–412 (1975).
8. Desaedeleer, G., Winchester, J. W., Pilote, J. O., Nelson, J. W. and Moffitt, H. A. Proton induced X-ray emission analysis of roadway aerosol filter samples for pollution control strategy. *In: Proc. Symp. Measurement, Detection and Control of Environmental Pollutants*. 233–44, IAEA, Vienna (1976).
9. Eddy, C. E., Laby, T. H. and Turner, A. H. *Proc. Roy. Soc.* Ser. A, **124**, 249 (1929).
10. Florkowski, T., Holyńska, B. and Piorek, S. X-ray fluorescence Techniques in analysis of environmental pollutants. *In: Proc. Symp. Measurement, Detection and Control of Environmental Pollutants*, 213–31, IAEA, Vienna (1976).
11. Hevesy, G. von *Chemical Analysis by X-rays and its Application*. McGraw-Hill Book Co. Inc., New York (1932).
12. Hooper, P. R. *Anal. Chem.* **36**, 1271–76 (1964).
13. Kaufmann, H. C. *et al.* REX, a computer programme for PIXE spectrum resolution of aerosols. *Advances in X-ray Analysis* **19**, Plenum, New York (1976).
14. Kemp, J. W. and Andermann, G. 'Refinements in X-ray Emission Techniques'. Pittsburgh Conference on Analytical Chemistry and Applied Spectroscopy, Pittsburgh, Pa. (1956).
15. Liebhafsky, H. A., Pfeifer, H. G., Winslow, E. H. and Zemany, P. D. *X-rays, Electrons and Analytical Chemistry*, Wiley, New York (1972).
16. Rank-Hilger. *Flurovac-Fluroprint X-ray fluorescence spectrometer*, 4–11, Westwood, Margate (1976).
17. Shenberg, C. and Boazi, M. Rapid qualitative determination of main components in archaelogical samples by radioisotope excited X-ray fluorescence analysis. *J. Radioanal. Chem.* **27**(2), 457–63 (1975).
18. Spatz, R. and Lieser, K. H. Determination of trace elements in dust samples by X-ray fluorescence analysis with radionuclide excitation with special regard to matrix effects (in German). *Z. Anal. Chem.* **280**(3), 197–200 (1976).
19. Tölg, G. Extreme trace analysis of the element. I. Methods and problems of sample treatments, separation and enrichment. *Talanta* **19**, 1489–521 (1972).

20. Vis, R. D. and Verheul, H. The capabilities of proton induced X-ray fluorescence in analytical chemistry. *J. Radioanal. Chem.* **27**(2), 447–56 (1975).
21. Vose, P. B. and Koontz, H. V. *Hilgardia* **29**, 575 (1960).
22. Whittig, L. D., Buchanan, J. R. and Brown, A. L. *Agr. Food Chem.* **8**, 419–21 (1960).
23. Wobrauschek, P. and Aiginger, H. X-ray total reflection fluorescence analysis. *In: Proc. Symp. Measurement, Detection and Control of Environmental Pollutants, 187–96, IAEA, Vienna (1976).*
24. Bertin, E. P. *Principles and Practice of X-ray Spectrometric Analysis*, 2nd Ed., pp. 1079, Plenum Press, New York and London (1975).
25. Müller, R. O. *Spectrochemical Analysis by X-ray Fluorescence*. pp. 326, Plenum Press, New York (1972) (original in German, 1967).

CHAPTER 10

Autoradiography

AUTORADIOGRAPHY has been known to science as long as radiation itself, because it was due to Becquerel finding in 1895 that uranium ore had fogged photographic plates that radiation was discovered. The word ''autoradiography'' dates from Lacassagne in 1924.

The principle is basically simple: it has been shown that ionizing particles have the same photochemical effect on a photographic emulsion as do light rays. Thus ionizing radiation induces a latent image in photographic emulsion which on development is revealed through developed silver halide grains. The number that are developed per unit area conform to the position and intensity of the radiation. If therefore a specimen containing a radioisotope tracer, such as a small plant for example, is closely applied to a photographic film or plate and left for a period of time and the emulsion subsequently developed and fixed in the normal manner, the silver halide grains which have been sensitized by the radiation will indicate the position of the radiation. The degree of darkening will also reflect the intensity of the tracer and hence the concentration of the substance being traced. Very many specialized texts have now been published [(2,5,9,10,21,22)].

Autoradiography is basically inexpensive, requiring very little equipment, but its real value is that it permits ready location and visualization of the movement and concentration of the tracer. If similar studies are undertaken by dividing the specimen and counting its component parts, very much greater work and time is involved, and moreover the same degree of fine detailed examination is not practicable. Consider the distribution of substance in a leaf for instance: autoradiography will instantly tell us whether the main site of deposition is in the veins or the lamina, and whether at the edge of the lamina, or irregularly or uniformly distributed throughout the leaf.

Because autoradiography integrates all the radiation being emitted from the experimental object it is especially useful for very low levels of radioactivity, and when it is desired to identify accumulation or reaction sites.

Macroautoradiography is used to denote the autoradiography of relatively large objects as in the examination of whole plants, roots etc., the identification of labelled spots on paper chromatograms or thin layer chromatograms (TLC), or for the localization of labelled ions in frozen soil sections, for example. In the case of microautoradiography we are concerned with the cellular level and the localization of the isotope label in groups of cells, and within individual cells. Refinements of technique have now made these methods suitable even for studies combined with electron microscopy.

Although firmly scientifically based there is a good deal of practical skill and art involved in making consistently good autoradiographs, and scarcely two workers do everything in identical manner, especially with micro-autoradiography. Nevertheless, there are some general principles and considerations basic to all procedures.

RESOLUTION FACTORS IN AUTORADIOGRAPHY

Resolution in autoradiography is customarily defined as the distance between the point of maximum density and that of half-density. Resolution is greatly influenced by type and energy of radiation, emulsion thickness, the thickness of the specimen, the width of the gap between the specimen and the emulsion, and potential scattering from film backing.

γ-rays give very poor autoradiographs because they have very low linear-energy transfer, having no mass and ionizing to only a small extent in the course of a long path. Thus they pass through thin biological materials largely without interaction, except possibly for a weak image arising from secondary electrons.

α-particles and low energy β-emitters, such as ^{14}C, ^{45}Ca, ^{35}S and ^{3}H, have much higher linear-energy transfer and are consequently very suitable. Although high energy β-emitters produce autographs, the quality is rather poor and diffuse due to the comparatively long path lengths, which means that the image in the emulsion may be produced relatively far from the radiation source. Reasonable macro-autoradiographs can be produced however with ^{32}P if comparatively low activity and relatively long exposure times are used.

The thinner the emulsion then the better the resolution. Most X-ray films have emulsion about 10–30 μm thick and with a grain size of about 5000 Å. This can give a resolution of 20–30 μm. Stripping films have much thinner emulsions of about 5 μm thick and much finer grain size can give resolution of 10–15 μm. The thickness of emulsions which are used to coat histological specimens clearly varies according to the technique of the investigator, although the ideal is to achieve a uniform monolayer of silver halide crystals. Grain size may be as small as 300–700 Å, though normally about 1200 Å is in more common use. Emulsions are a compromise between fine grain to improve resolution and high sensitivity to reduce exposure time, just as in normal light photography.

Good resolution is critically dependent on the thickness of the specimen, and the finer the detail that is sought then the thinner the specimen must be. Otherwise the radiation arising at different depths in a thick specimen makes it impossible to achieve a true reproduction, due to scattering of the radiation. In the case of macroautoradiography this means that one can achieve good autographs of leaves and thin stems such as of grasses, young bean plants etc., and of young roots of plants grown in solution culture. Thick material will only give a hazy, imprecise blackening of the emulsion. For the same reason the specimen must be closely and uniformly applied to the film.

With microautoradiography the same considerations apply, but with even greater force. Thus sections for light microscope autoradiography should be no more than 0.5 μm thick, but for electron microscope work they must be as thin as 600 Å. The use of emulsion directly applied to the specimen gives better resolution than use of stripping film because of the closer application and thinner emulsion.

DIFFICULTIES IN AUTORADIOGRAPHY

The four main difficulties in autoradiography are artifacts, movement of the tracer after completion of the experiment, leaching of the tracer, and in fine microautoradiography the presence of unincorporated tracer. These problems are usually overcome with careful work and attention to detail.

Artifacts, that is some sort of image or impression on the emulsion that is not due to radiation, may arise through excessive pressure or chemical reaction of the plant with the emulsion. In fact, although they have been reported ([13,16]) they seem to be a good deal more rare than is sometimes suggested.

Artifacts due to pressure or plant secretions of a chemically active nature are easily checked by the simple means of putting a number of specimens, untreated with radioisotope, through the identical processes as the treated material. The possibility of chemical action can also be prevented by placing a piece of very thin "Saran wrap", "Mylar", Melinex or similar plastic sheet between the specimen and the emulsion.

The problem of tracer movement during drying and exposure of the film is harder to solve. Most plants are prepared for autoradiography by drying in the oven or freeze drying, although sometimes fresh material is applied direct to the film and placed in a deep freezer for exposure. Before any form of drying, plants are usually mounted on herbarium drying paper or blotting paper using a few strips of "Scotch" tape to hold them in an appropriate position. It is then easy to prevent gross movement of the radioisotope between root and stem, or between stem and leaves, simply by using a razor blade to cut through each petiole or leaf base, and through the base of the stem, thus isolating all the major parts and preventing movement from one to another during drying. Movement within broad dicotyledon leaves, which can be significant, cannot be controlled in this manner, and the only answer is freezing or freeze drying.

The leaching of tracer and the presence of unincorporated tracer are not normal problems with whole plants, but can be significant in the autoradiography of thin tissue sections for the study of transport of molecules within cells by the "pulse-chase" method, and in the localization of metabolite synthesis.

Handling errors common to any careless photographic processing may also occur: scratching, finger prints, stray light, non-uniform development, displacement of the film during exposure and extraneous dust or other debris. An effect of histological stains on the emulsion can occur with microautoradiography.

PROCEDURES FOR MACROAUTORADIOGRAPHY

Virtually all the best procedures for macroautoradiography of plant material are based, consciously or unconsciously, on the methods developed by Yamaguchi and Crafts ([6,29]), whose autographs of ^{14}C herbicides in plants have seldom been surpassed.

Application of the Label

Leaf application is common, particularly for the study of herbicide uptake and movement ([29]), or the mode and pattern of translocation of micronutrients ([12,27]). In such cases the label may be applied as a small drop or spot of about 0.04–0.05 ml, and retained within a ring of lanolin or modelling plasticine. Alternatively the active solution is sometimes applied with a spray or brush. In the case of spray application, suitable precautions must be taken against contamination of both the operator, other parts of the plant, and nearby plants. Any spraying operation with a radioisotope must be regarded as hazardous and to be avoided if possible. With leaf application it is usual to use some sort of ''spreader'', such as dilute detergent, to assist the solution to flow over the leaf surface.

The plasticine rings may be prepared by cutting discs with a cork borer out of a thin sheet of worked plasticine, and then cutting out a smaller disc from this. These rings are very effective and probably less trouble to prepare than lanolin rings.

Suitable levels of activity to apply as a spot or sprayed over an area of about 2 cm diam. vary from about 5–50 μCi per application. In general, lower levels are given if the isotope is thought to be readily taken up and if the experimental period is long. Higher levels are given if the experimental period is short, the material not taken up readily, if the radioisotope has low energy, and if the experimental plant is very large. Normally about 5–15 μCi will be adequate. it should however be noted that it is possible to use very low levels (0.05 μCi) of ^{14}C labelled herbicides and similar materials if very long exposure periods are given, 6 weeks or more for example. This procedure can give very fine autographs.

When the radioisotope label is given in solution culture for subsequent autoradiography, the activities used must be low (10–30 μCi/l) to avoid getting the plants too radioactive with subsequent indiscriminate fogging of the film. This is especially the case with autographs of roots, where both low solution activity and short experimental uptake times, with a maximum of 15 minutes, are essential if any detail is to be observed ([19]).

Preparations for Exposure

The autographs are prepared by placing in direct apposition to no-screen X-ray film. Most types used in medical X-ray work are suitable. A selection of what have proven to be the most popular films, emulsions and developers for autoradiographic work are given in Table 10.1.

TABLE 10.1
Film type and emulsion developers and fixers for autoradiography

Film or emulsion type	Developer	Fixative
MACRO-AUTORADIOGRAPHY		
1. *Film*		
Kodak or Ilford[1] no-screen X-ray film, standard 20 × 30 cm	Kodak D-19[2] 2–5 min at 20°C or Amidol	30% sodium thiosulphate[3] *or* Kodak FX-40 X-ray liquid fixer *or* Kodak F-5 for 10 min at 20°C
MICRO-AUTORADIOGRAPHY		
2. *Coating emulsion*		
Ilford L-4[4] Ilford K-5 Kodak NTB-3 (Kodak NTE Gevaert NUC-307)[7]	Microdol-X[5] or Kodak D-19 for 5 min or Amidol	2.5% sodium thiosulphate[6] 3 min or $Na_2S_2O_3 \cdot 5H_2O - K_2S_2O_5$
3. *Stripping film*		
Kodak AR-10[8] Kodak AR-50[9]	Kodak D-19b 10 min at 20°C	

Notes
[1]Red safe-light
[2]Good for general use for all purposes
[3]Very commonly used
[4]Probably best for general use, 1200 Å grain size. Ilford safe light 5902 (Yellow)
[5]Best for L-4, increases grain size
[6]Much more dilute thiosulphate appropriate for autography from electron microscope thin tissue sections
[7]Very good resolution but difficult to use, grain size only 300–700Å
[8]Fine grain stripping plate, best for general use, weak red safe light
[9]Faster than AR-10 but poorer resolution

The plants are harvested (it is usually not necessary to use gloves if a "barrier" cream is worn on the hands) and the part of the plant, leaf, etc., to which the radioactive drop was applied should be cut away and disposed of safely. If it is left it will only "fog" a large area of the film. Each specimen should then be carefully placed on a sheet of blotting paper or herbarium drying paper, with the leaves flat, and the root carefully arranged. Obviously the whole specimen must be arranged in such a manner that it will be covered by the X-ray film that is available. The specimen may be held in position by a number of small pieces of "Scotch" tape. These usually do not show on the autograph, and if they do they are easily recognized. It is then frequently desirable to use a razor blade to cut the petioles or leaf bases, and between root and stem, in order to prevent any subsequent movement of the label between major parts.

There are then three possible procedures (1) dry the material in the oven at 70°C (2) freeze dry (3) using a safelight, immediately place in contact with the film and place in a deep freeze for the subsequent exposure time. Preservation of the plants by oven drying is then done on separation from the film after the exposure period. Melinex is used to protect the emulsion from the fresh plant material.

TABLE 10-1 (cont'd)

Developers			
D-17 Developer		*D-19b Developer*	
Na_2SO_4 (anhydrous)	96.0 g	Elon	2.2 g
Hydroquinone	8.8 g	Na_2SO_4 (crystalline)	130.0 g
Na_2CO_3 (anhydrous)	48.0 g	Na_2SO_4 (anhydrous)	48.0 g
KBr	5.0 g	Hydroquinone	8.8 g
Dissolve in order,		Na_2CO_3 (crystalline)	130.0 g
make up to:	1 litre	Na_2CO_3 (anhydrous)	48.0 g
		KBr	4.0 g
Amidol Developer		Dissolve in order,	
		make up to:	1 litre
H_3BO_3	35.0 g		
Na_2SO_3 (anhydrous)	18.0 g		
KBr, 10% soln.	8 ml		
Make up to:	1 litre		
Add 4.5 g Amidol ($C_6H_{10}Cl_2.H_2O$) immediately before use			
Fixers			
Kodak F-5 fixer		*Fixer for emulsion*	
$Na_2S_2O_3.5H_2O$ (crystalline)	240.0 g		
Na_2SO_4 (anhydrous)	15.0 g	$Na_2S_2O_3.5H_2O$	330.0 g
Acetic acid (glacial)	17.0 ml	$K_2S_2O_5$	33.0 g
H_3BO_3 (crystalline)	7.5 g	Made up to:	1 litre
Potass. alum.	15.0 g		
Dissolve in order,			
make up to:	1 litre		

The first alternative is obviously the easiest although technically the worst. Rapid drying will reduce the chance of label movement, therefore material should be placed in a pre-heated forced circulation oven, and care taken not to overfill.

Freeze drying requires special equipment. The arrangements of Crafts and Yamaguchi ([6,29]) have proven very suitable. Vacuum tanks with an internal size of about 35 × 45 × 10 cm are fabricated from welded mild steel plate. One end is left open and is flanged, to which a cover of aluminium plate is bolted, with a rubber gasket to provide a vacuum seal. The tanks are provided with an outlet pipe and a tap to release the vacuum. Three or four such tanks can be contained in an upright freezer with the open ends facing the door, and connected to a main vacuum pipe which is passed through the freezer wall to a two stage vacuum pump with a free air displacement of 125–150 litres/min. Moisture cold traps have to be placed somewhere in the pumping system to protect the pump, and probably the easiest arrangement is to have a single small trap for each vacuum tank. These can be contained in Dewar flasks containing a mixture of ethyl-alcohol and dry ice at approximately −70°C. The deep freeze is run at −5 to −10°C, vacuum is about 0.001 mm Hg, and the drying time 4–7 days.

In order to get immediate freezing of the specimens at harvest Levi ([14]) has utilized a system whereby envelopes of aluminium foil are made and filled with finely powdered dry ice. The plants are then placed between sheets of blotting paper and then stacked sandwiched between foil envelopes. Waxed paper is placed between blotting paper and the foil to prevent sticking. After harvesting is complete and all the plants are uniformly cooled they are transferred to the freezer tanks without the dry ice.

Plants which have been freeze dried are very brittle and must be allowed to take up atmospheric moisture before any further handling, or else they will be damaged.

Where the major interest of the experiment is in roots which have been in radioactive solution (wear gloves!), they should be drained and washed in five or six changes of deionized water. According to the purpose of the experiment it may also be useful to place them for a short time in an unlabelled exchange solution of the same ion. The roots can then be spread by flotation on stiff paper in water ([19]) and then blotted dry with soft tissue such as Kleenex. Roots cannot be oven-dried and still give meaningful autographs, so the specimens are placed in direct contact with no-screen X-ray film, or with a sheet of Melinex (cellophane) in between, and placed in a deep freezer at −10°C or lower, giving an exposure time of 10–14 days. If the activity is so great that this exposure time is too long, then the activity should be subsequently reduced if the best quality autograph is to be obtained.

Exposure of the Films

This stage of the work must be carried out in the darkroom with a red safelight. Each of the dried plants is left on its drying/mounting paper and placed against the emulsion side of a sheet of X-ray film. The specimen may be placed directly on the film or the mount may be previously covered with a single sheet of Saran wrap or Melinex. This has the advantage of preventng any possible chemical artifacts from the plants and also keeps all the specimen together. However, there is a loss of radiation intensity of about 20% with Melinex and as much as 40–50% with Saran wrap. Direct placement without any barrier is probably most convenient for the majority of dry material, checking for artifacts by processing unlabelled plant material simultaneously, and using Melinex only where fresh material is applied for exposure in a freezer.

Film may be permanently numbered at this stage by using a special ball-point pen with ^{14}C labelled ink.

Some workers then place the sheet of film together with the applied specimen back inside the individual envelope from which the X-ray film was taken, having discarded the protective piece of cardboard. The envelope is then sealed up again with adhesive tape and can if necessary be handled in daylight, while another possible advantage is that it reduces the chance of specimen movement. In general though it seems an unnecessary complication of handling for most situations.

A stack of X-ray film + specimen is then built up, with each specimen being backed by a 1.5 cm thick foam rubber or polyethylene foam plastic sheet of the same

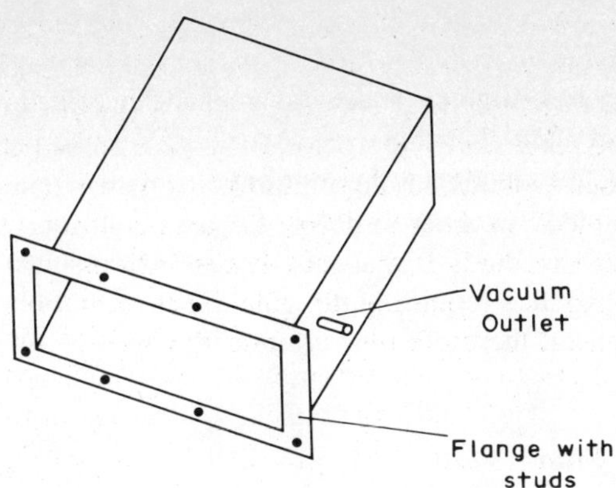

FIG. 10.1 Vacuum tank welded mild steel plate, used for freeze-drying of specimens for macroautoradiography. Size about 35 × 45 × 10 cm.

size as the film. This serves both to ensure close contact of specimen and emulsion without causing pressure artifacts. The film in turn should be backed by a 0.25 cm thick aluminium sheet of the same size. Alternatively, hardboard completely covered with aluminium foil may be used instead. These boards serve both to keep the specimens flat and ensure uniform pressure, but also make certain the β-radiation from one specimen cannot affect any film other than the one with which it is in contact. Thus a stack is built up: hardboard/film/specimen/foam rubber/hardboard etc., ending with a board. Then a strap is tightened lightly around the stack, sufficient to ensure pressure but not too tight, and the stack is placed in a light tight box or wrapped up securely in black plastic sheet.

During the exposure period the stack may be placed out of the way in a cupboard, often being left in the dark room, care being taken that it is neither near to a source of radiation nor in a position that the specimens might spoil unused X-ray film. Where fresh specimens have been used then every effort should be made to work quickly, and the stack is then placed in a deep freezer.

For an entirely new experimental procedure and/or radioisotope, the time required for exposure is usually found experimentally by developing one or two autographs at intervals of a few days, the first being taken at five days. When removing film for test development take care not to alter the positions of the remainder in the stack. The number of film sheets exposed should allow for test development, and in any case with all autoradiographic work it is very desirable to have quite a number of autographs for any treatment to ensure good interpretation. Subsequent repeat or similar experiments will enable both the initial radioisotope dose and the exposure time to be estimated more accurately. Over-exposure should be avoided as it gives loss of resolution.

One or two semi-quantitative procedures can also be used for calculating exposure time, based on the fact that it will take 5×10^6 or 10^7 β-particles to fall on each cm^2 of emulsion to produce sufficient darkening ([11]). Alternatively, as pointed out by

Comar ([7]), if the test object records 100 c.p.m. when a G-M tube is held at 2–3 cm distance then the exposure time will be 10–14 days. Counts are taken of the specimen using an end-window G-M tube and compared with a β-standard of known disintegration rate. The time taken for 5×10^6 or 10^7 counts to be accumulated can then be calculated. For both methods an aluminium plate with a hole of known area can be placed over the subject, in order to define the area contributing to the count. The great drawback of such methods is that they necessarily assume uniformity of radiotracer distribution, and also require quite wide experimental objects such as leaves to be effective. In general therefore they are not very useful.

Development of the Film

All working should be carried out using the recommended safelight if known, normally weak red, e.g. Wratten No. 1. If there are specific instructions for developing the film, then clearly these should be followed. In most cases a suitable standard procedure is:

1. develop in D.19 or Amidol developer for 3–5 minutes at 20°C with occasional gentle agitation.
2. rinse for 10–15 sec in each of two changes of deionized water.
3. fix in 30% sodium thiosulphate of F-5 fixer for 10 minutes at 20°C.
4. finally wash in running water for 30 minutes, and then hang up to dry.

At all times take care to avoid finger prints, scratches, debris etc. getting on the prints. In order to achieve uniformity of development, and hence densities which are comparable between autographs, it is essential that all film should be treated in exactly the same manner. Hence the developer should be fresh, the temperature carefully checked and ideally, controlled, and the times rigidly adhered to. When comparability is not important, it should be noted that light over-development may be used to compensate partly for low density in an autograph resulting from inadequate exposure. Old X-ray film shows background fog due to cosmic radiation, and if there is any doubt about the freshness of the film it should be checked. Film may be stored at room temperature up to 20°C, but for preference is kept at about 5°C.

Photographing autoradiographs

Some workers have photographed autographs by placing them on an opaque glass plate uniformly lighted from behind. Equally good results, with possibly better contrast, are obtained by placing the autoradiographs against a bright white surface and photographing with the use of side lighting.

Methods for Soils

It is sometimes useful to take autoradiographs of soil faces, in order to study ion migration, the location of roots of radioisotope injected plants, and zones of nutrient depletion ([1,8,17,20]).

Plants can be grown in suitable sized boxes, one of the sides being hinged to permit access to the soil face. If the hinged side is so constructed that it slopes inwards towards the base this will encourage root growth in close apposition to the side. More elaborate boxes may have glass sides which can be covered with a board to keep out the light. The advantage of glass sides is that growth of roots may be inspected without disturbing the system.

After a period of time has elapsed for "settling" of the soil and for root growth, usually the sides of such boxes may be opened for short periods without too much risk of the soil falling out, or they may be leaned over in the opposite direction as a precaution. If desired a sheet of thin plastic may be placed and fixed over the opening side of the box before it is filled with soil, to provide protection later for the X-ray film.

For exposure, the individual envelopes of X-ray film should be first opened in the darkroom to check the position of the film in relation to the cardboard backing, and to ensure that the emulsion side is facing outwards. All packets are arranged in a uniform manner and resealed with adhesive tape. The side of the growth box is then

FIG. 10.2 A glass-sided growth box, with removable wooden covers, suitable for root and ion movement studies in soil.

opened and the film packet, emulsion side towards the soil and protected by a thin Saran wrap or other plastic sheet, is placed in the desired position and the side closed.

Keeping the film in the packet protects it and avoids problems from light. The levels of radioactivity present in such experiments are much greater than we have hitherto been discussing for plant autoradiographs. Hence the reduction in activity reaching the film due to the envelope is not a serious matter. It will be appreciated that such soil autographs are concerned with major localization of the tracer and are relatively crude in both technique and imaging. In view of the higher levels of activity used only short exposure times, usually not more than 24 hours, are necessary. Develop in the same manner.

For certain studies more precise images can be obtained by freezing the soil in a deep freezer or cold store, subsequently sectioning the frozen block, in the region of interest, by means of a bandsaw. A uniform surface is obtained which can be applied direct to X-ray film protected by Melinex.

Methods for Chromatography

Autoradiography is used for localizing the radioisotope label on paper chromatograms and on thin layer chromatographic plates. Procedures are quite straightforward, but care has to be taken to ensure that subsequent accurate registration of the developed film and the chromatograph is possible. Film for TLC plates can previously be cut to exact size in the darkroom under a safelamp.

As the tracers most commonly used are ^{3}H and ^{14}C, exposure times of a week or so are common. Film may be developed in the routine manner given above.

In order to cut down the inevitable delay in getting results from autoradiographs of TLC chromatograms, Panax have developed a "spark imaging" system which can produce accurate and permanent records in as little as 20 minutes, Fig. 10.3.

With this equipment the chromatogram is placed on a tray inside a spark chamber, which has a counting gas mixture of argon and 10% methane. The electrode system is mounted on a transparent plate through which a Polaroid film is exposed to the spark pattern arising from ionisation of the gas by the β-particles. The individual flashes produced over the exposure period are integrated by the Polaroid film. Development takes place within the camera in the normal manner, and when the print is removed it shows the exact position of the labelled areas.

A second part of the instrument has an arrangement whereby the TLC plate can then be placed as the screen of an episcope, onto which is projected an image of the developed film, by means of a lens system and high intensity lamp. As the projected image is the same size as the TLC plate, it is then possible to mark exactly any areas of the plate desired for further study.

PROCEDURES FOR MICROAUTORADIOGRAPHY

Microautoradiography permits the study of cellular function at the cellular or subcellular level. ^{35}S, ^{14}C and ^{3}H are the most commonly used isotopes, together with ^{45}Ca and ^{32}P. Isotopes of high energy are not suitable.

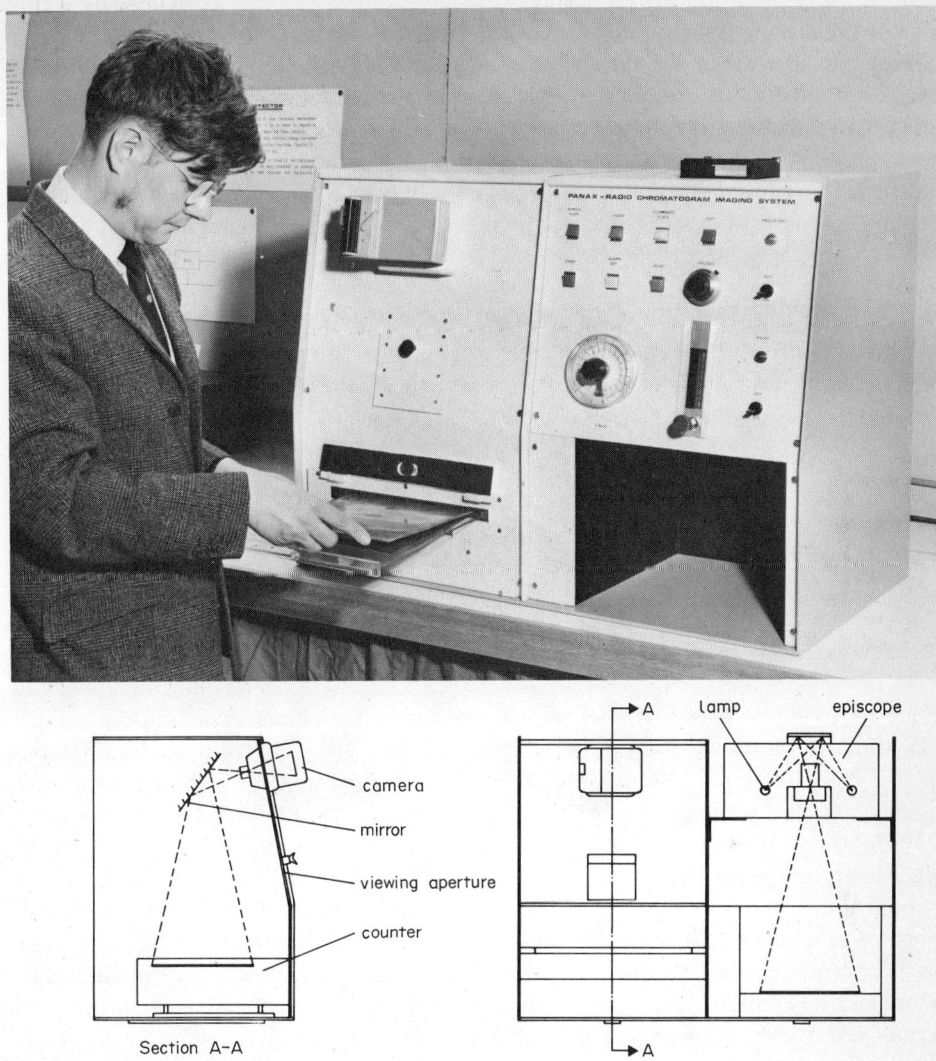

FIG. 10.3 Panax beta-graph for rapid spark-imaging of radiochromatograms on Polaroid film, with subsequent accurate location of radioactive areas (Panax Equipment Ltd.).

There are many varied applications of the technique. Localization of metabolite synthesis within the cell can be studied by placing the tissue in a solution of labelled precursor for an appropriate time, found by trial and error, with subsequent fixing and washing to remove unincorporated label. The autograph will then indicate the site of synthesis. Enzymes may be localized in cells by incubating the tissue with a radioactive specific enzyme inhibitor. As the enzyme and labelled inhibitor combine the enzyme can be located by means of the radioactivity indicated by the autograph.

Pulse-chase experiments permit the study of the movement of macromolecules within cells or tissues. First the tissue is placed in labelled precursor solution, the "pulse", and after an interval of time the labelled precursor is washed out and replaced by the unlabelled "chase" precursor. Time is allowed for the transport of the labelled molecule before fixing and autographing.

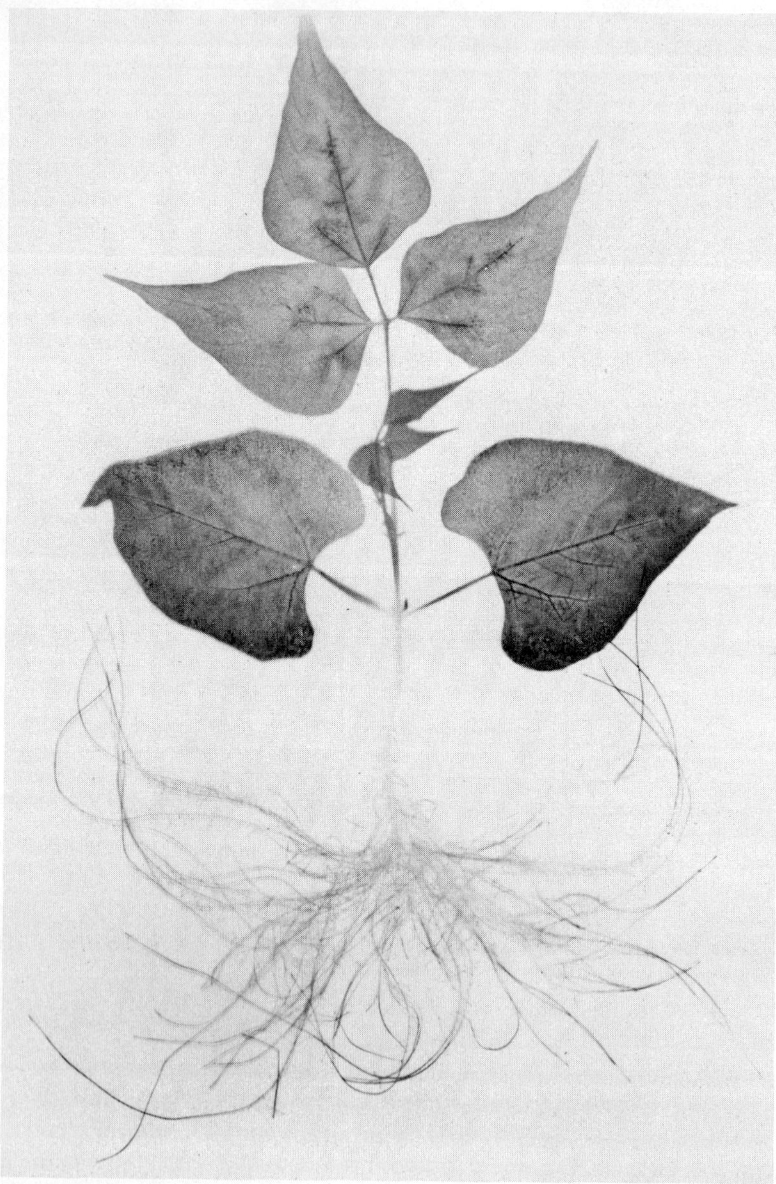

FIG. 10.4 Autoradiograph, showing uniform distribution of sulphur labelled with ^{35}S, in a young bean plant. Note detail of the roots (due to T. Muraoka).

FIG. 10.5 Autoradiograph of older bean plant labelled with ^{35}S, showing a probable artifact in the older leaves due to movement of the label during drying (due to T. Muraoka).

Application of the Label

Material for microautoradiography may arise in two ways: either the specimen is part of a larger organism that has been treated with a radioactive label by injection, surface application or through the roots, or else it is a tissue that has been placed in the labelled solution for some specific purpose. In either case the level of activity used must be fairly high, because of the thinness of the sample sections and the very small amounts of radioactivity finally present.

In the first case the level of activity given may well depend on the requirements of the whole organism, particularly with reference to possible radiation damage in relation to the length of the experiment. These considerations have already been discussed elsewhere, and we merely note here that when samples are to be excised

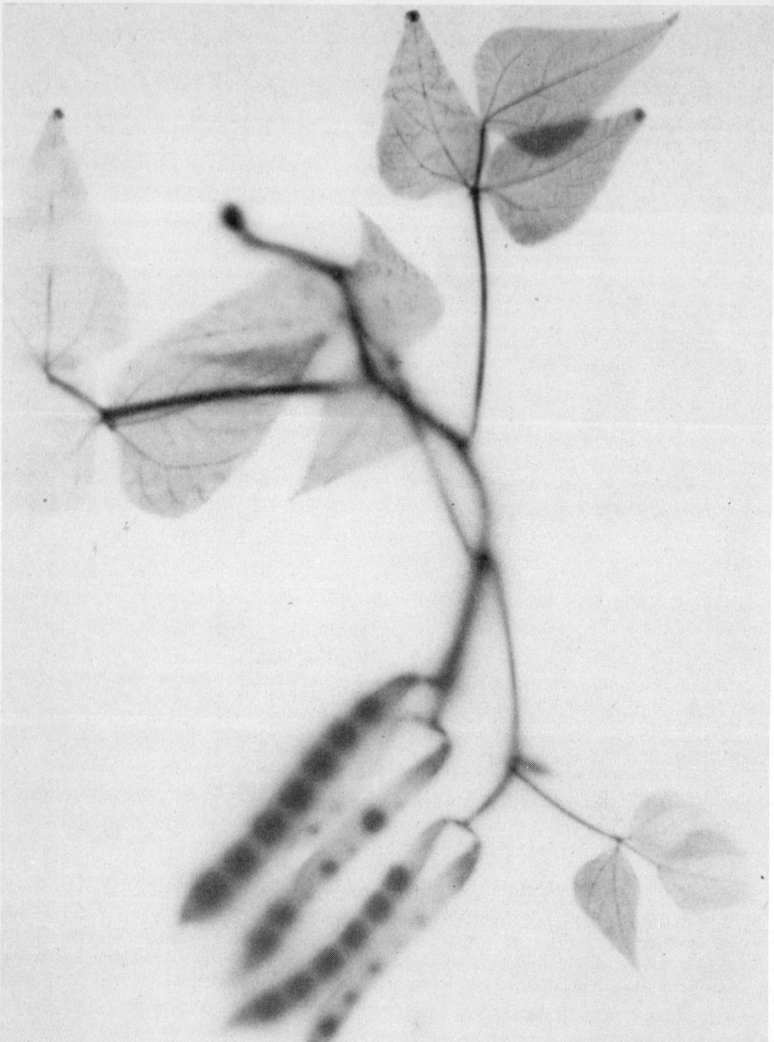

FIG. 10.6 ^{32}P labelled bean *(Phaseolus)* plant about 70 days old, showing preferential distribution of phosphorus to the seed (due to T. Muraoka).

for microautoradiography the highest levels of radioactivity consistent with normal physiological function should be used.

When an isolated tissue is incubated in a labelled solution a high level of both actual and specific activity is necessary, because of the short period of absorbtion or reaction. Specific activities of 20–100 Ci/m mol should be used when available, and total activity may be as much as 100–250 μCi/ml or per g for absorbtion periods of 0.5–4 hours. Clearly, localization experiments must have high activity for short periods as material will be transported from the synthesis site in long experiments.

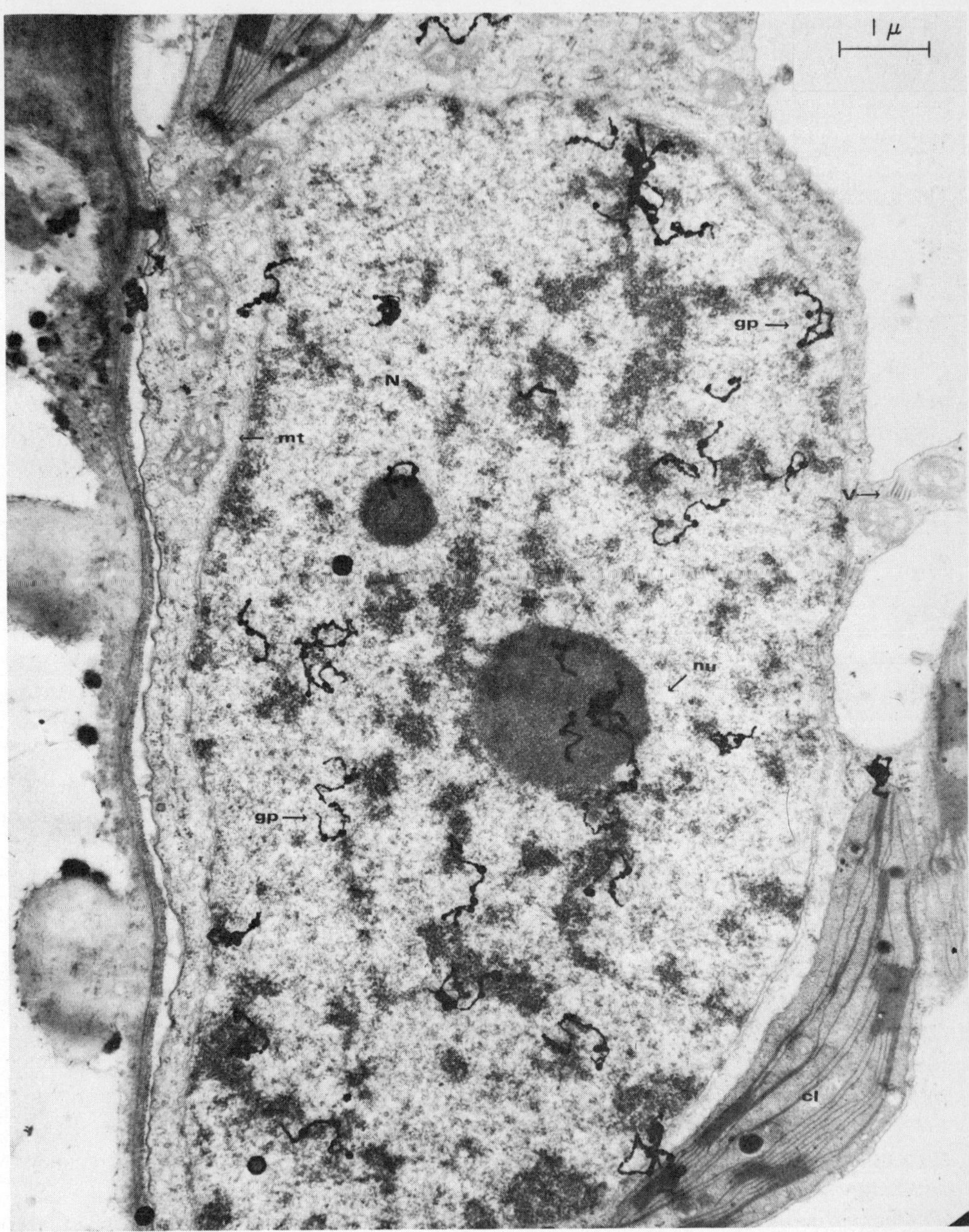

FIG. 10.7 A high grade electron microautoradiograph showing the distribution of marked viral-RNA inside the nucleolus and nucleus of a virus infected tobacco leaf cell treated with ^{3}H-uridine. *mt:* mitochondria; *gp:* silver grains; *cl:* chloroplast; *nu:* nucleolus; *v:* virus (due to Dr. Darcy M. Silva).

Methods

Three basic methods of microautoradiography are in common use: (1) direct apposition by mounting (2) application of stripping film, and (3) coating with liquid emulsion. Any of these methods may be used with the thick sections for light microscope work, but for electron microscopy with thin sections only the latter is suitable. The following account should be read "as a whole" because not all details are repeated for every method.

Direct Apposition by Mounting

This method is simple but crude, and is largely obsolete as the other methods are better. As the sections are directly mounted on the emulsion of photographic film or plate there is good contact and registry, but the grain size is comparatively large and the emulsion comparatively thick. Further disadvantages are the possibility of chemical effect on the emulsion due to the direct contact with the specimen and/or stain. This is also a general problem of the other methods.

Procedure

Typically, specimens are fixed (e.g. in Carnoy solution: absolute ethanol, chloroform, glacial acetic acid, 6:3:1), embedded in paraffin wax in the classical manner, and sections of 5 μm thickness are cut. The sections are floated on water at 35–40°C to remove wrinkles. They are then transferred to a bath of water at lower temperature (about 18–20°C) and then, using a safelight, a piece of no-screen X-ray film of appropriate size, or film plate, is dipped into the water bath emulsion side up. The film is slid under the section, which is held against the film by means of a mounted needle, as the film is carefully removed. As the water drains off the section remains mounted on the emulsion. After being allowed to dry, the film with mounted section is placed in a light-tight container in a refrigerator at 40°C for the required exposure period of 4–10 days.

After the exposure is completed the paraffin wax should be removed, under safelight conditions, by dipping the film in 2 changes of Xylol for 2 minutes each, then allowing to drain and dry completely (If embedding has been done in the more modern methyl or butyl methacrylate then subsequent staining can be done without further manipulation). Develop in Kodak D19b for 2–5 minutes at 20°C, rinse in two changes of water for 15 sec each, then fix in 30% sodium thiosulphate for 10 min, and finally rinse by dipping 15–20 times in distilled water. Normal lighting may now be used and the tissue can be stained.

Staining in basic fuchsin (as a 0.1–0.5% aqueous solution with 0.5% $Na_2S_2O_5$) for 5 min is a simple and useful general purpose procedure and especially for nuclear material, followed by differentiation in distilled water or sulphite water (5 g $Na_2S_2O_5$ in 1 l) for 10 min followed by washing. The preparation is dehydrated in successive increasing concentrations of ethanol, finally in toluene and mounted in Canada balsam.

An alternative for plant material is to stain in fast green or the classical safranin and fast green double stain when there is particular interest in the anatomy of the specimen. Staining should be kept very light. Another common staining alternative is methylene blue/Azure II (0.25 g methylene blue, 0.25 g Azure II, 1 g Na-tetraborate in 100 ml).

Alternative Procedures

Studies involving the localization of synthesis in a cell will require removal of the unincorporated label. After fixing in Carnoy or similar solution the tissue can be rinsed with several changes of 5% trichloracetic acid over a period of 1 hour at 4–5°C. In other studies it is necessary to prevent leaching e.g. of inorganic $^{32}PO_4$. The specimen can be initially frozen with "dry ice" CO_2 and sectioned with a freezing microtome, the frozen sections being transferred straight to the film emulsion (safelight!). The mounted specimen is then kept in the deep freeze for the period of development. Processing can then be carried out in a similar manner to that already described. Considerable attention has been given to the problem of autoradiography of diffusible substances ([22]).

Other procedures keep the specimen and film separate, both to reduce the problem of leaching, and to permit staining of the section without affecting the emulsion. However, in this method both contact, resolution and registry are all poor. The specimen is fixed, dehydrated, embedded and sectioned in the normal manner and sections are mounted on albumen-coated, or collodion (1% nitrocellulose in amyl acetate) coated microscope slides. The paraffin wax is removed as previously, and then after passing through 2 lots of absolute ethanol, the slides are coated by dipping in 1% celloidin twice, allowing time for drying in between. After drying, each slide is placed in apposition to the emulsion side of a piece of no-screen X-ray film cut to appropriate size. This can be backed up by another microscope slide, the whole assembly being clamped together with a spring clip. After exposure the specimen and film can be separated for individual processing. Before staining the section, the celloidin is removed by amyl acetate (5 min) and successively more dilute changes of ethyl alcohol.

Application of Stripping Film

The stripping film technique, originally devised by Pelc ([18]), has the advantage of quite good contact, resolution and registry, although it has lost ground to the emulsion coating method, which is now generally regarded as being easier and less troublesome. The basic principle is that the emulsion is stripped from its glass plate and a piece is flattened over the section on a glass slide. As in the previous procedures there can be leaching losses during processing, and the emulsion base can interfere with staining.

Procedure

Tissue sections are prepared as described in the previous section and are picked up from a distilled water bath at 20°C onto slides previously coated with albumen or collodion. After draining and drying, if embedded in paraffin wax, this is removed from the preparation by dipping in 2 changes of xylol. At this point the sections may be stained, if it is unlikely for there to be a loss of radioactive substance by leaching. After staining, cover with 1.0% celloidin for protection, as before. If staining is not carried out at this stage then a celloidin coating should not be applied, as staining must be done later through the film base following exposure. Work is then carried out under a safelight.

The Kodak AR10 stripping plate has an emulsion layer of 5 μm supported by a gelatin base 10 μm thick on glass. A razor blade is used to outline on the plate pieces of emulsion strip of appropriate size to cover the preparation. Using the razor blade a corner of a piece of film is lifted up and slowly stripped off the plate. If done too quickly static electricity may cause blackening of the emulsion. The strip is then floated emulsion side down in a bath of distilled water at 22°C for 3–4 minutes, to allow it to expand completely. A slide with mounted specimen is slid underneath the floating film, and is then brought out of the water at an angle, so that the film applies itself closely to the slide. Some workers use a piece of film slightly larger than the slide so that the ends can be folded over the edge of the slide to give even better adhesion and location. The film and preparation is dried slowly and then placed in a box, containing silica gel as a drying agent, for exposure at 4°C. After exposure, develop in D19b developer at 20–22°C for 10 min. Wash by dipping 10–15 times in a water bath, then fix for 10 min in 30% sodium thiosulphate, followed by thorough washing in water for 15 min.

If the sections were not stained earlier, then they may now be stained at this stage, through the emulsion base. Alternatively, unstained sections can be examined by phase contrast microscopy. Mount the preparations in Canada balsam under a cover slip.

In order to protect the sensitive emulsion from the tissue it is also possible to mount stripping film so that the relatively impermeable base is next to the section. This however is undesirable because of the poorer resolution and registry due to the 5–10 μm thickness of the gelatin base interposed between the specimen and the emulsion. Moreover, the activity reaching the emulsion from a ^{3}H-label is seriously reduced.

Coating with Liquid Emulsion

The preparation is covered with a heated liquid emulsion, which hardens on cooling to form a permanent bond with the section. The use of this coating procedure was first described by Belanger and Leblond ([3]) at a quite early stage of the development of microautoradiographic technique, but was not immediately popular. The availability of better emusions and the development of improved technique and understanding of the method now make it the procedure of choice for fine detail studies. It is essential for the still-growing area of electron micrography work, for which special methods have been developed.

The advantage of liquid emulsion coating is that it gives very good contact and good registry, with resulting excellent resolution. This makes possible the best correlation between radioactivity and histological detail. The main problems are obtaining a uniform thickness of emulsion, potential leakage of radioactivity into the liquid emulsion, and the fact that handling liquid emulsion tends to subsequently increase the background fog.

Procedure

Sections are mounted on microscope slides previously "subbed" with albumen or collodion. Staining may be done at this stage, if loss of radioisotope is not considered likely. In this case, or if no staining is to be done at all, the preparations may then be coated with 1% colloidin to protect the tissue section from photographic fluids. A number of suitable emulsions are available of which Ilford L-4 is one of the most useful, being suitable for either light or electron microscopy. Eastman Kodak NTB-3 has also been widely used, but it tends to have a high background.

Work must now be done under a safelight (Ilford S902 yellow for Ilford L-4). Preparations may either be dipped in the liquid emulsion or coated. Dipping is a rather bad practice as it is not possible to control the thickness of the emulsion, there is the possibility of cross contamination by radioactivity of the emulsion and other preparations, and it is also wasteful of emulsion. Therefore, the coating procedure is described here: (1) Put 5 or 10 ml of distilled water into a 50 ml beaker and add an equivalent amount of emulsion gel. (2) Heat to 40°C on a temperature controlled hot plate or water bath, stirring with a glass rod. (3) The final thickness of emulsion that is considered desirable varies between workers from 10 μm to 50 μm. The lower limit is generally to be preferred. Now, as the final thickness of the cool, gelled, emulsion is half that in the warm liquid diluted state, it follows that an initial coating of 20 μm must be applied. A calculation shows that about 0.25 ml of liquid emulsion spread over an area of 8 cm^2, i.e. 2 × 4 cm (adequate for most preparations) will give this thickness. (4) Mark a small Pasteur pipette with a rubber bulb to deliver 0.25 ml. Apply this amount of liquid emulsion in a line across the slide at one end of the preparation. (5) Then, holding the slide at an angle and using either the pipette or a small glass rod, smear the emulsion over the section as evenly as possible, preferably with a single sweep of the rod.

As an alternative to spreading with a glass rod, a small spreader can be made from polythene or perspex sheet, cut with "projections" to provide a constant clearance, and hence thickness, between the slide and the spreader when it is pulled along the slide on a flat surface. The slides are then put to dry in a flat position for 2–4 hours at 18°C. They may then be placed in a light tight box with silica gel, for exposure in a refrigerator at 4–5°C.

Exposure may take 2–14 days, the exact time being found by developing test slides at intervals. Develop in Amidol for 3–5 min at 22°C, rinse in two changes of water for 15 sec each and fix in $Na_2S_2O_3/K_2S_2O_5$ fixer (see Table 10.1) for 3–5 min. The exact time depends on the thickness of the emulsion and must be found by trial. Finally wash in distilled water by dipping for 15–20 times. Unstained sections may then be stained through the emulsion.

Electron Microscopy Autoradiography

Highly refined development of the liquid emulsion method is used in the preparation of electron microscope autoradiographs. The techniques are highly specialized, and all that will be attempted here is to give a general description of the procedures adopted. Detailed consideration of the problems and methods can be found in Saltpeter ([24,25]) and Stevens ([26]), while a useful practical working account has been given by Bienz and Bienz ([4]).

Fixing, Dehydration and Staining

As acetic-alcohol fixation damages the ultrastructure for electron microscope examination it is necessary to use a milder fixing solution such as 5% glutaraldehyde in pH 7.2 phosphate buffer or alternatively in pH 7.2 Na-cacodylate buffer (13.375 g Na-cacodylate.$3H_2O$ in 500 ml distilled water; adjust pH with HCl) for 60–90 min at 4°C. Glutaraldehyde reacts with amino and polyhydroxyl alcohols in tissues and the crosslinks bind the ultra-cellular structure and provide good images without artifacts (Sabatini *et al.*) [23]). If there is unincorporated label to be removed this is done by then washing in several changes of the same buffer over a period of 24–48 hours. The tissue is then post fixed for 60 min in 2% osmium tetroxide made up in the same buffer.

The tissue is placed in 75% ethanol containing 2% uranyl acetate as a block stain for 120 min; then transferred to two changes of 90% ethanol for 10 min each; three changes of absolute alcohol for 20 min each; finally into 1:1 propylene oxide/Epon 812 for 120 min.

Embedding and Section Cutting

The tissue is put in a small amount of Epon 812 epoxy embedding resin in a silicone rubber embedding mould (or alternatively in a gelatine or polyethylene capsule). The tissue is left for 10–15 min to allow the propylene oxide to evaporate and the mould is then filled completely with Epon. The resin is cured at about 45°C for 24 hours and 60°C for another 48 hours.

Sections have to be cut of between 600 and 800 Å thickness, the thinner sections being for potential magnifications greater than 10,000. The thinner the section the longer the exposure subsequently required, so comparatively thick sections are used when possible. The sections are picked up from the distilled water in the collecting "boat" by means of a hair loop, and transferred to carbon coated grids in the normal manner for electron microscopy.

Coating with Emulsion

Work must now be carried out under a safelight. The grids may be dipped in emulsion as previously noted for slides, but coating with a pre-gelled film of emulsion

(Stevens) ([26]) seems the method of choice in most cases. A stainless steel loop of about 15 mm diameter is dipped in Ilford L-4 emulsion (diluted 1:1 with distilled water at 40°C and allowed to cool to about 32°C), the surplus is wiped off and the loop rested on a support while the emulsion gels. In the meantime the grids have been mounted by means of double-sided scotch tape onto the ends of glass rods placed in holes in a wooden block stand. Then the loop is drawn over such a mounted grid, placing the monolayer emulsion film in close apposition to the section. The stand can then be put in a light tight box with silica gel for exposure at 4–5°C.

Development

The grids are developed by floating upside down on Microdol-X in a watch glass for about 4 min at 22°C; rinsed in two changes of distilled water for a few seconds in each; fixed in 2–5% sodium thiosulphate for about 3 min; finally washed in three or four changes of distilled water for 10 min each. As the developer sometimes removes uranium acetate from DNA related structure, giving a bleached condition, it may be desirable to restain in 2–5% uranium acetate for 5 minutes.

REFERENCES FOR FURTHER READING

1. BARBER, S. A. and Ozanne, P. G. Autoradiographic evidence for the differential effect of four plant species in altering the calcium content of the rhizosphere soil. *Soil Sci. Soc. Amer. Proc.* **34**, 635 (1970).
2. BASERGA, R. and Malamud, D. *Autoradiography Techiques and Applications*. Harper & Row, New York (1969).
3. BELANGER, L. F. and Leblond, C. P. A method for locating radioactive elements in tissues by covering histological sections with a photographic emulsion. *Endocrinology* **39**, 8–13 (1946).
4. BIENZ, K. and Bienz, D. Course in electron microscopy and autoradiography. Mimeo. pp. 24, CENA, CP 96, 13400 Piracicaba, SP, Brasil (1974).
5. BOYD, G. A. *Autoradiography in Biology*. Academic Press, New York (1955).
6. CRAFTS, A. S. and Yamaguchi, S. The autoradiography of plant materials. *Calif. Agr. Expt. Sta. Manual* 35, 143 pp. (1964).
7. COMAR, C. L. *Radioisotopes in Biology and Agriculture*. p. 481, McGraw-Hill, New York (1955).
8. EVANS, D. T. and Syers, J. K. An application of autoradiography to study the spatial distribution of ^{32}P-labelled orthophosphate added to soil crumbs. *Soil Sci. Soc. Amer. Proc.* **35**, 906 (1971).
9. GAHAN, P. B. *Autoradiography for Biologists*, Academic Press, New York (1973).
10. GUDE, W. *Autoradiographic Techniques*, Prentice Hall (1968).
11. HELLER, D. A. The Radioautographic Technique. *Adv. Biol. Med. Phys.* **2**, 133–70 (1951).
12. KOONTZ, H. V. and Biddulph, O. Factors affecting absorbtion and translocation of foliar applied phosphorus. *Plant Physiol.* **32**, 463–70 (1957).
13. LEVI, E. An artifact in plant autoradiography. *Science* **137**, 343–44 (1962).
14. LEVI, E. Handling plants for macro-autoradiography. Proc. Symp. *Isotopes in Weed Research*, 1965. 189–94 IAEA, Vienna (1966).
15. LEWIS, D. G. and Quirk, J. P. Diffusion of phosphate to plant roots. *Nature* **205**, 765–766 (1965).
16. MILIKAN, C. R. Radioautographs of manganese in plants. *Aust. J. Sci. Res.* **4**, 28–41 (1951).
17. NEILSON, J. A. Autoradiography for studying individual root systems in mixed herbaceous stands. *Ecology* **45**, 644 (1964).
18. PELC, S. P. Autoradiographic Techniques. *Nature* **160**, 749–50 (1950).
19. RANDALL, P. J. and Vose, P. B. Effect of aluminium on uptake and translocation of phosphorus-32 by perennial ryegrass. *Plant Physiol.* **38**, 403–49 (1963).
20. RENNIE, D. A. and Bole, J. B. The migration and mechanism of phosphorus movement from ammonium

phosphate pellet using autoradiographic techniques. 9th Int. Congr. *Soil Sci.* **7**, 775 (1968).
21. ROGERS, R. *Techniques of Autoradiography*, Elsevier (1967).
22. ROTH, L. and Stumpf, O. *Autoradiography of Diffusible Substances*, Academic Press (1969).
23. SABATINI, D. D., Bensch, K. and Barnett, J. *J. Cell Biol.* **17**, 19 (1969).
24. SALTPETER, M. M. and Bachman, L. Autoradiography with the electron microscope. A procedure for improving resolution, sensitivity and contrast. *J. Cell Biol.* **22**, 469–77 (1964).
25. SALTPETER, M. M. *In: Methods in Cell Physiology* (D. M. Prescott, Ed.) Academic Press Vol. II, 229 (1966).
26. STEVENS, A. R. *In: Methods in Cell Physiology* (D. M. Prescott, Ed.) Academic Press Vol. II, 255 (1966).
27. VOSE, P. B. The translocation and redistribution of manganese in *Avena*. *J. Expt. Bot.* **14**, 448–57 (1963).
28. WALLACE, A. Retranslocation of ^{86}Rb, ^{137}Cs and K to new leaf growth in bush beans. *Plant Soil* **29**, 184 (1968).
29. YAMAGUCHI, S. and Crafts, A. S. Autoradiographic method for studying absorbtion and translocation of herbicides using ^{14}C labelled compounds. *Hilgardia* **28**(6) 161–91 (1958).

CHAPTER 11

Isotopes In Soils Studies

I. SOIL CHEMISTRY, NUTRIENT MOVEMENT AND AVAILABILITY INTRODUCTION

RESEARCH workers in soil chemistry and soil fertility were among the first to utilize isotopic methods when they became available. Such techniques have now very greatly advanced our knowledge in regard to measuring the magnitude of nutrient reserves that can replenish the soil solution during crop growth, and the factors governing the rate of movement of nutrients into the soil solution from the solid phase. The number of research papers reporting the use of radioisotopes in soils studies is now so great that a complete review is no longer practicable.

Contemporary work can be found in such journals as *Soil Sci. Soc. Amer. J. Soil Science, J. Soil Sci., Plant and Soil*, etc. What is attempted here is to outline the situations where isotope methods have played a significant role, and to detail a number of specific techniques which have proven to be of general interest or applicability.

A soil is a multi-component system in which there is a continual tendency to attain equilibrium between the solid (soil minerals and organic matter) and liquid (soil solution) phases. Equilibria also exist between different components of the solid phase, such as fixed potassium and ammonium in the interlayer positions of clay minerals. Figure 11.1 represents the situation in a normal plant system, showing the series of equilibria which govern the movement of nutrient ions from soil solid to plant root.

The concentration of an ion in the soil solution is determined by the dynamic equilibria between solid and solution phases, the ion being continually dissolved from, and subsequently adsorbed on the surface of the solid phase via the solution phase. When a plant is grown in the medium and absorbs the ion from the soil solution this equilibrium will be altered, and the concentration of the ion in the soil solution lowered. If uptake by the plant continues without replenishment of the soil solution then ultimately the rate of uptake of the ion will become reduced. However, at least for a time, there will be a tendency for the ion in the soil solution to be replenished from the solid phase, the concentration of the ion in the solution being controlled by the plant uptake and the rate of dissolution or desorbtion from the solid phase. When plant growth, and hence demand for the nutrient ion, exceeds the capacity of the solution phase to supply it, then plant deficiency will occur. Such deficiency may be temporary, following a period of very rapid growth, or may be permanent if the soil solid phase is itself deficient of the ion or the ion is "unavailable" due to unfavourable pH.

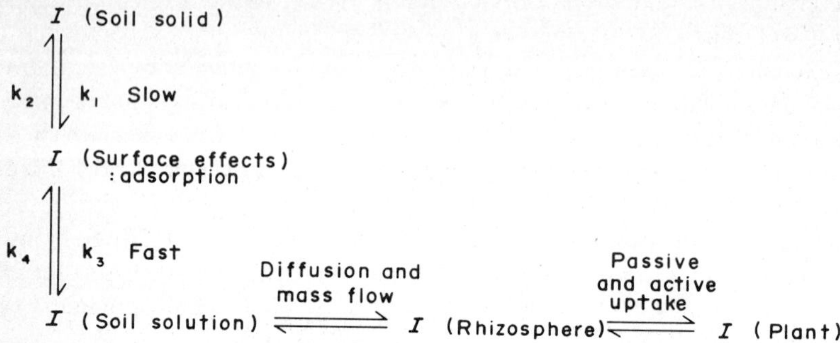

FIG. 11.1 The movement of a nutrient soil, *I*, from soil solid to plant.

The movement of ions, e.g. phosphate and cations, from the soil solids is relatively slow. It is a product of the weathering of soil minerals and microbiological activity. The second process is much faster and is primarily concerned with desorption of the solid-adsorbed nutrients, when the concentration in the soil solution decreases due to plant uptake, or leaching. The reverse adsorptive processes may also occur following the addition of nutrients as fertilizer.

It is of course relatively easy to model the movement of an ion from soil solid to plant as being a series of reactions each with its own rate constant, but it is quite another matter to specifically evaluate the different processes involved. Such things as dissociation equilibria and diffusion coefficients may be quite well known for pure solution, but the situation in the soil is vastly more complex, where rates of solubilization, ion exchange equilibria and surface or bulk diffusion coefficients are much harder to determine. Without the availability of either carrier-free isotopes or isotopes of very high specific activity, the considerable progress that has been made would scarcely have been possible.

Isotope based studies such as those in references 8, 9, 10, 11, 21, 24, 27, 33, 40, 51, 57, 60 65, 66 and 74 greatly enlarged our knowledge of the exchange mechanisms of cations and anions in the soil system. Inevitably, most procedures have been primarily of a research nature, developed for specific circumstances, but the determination of total cation exchange capacity by a radioisotope method has more general application.

DETERMINATION OF THE CATION EXCHANGE CAPACITY OF A SOIL

The cation exchange properties of a soil are due to the negatively charged particle surfaces of the clay and organic matter colloid complexes. The positively charged nutrient cations are thus held on the surface of these particles. It follows that the cation exchange capacity of a soil is the quantity of positive ions needed to neutralize the negative charge of a unit amount of soil and is usually expressed as milliequivalents per 100 g soil. The CEC is equivalent to the sum of the total exchangeable cations and the exchangeable hydrogen and it may therefore be obtained by calculation from these two determinations or it may be obtained experimentally by direct means. The CEC of a soil is important because it is a direct measure of the capacity of a soil to retain nutrients in a form available to plants but not susceptible to heavy losses from leaching by rainfall. The CEC thus distinguishes between potentially fertile soils with adequate clay and organic matter content, and infertile sands. CEC may be as low as 2 meq/100 g soil for a sandy soil with 70% sand to 20–35 meq/100 g soil for clay and clay loam soils with 40–60% clay. Very heavy clays and highly organic soils may give higher figures than this.

As a determination, the cation exchange capacity of a soil is an equilibrium measurement of cations adsorbed when a solution of a simple ion is percolated through the soil. Being an equilibrium measurement, the value varies to a certain extent with the cation employed for determination, the pH and the concentration. A number of radioisotope methods have been described ([12,47]) the one given here is that of Beek ([7]), in which the soil is first washed with ethanol to remove soluble salts, then mixed with inert sand and transferred to a percolation tube, which is then saturated with a solution of strontium acetate. Following removal of excess salt the sample is dried, and an aliquot of strontium acetate solution of known concentration labelled with ^{85}Sr is added to a known amount of soil. When equilibrium has been attained the activity of the supernatant liquid is determined and the cation exchange capacity calculated from the decrease in activity.

Experimental

1. 5 g of ethanol-washed soil are weighed out and 10 g of inert sand are added and thoroughly mixed. The mixture is transferred quantitatively to a percolation tube.
2. Vacuum is applied to the tube for 10–15 min and 100 ml of 0.5M strontium acetate solution of pH 7.5 is percolated.
3. Remove excess salt by percolating with 200 ml of 0.0025M strontium acetate solution, pH 7.5, followed by 100 ml of ethanol
4. Dry at 105°C, grind with pestle and mortar and weigh out 3 g into a 20 ml plastic bottle (note: of this 3 g mixture, 1 g is soil). Add 10.0 ml of 0.01M strontium acetate solution, labelled with about 0.1 μCi ^{85}Sr/ml.
5. Shake for 30 min, then centrifuge for 15 minutes at 3000 r.p.m. Then pipette duplicate aliquots of 3.0 ml into counting tubes. Also take duplicate 3.0 ml aliquots of the labelled strontium acetate solution that was added initially.
6. Determine the activity by counting with a β-counter.

Calculation

The CEC is expressed as meq/100 g soil.

$$\text{CEC} = 100 \times V_{as} \times \frac{(A_{as} - A_e)}{A_e} \times N_{as}$$

where, N_{as} = normality of the labelled strontium acetate solution
V_{as} = volume of the labelled strontium acetate solution
A_{as} = activity (c.p.m.) of the initial labelled stronium acetate solution
A_e = activity (c.p.m.) of the supernatant liquid.

NUTRIENT MOVEMENT AND SUPPLY

Nutrients in the soil solution reach the root surface (rhizosphere) by means of mass flow and diffusion processes ([3]) through spaces in the soil matrix, and also directly by root interception and contact. Mass flow of soil solution occurs due to water movement induced by the uptake of water by transpiring plants, and from rainfall, irrigation and drainage. Diffusion takes place when the soil solution is nominally stationary. In these conditions ion concentration gradients will be set up due to root and microbial activity removing ions from the soil solution at the root interface, thereby setting up a concentration gradient between the soil solution close to the root and soil solution further from the root. Isotope studies confirmed that if such a gradient is maintained, ions will diffuse towards the root ([17,36,45]).

Probably a majority of diffusion studies ([17,36,64]) have relied on the transient state method, where two samples, one isotopically labelled, are brought into contact and then the arrival of radioactivity in the unlabelled sample (or the reduction in radioactivity in the labelled sample) is determined as a function of time. Steady state method has also been used ([44]) where essentially the soil is equilibrated with a solution of the test ion and then arranged so that it is between two volumes of the same solution, one being labelled with a radioisotope. Again, the accumulation of radioactivity in the unlabelled solution is determined as a function of time.

Clearly, mass flow and diffusion processes can occur simultaneously ([41]), but their relative importance varies with the nutrient under consideration and the soil water conditions. If the water content of the soil is above wilting point, reduction in water content due to plant transpiration will increase the relative importance of ion movement due to mass flow. This is because as water content decreases, salt concentration will increase, and thus increase the amount of ions carried by mass flow because the amount of water transported will tend to be the same. Diffusion is likely to decrease significantly with severely reduced water content ([22,23,26]).

Isotope studies have shown that mass flow can supply large amounts of Ca, Mg and NO_3, and in certain soils also K. Root interception is also apparently important for Ca and Mg. On the other hand diffusion is usually an important mechanism in P, K and Mo supply ([4,5,6,17,34,45]). ^{35}S studies show that SO_4 movement is largely influenced by mass flow [1]. Contact is reported to be a major factor in the supply of Fe and Mn [20,46].

It has been known since the last century that not all of a nutrient element in the soil can be used by the plant. Since Dyer's use [13] of 2% citric acid as a means of imitating what was believed to be the roots' capacity to dissolve mineral elements, there is a long history of attempts to develop extraction procedures, particularly for phosphorus, which would permit estimation of the amount of element available to the plant and hence a comparison of the fertility of different soils by correlation with known crop response to fertilizer. Such methods have, from the practical point of view, been more or less successful in many cases but they have almost universally lacked a sound scientific basis.

Most effort and theory has been put into the mechanism of phosphorus supply. This is because of the complexity of its soil chemistry, the ready access to ^{32}P, and because of all the macroelements important in plant nutrition P has the lowest concentration in the soil solution, and its constant renewal from the solid phase is of critical importance. Thus Stout and Overstreet [25] calculated that in one experiment the soil solution P must have been replenished from the soil at least ten times a day. Collated data from many experiments showed [50] that in 25% of the samples P in the soil solution did not exceed 0.03 p.p.m. and in only 8% did it exceed 0.5 p.p.m., while other elements such as K, Ca, Mg, and Na may have concentrations a hundred or even a thousand times higher.

As we have noted the supply and transport of plant nutrients in the soil is now regarded as a succession of rate processes, each with its rate constant and each potentially rate limiting. This dynamic approach now largely influences present thinking on the availability of plant nutrient elements, and has lead to the concept of "capacity" and "intensity" factors governing nutrient supply. Thus the ability to renew the nutrient in the soil solution is a capacity factor, and Shapiro and Fried [56] developed the relationship for phosphorus.

$$\text{soil } P = -K \frac{\text{soil} - P}{P} + \Sigma \text{ soil}$$

where P is the concentration in the soil solution (an intensity factor); K is the apparent dissociation for the system at a specified pH; soil-P is the amount of phosphorus adsorbed per gram of soil (the P capacity factor); while Σ soil is the amount of phosphorus adsorbed per gram of soil if all the adsorption sites are saturated with P.

The long term evaluation of nutrient supply must also take into account a third parameter, a quantity factor, which represents the amount of nutrient potentially available during the growing period.

In practice we are presently limited to determining the amount of labile-P, i.e. the adsorbed and solution-P, which has been shown to give reasonable correlation with plant production [58]. Conventional soil extraction techniques for P although given a measure of the solid-adsorbed P are not specific for P (or any other element), moreover they are likely to cause re-distribution of the forms in which P is held. Isotope techniques make it possible to determine the exchangeable-*P* adsorbed on the solid surface without essentially altering the soil. If a known amount of an isotopic tracer is added to a system of a solution and a sparingly soluble solid then determination, at quasi-equilibrium, of the distribution of the label between solid and solution permits the calculation of the amount of surface ions maintaining the concentration of the solution. Thus for *P* we have the exchange equation:

$$\frac{P\text{ (solid surface)}}{P\text{ (solution)}} = \frac{{}^{32}\text{P (solid surface)}}{{}^{32}\text{P (solution)}}$$

This application of isotope dilution principle is the basis of a number of methods of determining available phosphorus; the estimation of 'E' value being a direct laboratory determination of isotopically exchangeable phosphorus.

Using exactly the same principles it was found ([48,49]) that ^{45}Ca could be used to determine the adsorbed calcium in calcareous and other soils, while isotope techniques permitted the determination of the labile pools of manganese, iron and zinc ([37,53]).

DETERMINING THE AVAILABILITY OF NUTRIENT ELEMENTS IN SOIL: 'A', 'L' AND 'E' VALUES

The A-value

We have briefly considered the use of isotopes in research on nutrient ion exchange kinetics, and especially with reference to the rates at which element reserves are able to exchange with soil solution elements and for evaluating the rate processes concerned with plant nutrient uptake from soil. Many are techniques for basic research, but a number of procedures have been developed to provide practical methods for expressing the availability of a soil nutrient.

These methods, all examples of the principle of isotope dilution, were first applied to the determination of plant-available soil phosphorus, but the principles employed have also been applied to a number of other elements. The so-called 'A', 'L' and 'E' values ([16]) have been particularly used for comparing the phosphorus supplying power of different soil types, and can provide a defined reference base for comparing different chemical extraction methods for routine large-scale determinations of P-availability.

Of the three commonly determined values, the A-value has achieved the widest acceptance and use. It is derived from the basic isotope dilution equations (page 131). Fried and Dean ([14]) pointed out that an available soil nutrient could be measured relative to a standard provided the available nutrient was adequately defined, and if one made the assumption that when two sources of nutrient are present in the soil the plant will absorb from each of these sources in proportion to the respective quantities "available". From this it followed that the "amount of available nutrient in the soil can be determined in terms of a standard, provided the proportion of the nutrient in the plants derived from this standard is determined''.

Assuming then that plants take up nutrients from two different sources in direct proportion to the amount available, the A-value was developed as the expression:

$$A = \frac{B(1-y)}{y} \tag{1}$$

where, A = amount of available nutrient in the soil
B = amount of fertilizer nutrient (standard) applied
y = proportion of nutrient in the plant derived from the standard.

In practice we use an isotopically labelled standard because it gives a direct measurement of the proportion of nutrient absorbed that was derived from the standard fertilizer. Note, however, that the use of an isotope is not an essential part of the concept, and one could for example estimate the amount of nutrient taken up by the plants from the standard by assuming the amount of nutrient taken up by the plant is linear to the amount present in the medium. Then the amount taken up from the standard is found by deducting the value obtained for a zero addition standard check treatment from the value obtained with added standard. Indirect methods such as this, normally imply using more than one fertilizer level and growing the zero check plants under conditions of nutrient stress. In contrast, the isotope procedure gives a direct measure, requires only one level of fertilizer, and normal fertility conditions.

It has been found that determination of A-value is largely independent of rate of fertilizer application, size of test pot, or the growth of the test crop. In other words, for a specific soil, crop and growing situation, the A-value is a constant. Objections have been made ([54]) concerning the validity of applying simple isotope dilution principles to soil-plant systems, in view of the possibility of considerable isotopic exchange taking place between the labelled P-fertilizer and soil-P when the fertilizer is mixed uniformly through the soil. This is indeed an obvious difficulty with A-value experiments, and experimental procedures must be so designed as to minimize the degree of isotopic exchange reaction between fertilizer and soil. Ideally this requires using a test crop with rapid germination and growth, and having as little mixing between fertilizer and soil as possible consistent with permitting free contact between fertilizer and roots, i.e. the labelled fertilizer is best applied in a band. Good experimental evidence supports the effectiveness of band placement in minimizing exchange ([52]).

Although the A-value was primarily developed with a view to determining the availability of phosphorus in soil, the principles employed can in fact be applied to a number of nutrient elements including N (using ^{15}N label), S (using ^{35}S), Zn (with ^{65}Zn) and Ca (using ^{45}Ca). It will be appreciated that the use of A-values is not confined to estimating availabilities of nutrients in soil, but can also be applied to evaluate the relative availability of different phosphorus sources ([15]) on the same soil, and to determine the residual value of applied phosphorus ([18,43]). The principle of A-value determination has also been extended for field use (see Chapter 12).

Determination of A-value

Experimental

Considerable variation is possible, but the essential features are to use a fast growing test crop and to apply the labelled fertilizer so as to reduce interaction between soil and fertilizer as much as possible. The soil containers may conveniently contain between 100 and 1000 g of soil, and it is usually appropriate to give a P application of 20 mg P/100 g soil. Barley, oats and ryegrass are suitable test crops.

1. Weigh out four or six replicates of 500 g of soil into suitable containers.
2. Remove the top 2–3 cm of soil and place on one side.
3. Apply 1.15 g of ^{32}P-superphosphate fertilizer (i.e. 20 mg P/100 g soil) uniformly over the exposed surface. To achieve good distribution of the fertilizer it may be useful to first mix it with a minimal amount of fine dry soil.
4. Sow the test crop e.g. barley, in the fertilizer band and cover with the soil previously removed.
5. Place each soil container in an appropriate sized saucer and irrigate through the bottom.
6. Place the containers in a growth room or glasshouse and allow to grow for 2–4 weeks; daily making up the losses of irrigation water due to evaporation and transpiration.
7. Note: as an alternative to sowing the seeds in the band, fine seeds such as ryegrass may be sown on the surface of the soil. Another alternative to encourage fast plant growth and a short experiment, is to germinate the seeds on a disc of nylon net or glass fibre the same diameter as the soil container. After germination a root mat is soon formed which can then be placed in contact with the freshly prepared soil and gently pressed down with a glass rod. Root growth is then very rapid, and accordingly this is a preferred method.

Analysis

1. *Reagent and Standards*

Vanado-molybdate: Dissolve separately in water 20 g ammonium hepta-molybdate and 1 g ammonion metavanadate. Mix, add 140 ml conc. nitric acid, dilute to 1 litre.

Standard solution: Dissolve 0.8784 g of potassium dihydrogen phosphate in distilled water and dilute to 1 litre. Then 5 ml contains 1 mg P.

2. *Procedure (plant material)*

Note that as *specific activity* is being determined the actual weight of the sample does not need to be known.

1. Harvest a sample of plant material which will not exceed 1 g dry weight.
2. Digest the material in conc. H_2SO_4 and H_2O_2 (5 ml of conc. H_2SO_4 for each gram dry matter and the minimum amount of H_2O_2 needed for obtaining colourless solutions). Make up to 50 ml.
3. Pipette 10 ml of plant digest into a 100 ml volumetric flask.

4. Add 10 ml of vanado-molybdate reagent.
5. Shake and make up to volume.
6. Allow to stand for 35 min to enable full colour development.
7. Prepare a set of standards simultaneously, containing 0.2, 0.5, 1, 1.5, 2.0 mg P (1 ml, 2.5 ml, 5 ml, 7.5 ml, 10 ml of standard solution) in 100 ml flasks. Include a blank with reagents above, and proceed as above.
8. Take readings on the spectrophotometer at a wavelength of 420 nm against a reagent blank.
9. Plot readings of the standards against mg P to obtain a calibration curve. Determine the amount of P in a 10 ml aliquot of digest.
10. Take another 10 ml aliquot of plant digest and count by Cerenkov or M6H or MX124 liquid GM counter to determine ^{32}P activity. (It may be necessary to adjust size of aliquot according to the activity). Calculate specific activity e.g. c.p.m./mg P.

Fertilizer

11. To obtain a standard fertilizer solution, weigh out 0.23 g of labelled superphosphate (i.e. about 20 mg P) into a 200 ml volumetric flask, add 20 ml of dilute nitric acid (dil. 1:2) and dissolve. Make up to volume. Filter if necessary. Note: It is not necessary to take an amount of fertilizer as standard, equivalent to the amount applied to the soil. In this case it just happens to be a convenient amount.
12. Take 10 ml aliquot of fertilizer solution and determine P content as for plant digest.
13. Determine ^{32}P activity in suitable aliquot of fertilizer solution. Calculate specific activity e.g. c.p.m./mg P.
14. Calculate A-value as described.

Calculation of A-value

We have already noted that inherent in the A-value concept is the assumption that if a plant is supplied with two sources of a nutrient it will utilize them in direct proportion to the availability of the element in each. This we have noted in Chapter 12 is also the basic assumption of labelled fertilizer studies, in which the specific activity of the plant divided by the specific activity of the labelled fertilizer equals the fraction of the nutrient in the plant which was derived from the labelled source (fertilizer):

$$\frac{\text{Specific activity of plant}}{\text{Specific activity of fertilizer}} \times 100 = \begin{array}{l}\text{percent nutrient in} \\ \text{plant derived from} \\ \text{the fertilizer}\end{array} \qquad (2)$$

In relation to the concept of A-value this may be further developed as:

$$\frac{\text{Specific activity of plant, } S_p}{\text{Specific activity of labelled standard nutrient, } S_f} \quad \frac{B}{A + B} \tag{3}$$

where A = amount of nutrient available in the soil
B = amount of labelled nutrient added as standard

Rearranging equation (2), we get:

$$A = \frac{B[1 - (S_p/S_f)]}{S_p/S_f} \tag{4}$$

Note that S_p / S_f is the fraction of the nutrient in the plant derived from the labelled standard nutrient, thus $1 - S_p / S_f$ is the fraction derived from the soil.
Alternatively we may express equation (4) as:

$$A = (B) \times (S_f / S_p - 1) \tag{5}$$

In full:

$$\text{A-value} = \left[\begin{array}{c}\text{amount of labelled nutrient}\\ \text{added as standard}\end{array}\right] \times \left[\frac{\begin{array}{l}\text{specific activity in}\\ \text{labelled nutrient}\end{array}}{\begin{array}{l}\text{specific activity in}\\ \text{plant}\end{array}} - 1\right]$$

If for example we wish to express this in practical terms of A-value for phosphorus, then this will give us:

$$\text{A-value} = \left[\begin{array}{l}\text{mg P added as}\\ \text{fertilizer phosphate}\\ \text{per 100 g soil}\end{array}\right] \times \left[\frac{\begin{array}{l}\text{specific activity of}\\ \text{P in fertilizer}\end{array}}{\begin{array}{l}\text{specific activity of}\\ \text{P in plant}\end{array}} - 1\right] \text{mgP/100g soil} \tag{6}$$

or,

$$\text{A-value} = \left[\begin{array}{l}\text{mg P added as}\\ \text{fertilizer phosphate}\\ \text{per 100 g soil}\end{array}\right] \times \left[\frac{\text{\% P derived from soil}}{\text{\% P derived from fertilizer}}\right] \text{mgP/100g soil} \tag{7}$$

The L-value

The L-value (Larsen) ([28]), similar in concept to the method of Gunnarsson and Fredriksson ([19]), is a measure of the isotopically exchangeable fraction of phosphate in soil, and has been used to determine the quantity of soil phosphate in labile form ([30,31]). It has frequently been confused with the A-value from which it differs in concept and basic experimental requirements, though sharing the identical mathematical formula for calculating the results.

As originally developed ([28,29]) the test was carried out over a complete plant growth cycle and used the basic isotope dilution equation to determine the quantity of labile soil phosphorus associated with the isotopic dilution of the applied labelled phosphorus. Thus the value which has come to be known as the L-value is given by

$$L = w^*\left(\frac{S_1^*}{S_2^*} - 1\right) \tag{8}$$

where S_1^* and S_2^* are the specific activities of the applied phosphorus and the plant phosphorus, w^* the amount of ^{32}P-labelled phosphorus applied and L is the amount of isotopically exchangeable soil phosphorus. Successive determinations of specific activity of the test plants showed that the value was independent of the quantity of applied phosphorus, and ultimately became independent of time, indicating the attainment of isotopic equilibrium.

Subsequent work ([55]) demonstrated that the L-value was independent of the quantity of carrier phosphorus over a very wide range (up to 1000 times) and even the time of sampling provided that the ^{32}P-labelled phosphorus was extremely well mixed throughout the soil. The effect of mixing is to speed up the attainment of equilibrium between labelled phosphate and labile soil phosphorus, which is also encouraged by the use of "carrier-free" ^{32}P.

The attainment of isotopic equilibrium is an essential feature of the method, because it is only in this way that the L-value can equate to a definite amount of soil phosphorus. When equilibrium is not achieved, although it may be quite possible to use the procedure to give a valuable comparison of the available phosphorus in a series of soils, the value measured will not be a true L-value. The only certain means of ensuring that isotopic equilibrium has been reached is to take a series of harvests and plot the curve of L-value versus time ([31,32]). The only alternative is to take a single harvest when the test crop approaches maturity.

Determination of L-value

Experimental

Procedures adopted can be very similar to those described for A-value determination, with certain important differences. The amount of soil used should be comparatively large, 1 kilo or more, as the experiment must be carried out over a longer period to permit isotopic equilibrium to be attained. While a basic level of labelled phosphorus application of, say, 20 mg P/100 g soil is not inappropriate, the ^{32}P is better applied "carrier free". Equally, every effort must be made to mix the labelled phosphorus uniformly through the whole soil mass. This means that the ^{32}P is best applied in aqueous solution, about 150–200 ml containing 50 μCi ^{32}P, per kilo of soil.

It is best to take a number of harvests in order to confirm isotopic equilibrium. Analysis may follow the procedures given for A-level determination.

Calculation of L-value

The calculation of 'L-value' is exactly the same as for 'A-value', but recall that the concept, stated objective and ideal experimental conditions are different.

Thus the A-value has the objective of measuring the availability of a soil nutrient (often, but not necessarily phosphorus) relative to a standard fertilizer source. The L-value has the objective of measuring the total quantity of plant available soil phosphorus. It is the amount of phosphorus in the soil and in the soil solution which is exchangeable with orthophosphate ions added to the soil, assuming the attainment of isotopic equilibrium.

The ideal conditions for an A-value experiment require that the fertilizer should interact with the soil as little as possible, even though it is distributed through the soil to permit ready contact with plant roots. Ideally a short experiment will minimize the degree of reaction between fertilizer and soil. On the contrary, in the case of the L-value the ^{32}P is added with a minimum amount of carrier to facilitate exchange with soil phosphorus, and with the maximum amount of mixing. Additionally, a long experiment is desirable to allow attainment of isotopic equilibrium and a steady state value.

Achieving the ideal conditions for either value is, in practice, scarcely possible.

The E-value

The E-value of a soil is a measure of the amount of isotopically exchangeable phosphate of that soil. McAuliffe *et al.*, ([38]) used the principle of isotopic exchange for determinations of surface phosphate, as a more fundamental approach than that inherent in extraction methods. Later, following the utilization of L-value procedures, Russell *et al.*, ([54]) developed the E-value as its laboratory equivalent.

In this method a small sample of soil is equilibrated with a ^{32}P-labelled phosphate solution, the suspension being shaken for an arbitrary period of time. Although based on the same principle as the L-value method, in practice the E-value differs in two significant respects. First, complete isotopic equilibrium is not achieved because although most exchange takes place fairly rapidly, some exchange continues over a long period as shaking makes frsh exchange sites available, and by slow diffusion in the solid phase. Secondly, in a soil suspension no phosphorus is removed from the system while in the determination of L-value the plants grown in the soil remove phosphorus, and may give rise to greater dilution than due to isotopic exchange on its own. Results will also be influenced by temperature, which should be specified, and soil:solution ratio.

Dalal and Hallsworth ([131]) found that over a range of soils there was not very good correlation between E-values and L-values. E-values were comparable with L-values on low phosphorus fixing soils but were relatively much higher on high phosphorus fixing soils. They found that the inverse-dilution technique (E_{id} value) of Mekhael *et al.*, ([132]) gave results which were most comparable.

Like the L-value, the E-value measures labile P ([54]), made up of "surface-P" and "solution-P". Modification of the E-value determination has sometimes been made in order to measure these components separately.

Determination of E-value

Experimental

1. Prepare stock solution 'A' containing 100 μg P/ml by weighing out 0.4392 g KH_2PO_4, dissolving in distilled water and making up to 1 litre. Prepare working solution 'B' by taking 20 ml of solution 'A', adding 100 μCi ^{32}P and making up to 1 litre with 0.01 M $CaCl_2$ (the $CaCl_2$ is to represent the soil solution and to make filtering somewhat easier). Then stock solution 'B' contains 2.0 μg P/ml and 5 μCi ^{32}P/50 ml.

2. Weigh out duplicate 5 g samples of each soil and place in 100 ml stoppered conical (Erlenmeyer) flasks.

3. Add 50 ml of working solution 'B' to each flask, stopper the flasks and place on a shaker for 24 hours (for class experiments a 4 or 6-hour shaking period may be conveniently adopted with little practical loss).

4. At the end of 24 hours, filter or centrifuge the suspensions.

5. The ^{32}P activity in the working solution and the filtrates is determined. Pipette a 1 or 2 ml aliquot into a counting tube, dilute to volume and count by Cerenkov method. Alternatively, evaporate 1 ml of filtrate or stock solution in a planchet, dry, and count with G-M counter.

6. Determine the ^{31}P colorimetrically in 10 or 20 ml of filtrate, as described for A-value determination.

Calculation

As the laboratory equivalent of the 'L' value, it follows that the E-value is a straightforward application of the isotope dilution principle and the same basic equation holds, thus:

$$E_t = w\left[\frac{S_s^*}{S_f^*} - (1)\right]$$

where E_t is the E-value at a given temperature

S_s^* is the specific activity of the added phosphate solution

S_f^* is the specific activity of the filtrate

w is the amount of phosphate added per unit of soil.

In working terms, this becomes

$$\text{E-value} = \left[\begin{array}{l}\mu\text{g P added}\dagger\\ \text{per g soil}\end{array}\right] \times \left[\frac{\begin{array}{l}\text{specific activity of}\\ \text{added P solution}\end{array}}{\begin{array}{l}\text{specific activity of}\\ \text{P in filtrate}\end{array}} - 1\right] \mu\text{g P/g soil} \qquad (10)$$

†when 50 ml of solution with 2.0 μg P/ml are used with 5 g of soil, this is equivalent to 20 μg/g.

An alternative way of calculating E-value is:

$$\text{E-value} = \frac{c_i}{c_f} \times [{}^{31}P_f - {}^{31}P_i]\, \mu g\ P/g\ soil \qquad (11)$$

where c_i is the initial count rate per unit volume of the ^{32}P solution, and c_f is the final count rate per unit volume; and $^{31}P_i$ and $^{31}P_f$ are the initial and final amounts of ^{31}P in the test solution in μg P/g soil.

Determination by inverse isotope dilution method

In the method of Mekhael *et al.*, ([132]) as modified by Dalal and Hallsworth ([131]), two sets of soil samples are shaken in 0.01 M $CaCl_2$ with carrier free ^{32}P (0.5 μCi/g soil) for a week. The suspensions are then centrifuged and one set of samples filtered, and P and ^{32}P determined. The other set of samples has 0.5 ml of 20 and 200 μg P/ml added to exchange with ^{32}P on the solid phase during 24 hours shaking. These suspensions are then centrifuged, filtered, and P and ^{32}P determined.

ISOTOPIC TRACERS IN POT EXPERIMENTS

Isotopic tracers are frequently used in pot experiments concerned with crop nutrition, either because greater control of environment and experimental conditions is required than is possible in the field, or as an inexpensive preliminary to such experiments. Results obtained in a pot experiment often help to design a more useful and informative field experiment than would otherwise by the case. However, isotope experiments in pots are no different from other pot experiments. That is they may provide a valuable guide to field behaviour but great care must be taken in extrapolating results to field situations.

Such pot experiments may be useful for determining time of application, level of application, source of nutrient, interaction of nutrients, soil pH and effect of liming or other soil amendment, and when interpreted with caution the placement of fertilizer. All results ultimately need confirming in the field.

In the field soil applications have been primarily restricted to ^{32}P or ^{45}Ca because of considerations of isotope availability, half-life and cost. Approximate activity levels used are:

200 μCi ^{32}P per gram P (minimum activity, for long experiments use 500 μCi/g P)

and 500 μCi ^{45}Ca per gram Ca

and suitable levels of fertilizer application are:

25, 50 and 100 kilograms P per hectare (or lb/acre)

and 200–800 kilograms Ca per hectare (or lb/acre).

For pot experiments use 50–100 μCi activity per pot containing 1.5–2 kilos of soil. For these experiments, a wider range of radioisotopes are practicable, including ^{58}Co, ^{55}Fe, ^{59}Fe, ^{54}Mn, ^{86}Rb, ^{35}S and ^{65}Zn. ^{99}Mo unfortunately has a half-life of only 66 hours.

Common practice in pot experiments varies in regard to the method of calculating the amount of nutrient that is applied, whether it should be on an ''area'' basis or on a ''percent weight of soil'' basis. Frequently both are combined to take account of the average weight of soil in the plough layer. This assumes a plough layer 18 cm deep, weighing 2 million pounds per acre (2.242 million kilograms per hectare). Thus 2.2 kilos of soil in a pot would ''represent'' 1 millionth of a hectare.

Assessing the levels of trace elements for pot experiments is always difficult, and Table 11.1 may prove a useful starting point.

TABLE 11.1
Suitable additions of trace-elements for pot experiments

Element	p.p.m. low	air-dry medium	soil basis high	Common source
B *	0.005	0.025	0.125	$Na_2B_4O_7.10H_2O$
Cu	0.025	0.125	0.625	$CuSO_4.5H_2O$
Mn	0.025	0.500	0.625	$MnSO_4.4H_2O$
Mo	0.005	0.025	0.125	$(NH_4)_6Mo_7O^2_4$
Zn	0.025	0.125	0.500	$ZnSO_4$

*no radioactive isotope available.

II. THE STUDY OF SOIL ORGANIC MATTER INTRODUCTION

There has been growing awareness in recent years of the possibility of using isotope methods for soil organic matter studies, while the application of such techniques as radiocarbon dating and stable isotope ratios has opened up whole new fields of research.

The main areas where isotope techniques have found a unique role include the effects of organic matter on nutrient availability, the synthesis and decomposition of organic matter in soils, the chemistry of humic substances and the transformation of nitrogen in soils. Only some of these areas can be touched on, including the preparation of suitable labelled material.

PREPARATION OF ^{14}C AND ^{3}H LABELLED PLANT MATERIAL

For studies on the effect of microorganisms on the decomposition of organic matter and of the humification of plant and microbial tissue, the preparation of ^{14}C labelled experimental material is essential.

The normal procedure for producing uniformly ^{14}C labelled plants is to grow them in a closed system enriched with $^{14}CO_2$. A number of successful approaches have been adopted, based on sealed plant growth chambers of ranging complexity, from the very elaborate $^{14}CO_2$ growth chamber of Zeller *et al.*, ([129]) Fig. 11.2, to the simple polystyrene canopies of the Saskatoon group ([82,126]) Figs. 11.3 and 11.4. Essential features include complete sealing from the outside atmosphere, continuous air circulation, provision for producing, maintaining and monitoring the desired level of $^{14}CO_2$, and control of the ambient temperature ([67,84,108,113,118]). Supply of nutrients and watering must also be provided. Usually nutrient solutions are used.

Sauerback and Fuhr ([108]) have described many of the problems and parameters of ^{14}C plant labelling and is an essential reference. Normally closed systems will have 0.1% CO_2 by volume in the system, and this is produced by reacting the required amount of labelled carbonates, $Na_2{}^{14}CO_3$, with dilute acid to provide an activity of about 100 μCI ^{14}C per g ^{12}C. The final specific activity of labelled plant carbon will be approximately one third that of the $Na_2{}^{14}CO_3$, or a little less.

The possibility of radiation damage to the plants imposes a limit to the activity of ^{14}C that can be used in the system, but the objective is to obtain labelled plant material containing 25–100 μCi per g C. Material with this amount of activity can be used for studies on the chemistry of humic substances, although less activity may be suitable for studies where specific fractionation and separation is not necessary. Normally, fast growing material such as wheat or barley is used. Alternatively, turves of established grassland may be placed in the growth chamber after cutting, and the regrowth takes place in an atmosphere of $^{14}CO_2$.

As it is essential that the material should be uniformly labelled, that is the lignin and cellulose must have the same specific activity as the water soluble fractions, it will be realised that there are no short cuts to its preparation. Suggestions have been made to simplify methods by providing the ^{14}C label to the roots by means of labelled cyanamide fertilizer $H_2{}^{14}CN_2$ or as labelled sodium bicarbonate $NaH^{14}CO_3$. However, these possible methods raise problems of their own, and a sealed growth chamber would still be required to prevent loss of $^{14}CO_2$ through dark respiration, leading to radioactive contamination and loss of valuable label. Equally important is the very large CO_2 dilution that would occur in plants exposed to the normal atmosphere.

^{14}C is normally introduced into soil organic matter and humic substances by providing labelled plant material as a substrate for organisms already present in soil. Neutron irradiation has been used to label humic acids ([113]) but is not practical due to lack of specificity, the unknown nature of the products and the presence of radiation decomposition products.

Tritiated ^{3}H humic substances can be prepared by allowing plant material to decompose in soil to which tritium labelled water has been added. Alternatively the Wilzbach ([127]) gas exposure method can be used. In this method ([114]) of tritiating

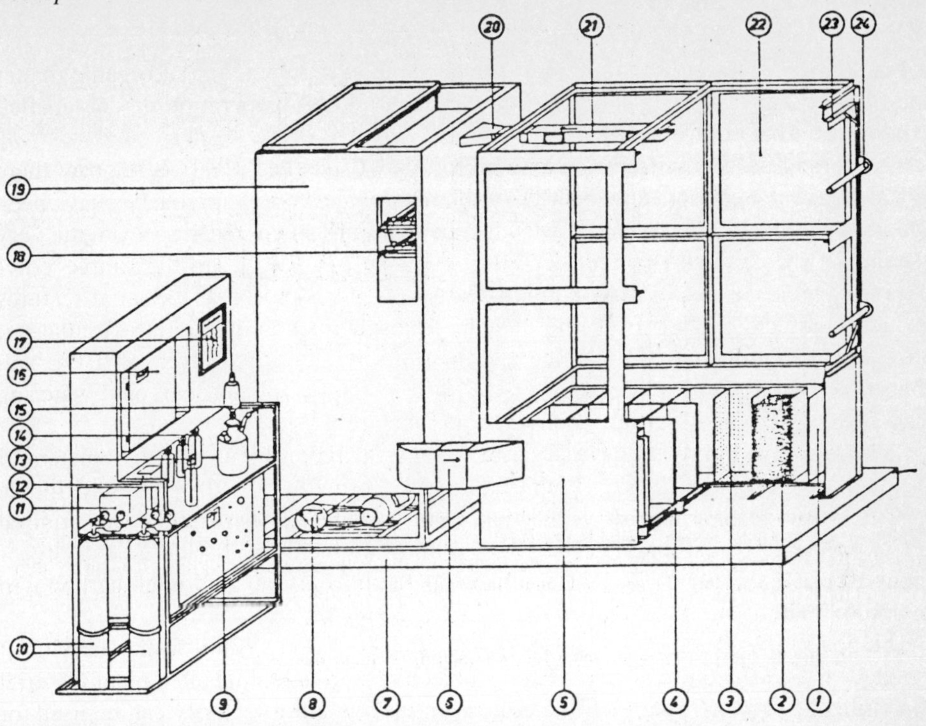

1. Nutrient tank 2. Graven bed 3. Plastic baskets with perforated walls 4. Rails for the nutrient tank 5. Growth chamber 6. Air connection to equipment 7. Steel-net reinforced concrete base 8. Peristaltic pump (low pressure air) 9. C^{14}-monitor 10. Gas-cylinders for IR-CO_2-analyser 11. Membrane pump for IR-CO_2-analyser 12. Membrane filter for IR-CO_2-analyser 13. Air-dryer for IR-CO_2-analyser 14. $C^{14}O_2$-absorption 15. $C^{14}O_2$-generation 16. IR-CO_2-analyser 17. IR-CO_2-recorder 18. Heaters for $C^{14}O_2$-chamber 19. Equipment chamber 20. Air connection between equipment and growth chamber 21. Air duct 22. Double glass panes 23. Door 24. Bars for handling the door

FIG. 11.2 Growth chamber for growing ^{14}C-labelled plant material (129).

compounds in a random manner, tritium gas and the humic substances are sealed together in an ampoule and are left to stand for a week or two. During this time exchange takes place and some of the ^3H atoms of the humic substances will be replaced by tritium. The hydrogen atoms which are most easily exchanged are determined by the configuration of the molecule and bond energy.

A major difficulty here is the problem of potential isotopic exchange, as discussed in Chapter 6, and the deuterium and tritium labels can be very easily lost. This is particularly the case if they are attached to -COOH, -OH or -NH_2 groups or to active carbon. In these cases the label can exchange relatively quickly with unlabelled water, and therefore in an aqueous medium the presence of the label will not necessarily be representative of the active molecule. Obviously there will be less danger of exchange when tritium is attached to sulphur groups like cysteine, H_2S or SH^-, or is part of a tightly bound organic molecule.

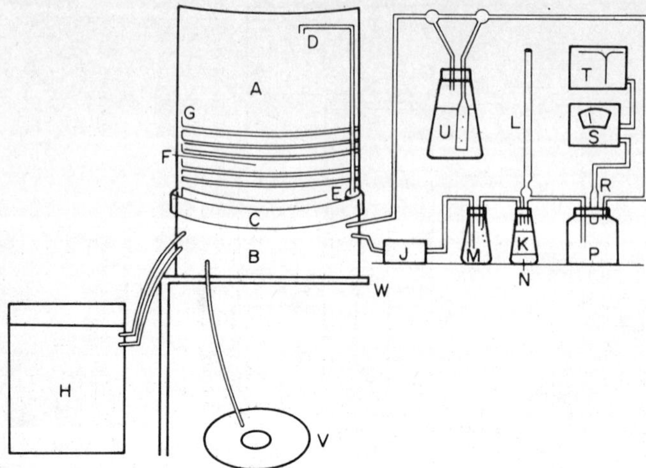

A canopy of cellulose acetate-butyrate; B metal base containing soil; C soft rubber gasket seal; D, E air inlet and outlet; F copper cooling coils; H refrigeration unit; J small air pump; K reaction vessel with 85% lactic acid; L burette with $Na_2{}^{14}CO_3$; M flow rate monitor; N magnetic stirrer; P counting chamber; R thin-window G-M tube; S scaler; T recorder; U liter flask with NaOH solution; V expandable tube to accept pressure; W bench.

FIG. 11.3 Relatively inexpensive equipment for ^{14}C-labelling of plant material ([82]).

ASSIMILATE TRANSFER AND ORGANIC MATTER PRODUCTION

Non-uniform labelling of growing plants in the field has been carried out by exposing them to $^{14}CO_2$ under a cellulose acetate-butyrate canopy, Fig. 11.4 ([126]). Here the objective was not to produce ^{14}C-labelled organic matter for other studies, but to determine carbon transfer through the plant-soil system by measuring carbon assimilation by the plants, the transfer of photosynthate below ground and the respiration of the roots. The overall objective being based on the concept that carbon flow can be used to characterize the soil-plant complex.

In order to study root formation and decomposition during plant growth, plants were grown for several months in a $^{14}CO_2$ atmosphere, with the shoots sealed from the roots ([112]). It was found that the labelled organic carbon in the soil at harvest was 20–50% higher than could be accounted for in the separated roots alone. Moreover, three times the remaining quantity of root carbon had been mineralized during growth. Four-fifths of the total soil $^{14}CO_2$-production resulted from microbial decomposition of root material. It was concluded that the plant roots are of much greater importance as a continual source of both soil humus and microbial energy than generally recognized, and that the residual quantity of roots at harvest cannot give a complete view of the role of roots in organic matter turnover and formation.

Plants were grown in a $^{14}CO_2$ atmosphere and a comparison made of the chemical nature of ^{14}C-labelled organic matter under sterile and non-sterile conditions ([98]). It

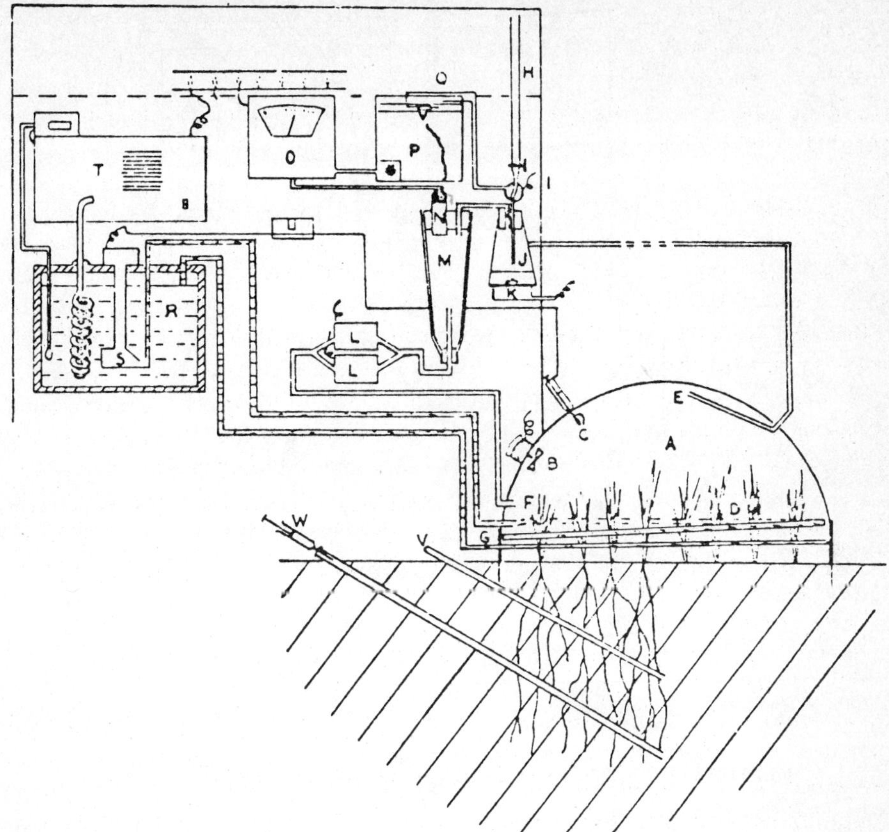

FIG. 11.4 $^{14}CO_2$-generating, temperature control equipment and photosynthesis canopy used in the field (126).

was suggested that a large part of the organic ^{14}C present in the soil resulted from autolysis of root tissue. About 15% of the ^{14}C in both sterile and non-sterile soils was present in humic acid fractions, and show significant formation of organic matter during root growth, much of it direct from root tissue without the mediation of soil microflora.

$^{14}CO_2$-studies such as described above will do much to improve our knowledge of root metabolism, breakdown, and the formation and maintenance of soil organic matter.

INFLUENCE OF SOIL ORGANIC MATTER ON NUTRIENT AVAILABILITY, PLANT NUTRITION AND METABOLISM, AND ON CHEMICAL RESIDUES

Organic matter is itself an important source of plant nutrients, but it also has a considerable effect on the availability of plant nutrients from the soil, and on plant nutrition and metabolism. It also has an effect on the exchange capacity of soils, the uptake of fall-out products such as cesium-137, and on the inactivation of herbicides and pesticide residues.

For many such studies we do not need isotopically labelled organic matter, as unlabelled organic matter is the variable and its influence on the uptake of isotopically labelled nutrients or compounds will be studied. Clearly the range of potential studies is very great and covers the whole area of the use of isotopes in soils research. Only a few typical examples of applications to organic matter studies can be given here.

^{32}P was used to determine the influence of soil organic matter on the loss of available phosphate in soil due to chemical reactions ([94]). The problem of the rapid removal of zinc from the available pool when zinc is added to high organic matter "Zn-deficient" soils has been studied using ^{65}Zn ([120]). It was possible to follow changes in the zinc distribution as reflected by different extractants, and in the organic matter residues. Adding organic matter to the soil resulted in more zinc going into the organic fraction, and zinc deficiency was caused during active decomposition of organic matter.

It was found that precipitated manganese isotopically exchanged with ^{54}Mn and was chelated by soil organic matter ([105]). A major conclusion was that the presence of organic matter is one of the most important factors responsible for retaining high concentrations of soil solution (chelated) manganese. Similarly a study ([75]) with ^{59}Fe and ^{54}Mn indicated that organic matter helps mobility of iron and manganese in calcareous soils. The chelating effect of organic matter has also been found to have an essential rôle in the migration of fission products ([70]).

References ([73,77,78]) are typical studies of the uptake of ^{14}C-labelled organic matter and breakdown products by the plant and their influence on plant metabolism, though the complexities of such biochemical systems do not now make this approach popular.

EFFECT OF MICROORGANISMS ON DECOMPOSITION OF ORGANIC MATTER AND PLANT TISSUE IN SOIL

^{14}C products have made it much easier to determine factors concerned with the turnover of organic matter in soil; the effect of adding fresh organic matter such as crop residues to the decomposition of native carbon soils; and the general rapid estimation of microbial activity.

Often the use of ^{14}C or ^{13}C labelled material is an essential way of tracing the decomposition of added organic matter. In so called "priming" studies ([69,71,81,107,121]) ^{14}C (or ^{13}C) labelled organic matter is incorporated in the soil, and $^{14}CO_2$ is evolved due to microbial decomposition. By comparing the specific activity of the $^{14}CO_2$ with that of the added organic matter it is possible to calculate what proportion of the mineralized carbon derived from the organic addition and what came from the decomposition of the native organic matter. These are typical isotope dilution procedures and have largely contributed to present views that the priming effect is largely a turnover phenomenon.

General microbial activity can be determined by a relatively simple process involving measurement of CO_2 evolution and gives a means of estimating the potential productivity of a given soil ([100]).

Jenkinson ([87]) demonstrated that fumigation with $CHCl_3$ of soils in which ^{14}C labelled plant material had been allowed to decompose, greatly increases $^{14}CO_2$ evolution on subsequent incubation. It was demonstrated that this method can be taken as a measure of biomass ([90]) and has been used by a number of workers ([93,123]), measuring the nett formation of CO_2 or mineral-N from decomposing cells following fumigation. $^{14}CO_2$ or ^{15}N are not essential to the principle of the method but clearly can often given more specific data. The method involves exposing the test soils to alcohol-free chloroform for 24 hr followed by removing the chloroform under repeated vacuum. Soils are then inoculated with a suspension of fresh soil in water and incubated for 10 days, with the CO_2-C evolved being collected. Biomass-C may then be calculated on the assumption that 50% of the carbon in killed biomass is mineralized to CO_2 by 10 days incubation after fumigation ([91]). When mineral-N is determined, biomass may be calculated assuming the C:N ratio of soil biomass to be 5:1 ([93]).

There are now many references concerning the use of isotopes in studies on the decomposition of organic matter, and references 76, 85, 101, 110, 117 and 119 illustrate the type of experimentation that has been made possible. The use of specifically ^{14}C-labelled lignins, coniferyl alcohols and humic acid type polymers has given information of their breakdown in soil ([99]). In a month 66% of $O^{14}CH_3$ compounds had evolved as $^{14}CO_2$ compared with less than 50% of ring ^{14}C and 2-^{14}C side chain carbons. It was found that carbon in the model humic acids was most resistant, a maximum of 14% being evolved as $^{14}CO_2$ after four weeks incubation.

Relatively little quantitative data on turnover of soil organic matter has been hitherto available. Tracer techniques with ^{14}C and ^{15}N have made it possible to determine readily the turnover time of the less resistant soil organic constituents ([83,89,103,104]), showing that there is an early rapid breakdown of plant residues when placed in the soil, but there is a resistant fraction of principally microbial metabolites and cell wall constituents, which stabilizes and remains over a period of years ([89]).

Determining the turnover-rate or mean residence time of the stable fraction has been made practicable by using the naturally occurring ^{14}C for radiocarbon dating. This technique is described in the last section, and it has completely altered our earlier ideas ([74]) based on the estimates derived from plant residue decomposition experiments. It may be combined with tracer techniques to give a complete picture of the mean residence time (MRT) of the different soil organic constituents ([104]).

Determination of Microbial Activity in Soil by Radiorespirometry

This straightforward, rapid procedure provides a measure of the microbial activity of soils. It may be used to compare soils as an estimate of their inherent fertility, or can be used to evaluate the possibly transient effects of the application of a pesticide or of some soil treatment.

The method is based on the principle that if a ^{14}C-labelled substrate is presented to the sample and the transformation rate is estimated by $^{14}CO_2$ evolution, then this will be a direct measure of microbial numbers and activity ([100,101]).

Clearly, many variations in detail are possible in the following method. Simple respirometer flasks are prepared, consisting of 50 or 75 ml conical (Erlenmeyer) or flat bottom flasks fitted with rubber stopper, into which is inserted a glass rod or a wire carrying a small glass or polythene well or vial. The vials can be attached to the rods with Araldite or similar adhesive, Fig. 11.5. In the vial is a strip of pleated filter paper to which 0.25 ml of phenethylamine is added as a CO_2 absorbent.

A 2 or 3 g soil sample is placed in each flask and moistened with enough water containing 0.5 ml of a ^{14}C-glucose solution to ensure that it is at optimum water content. The stopper is inserted and the sample incubated at 25°C for the desired time, when the reaction is terminated by adding 2–3 ml of 1N HCl. The filter paper and remaining phenethylamine are then transferred quantitatively to a scintillation vial, 10 ml of suitable scintillation mixture added, and then shaken well previous to counting. By comparison with a standard, calculate and express results as μ moles glucose consumed $h^{-1}g^{-1}$ oven dry soil. Expect typical results from 0.015 to 0.25 μmoles $h^{-1}g^{-1}$.

Normally duplicate samples will be prepared, with incubation times of 30, 60 and 90 minutes. At least initially, more than one substrate concentration should be used, e.g. 0.5, 1.0, 5.0 and 10 μmol glucose per 0.5 ml solution. The activity of the ^{14}C-glucose should be approximately 500–1500 d.p.s. As a measure of the dynamic biological activity is wanted it is undesirable to give too large an amount of substrate, as this may lead to unusual proliferation of soil flora much of which may be in the resting stage.

This method is reproducible, and provides a useful and accurate measure of soil microbiological activity. As we are concerned here with essentially enzymatic processes, and enzyme concentrations as reflected by microbial respiration, it is possible to apply Michaelis-Menten enzyme kinetics to the system.

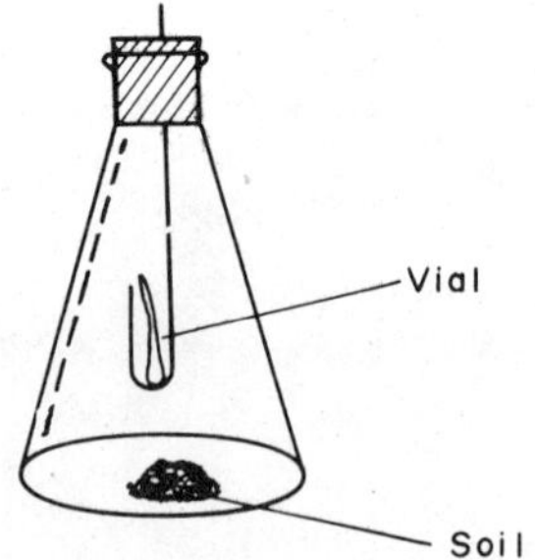

FIG. 11.5 Experimental set-up for radiorespirometry.

Theoretically this would permit calculation of a maximum reaction velocity (V_{max}) which should be proportional to total enzyme concentration and microbe population. Whether however this is a legitimate and meaningful mathematical manipulation for such a heterogenous system is a matter of opinion.

Determination of the Turnover Rate or Half-life of Crop Residues in Soil

We are interested here in determining the fractional rate at which the organic matter pool decomposes and is renewed. Clearly, under steady state conditions the amount lost will over a period of time equal the amount added.

Like many other biological processes the fractional rate of change is constant, that is it is exponential. Thus a semi-log plot of log residual added carbon versus time on a linear scale should give a straight line (Fig. 11.6). In fact due to the plant material decomposing into a number of chemically different components, the line will be made up of a number of sections, which may or may not be resolvable into individual rates ([79]) as in the manner of separating the half-lives in a mixture of more than one radioisotope.

If we can assume here that

$$2.3 \log \frac{Q}{Q_o} = -kt \tag{12}$$

$$\text{or } Q = Q_o e^{-kt} \tag{13}$$

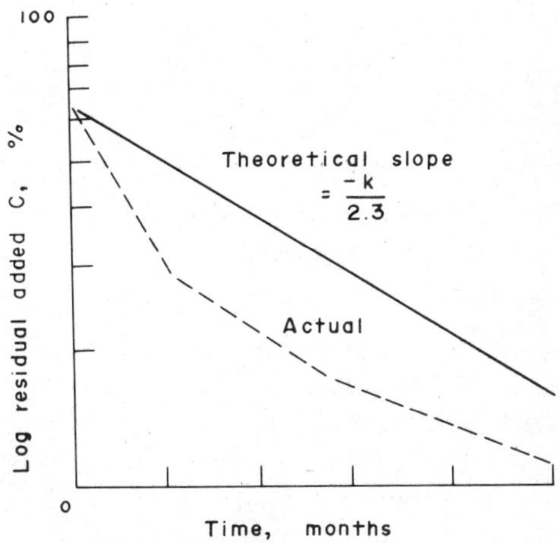

FIG. 11.6 Decomposition of added plant residues with time.

where Q_o = quantity of organic matter present at zero time

Q = quantity of organic matter present at time t

e = base of the natural logarithm = 2.7183

then from equation (12) the biological half-life

$$t_{1/2} = \frac{-2.3 \log \frac{1}{2}}{k} = \frac{0.693}{k} \qquad (14)$$

and the turnover rate constant,
$$k = \frac{0.693}{t_{1/2}} \qquad (15)$$

The principle of the method is that ^{14}C labelled plant material is added to microplots in the field. Then at successive periods of time, every few weeks or months, the plot is sub-sampled and the amount of ^{14}C from the labelled plant material still residual in the soil is measured as $^{14}CO_2$ ([79,89,103,117,130]). If desired the ^{14}C of any particular soil organic fraction can be estimated after fractionation by one of the existing schemes ([92,103,111]).

Preparation of microplots

The cost of the ^{14}C labelled material inevitably means that the microplots must be small, say 25–35 cm in diameter, though larger sub-plots of 1.2 m square have been used to more nearly simulate field conditions ([130]). It is convenient to drive metal or plastic cylinders of the appropriate diameter into the soil to a depth of 15–30 cm, to delineate the plot and prevent lateral movement of the labelled material and soil. The soil is usually removed from the cylinder to the desired depth in order to incorporate the finely chopped (1–2 cm) ^{14}C-labelled residues. Before returning the soil a stainless steel wire or plastic mesh screen may be placed at the bottom of the cylinder to define the working volume, without interfering with the growth of the longer roots. Three or four replications should be prepared.

Sampling

Sampling is carried out at the pre-determined time intervals over a period of months, say 1½, 3, 6, 9, 12 and 15 months, but the time scale will vary with the purpose of the experiments. The minimum size of corer is used necessary to get a representative sample of 10.0 g of soil from 3 or 4 cores per cylinder, these being subdivided into upper and lower layers.

Analysis

The large fragments of comparatively unaltered plant material are picked out from the soil, which is then dried and ground to pass an 80-mesh sieve. A sample is weighed out accurately, so as to contain 5–10 mgC (say 0.4–0.5 g soil), and transferred to the combustion flask. $^{14}CO_2$ is released from the organic matter by the Van Slyke-Folch ([125]) wet oxidation procedure, the $^{14}CO_2$ being absorbed in sodium hydroxide solution.

A system can be assembled as shown in Fig. 5.1(a) of Chapter 5. The determination should be carried out as described in Chapter 5, the CO_2 being collected in 15 ml of 0.1 N NaOH. An aliquot of the 0.1 N NaOH containing $^{14}CO_2$ is then placed in a scintillation vial with a scintillation mixture, for counting, as described on page 83. The typical example calculation given there may also be followed. From the aliquot transferred to the scintillation vial for counting it is now possible to calculate the specific activity of the labelled carbon from the organic matter. The specific activity of the original labelled material applied to the soil is of course already known from a similar determination and both are expressed as d.p.s./mg C.

Example: Consider a soil to which ^{14}C labelled material of a specific activity of 1500 d.p.s./mg C was applied to a plot at the rate of 10 mg C per g soil. After sampling, the weight of soil taken for combustion was 0.4 g and yielded 10.5 mg of C with a specific activity of 230 d.p.s. per mg C.

Then, the concentration of soil carbon is

$$\frac{10.5 \text{ mg}}{0.4 \text{ g}} = 26.25 \text{ mg C/g}$$

while the concentration of carbon from the labelled residues still in the soil will be

$$\frac{(26.25 \text{ mg C/g soil})\ (230 \text{ d.p.s./mg C})}{1500 \text{ d.p.s./mg C}} = 4.0 \text{ mg C/g soil}$$

But the ^{14}C labelled material was added at a rate of 10 mg C per g soil, therefore the soil now has

$$\frac{4.0}{10.0} \times 100 = 40\% \text{ of the original material.}$$

NATURAL RADIOCARBON DATING OF SOIL ORGANIC MATTER

The purposes of natural radiocarbon dating studies of soils are to evaluate the long term organic matter turnover of soils, and indirectly provide a means of estimating the basic nitrogen supplying power of soils.

Evidence from many experiments shows that when plant residues are added to soil, mineralization of readily decomposable plant compounds is completed in a relatively short time ([74, 89, 107, 110, 117, 121]). Following the short initial peak activity the rate of decomposition is greatly slowed and fractions such as cell wall constituents become relatively stabilized. Determinations of carbon balance in long term field studies and isotopic turnover experiments have indicated annual losses of organic matter between 1 and 5%, corresponding approximately to the stable constituents having half-lives of 5–25 years ([68, 74, 86, 88, 92, 116, 117]).

There is now evidence from radiocarbon dating that some of the persistent organic matter fractions remain in the soil for a considerable time, of the order of hundreds of years. Although radiocarbon dating has been applied for over twenty years to the study of buried soils in connection with studies of prehistory ([106]) the application to contemporary surface soils is comparatively recent ([72, 88, 97, 103, 115, 128]), now making it feasible to estimate the turnover rate or mean residence time ([80]) of these very long lived fractions.

Radiocarbon dating estimations depend on the fact that high energy neutrons from cosmic radiation react with nitrogen to give ^{14}C. Oxidation of the carbon produces $^{14}CO_2$ which ultimately equilibrates with the stable ^{12}C and ^{13}C isotopes and is incorporated into plants by photosynthesis, and hence into soil organic matter.

Two assumptions are involved ([95,96]): that when plant or animal material dies the carbon atoms of the dead matter do not exchange with external carbon such as CO_2 of the atmosphere, and secondly the specific activity of ^{14}C in the biosphere has remained at a constant value up until the time of the first nuclear explosions. However, ^{14}C is decaying constantly with a long half-life of approximately 5700 years, and it has been determined that the steady state specific activity, S_o, is about 0.25 d.p.s. per g of carbon.

It follows that if plant material died $\bar{t}$ years ago ($\bar{t}$ = 0) it had then a specific activity of ^{14}C equal to S_o. If organic matter from this material is isolated today ($\bar{t}$ years afterwards) and the present specific activity S is determined, then $\bar{t}$ can be calculated from the decay equation

$$S = S_o \left(\frac{1}{2}\right)^{\frac{\bar{t}}{t_{1/2}}} \tag{16}$$

where $t_{1/2}$ is the half-life of ^{14}C, S_o the specific activity of a standard reference, and t the mean residence time of the organic C in the soil. Solving equation (16) for $\bar{t}$, the unknown mean residence time of a sample will be

$$\bar{t} = 19{,}000 \log \left(\frac{S_o}{S}\right) \text{years} \tag{17}$$

where S_o is 0.950, being the activity of the National Bureau of Standards (U.S.) oxalic acid standard.

Techniques

The very low levels of naturally occurring ^{14}C, itself a very weak β-emitter, give rise to considerable problems of technique to get a representative sample and especially to have sufficient activity for measurement. As the activity is low, the CO_2 evolved is usually converted into acetylene (C_2H_2) for gas proportional counting ([124]), or else the acetylene is converted into benzene ([115]) to which is added a toluene based scintillation mixture for liquid scintillation counting. The activity is so low that all preparation and counting must be done in a building where no other radioisotope work is being carried out.

Calculation will show that the usual methods adopted for liquid scintillation counting of ^{14}C cannot be utilized, as the normal miscible CO_2-absorbing solutions cannot absorb the volume of $^{14}CO_2$ necessary. Thus, if 5 g organic matter are taken for determination this will yield approximately 3 g carbon, in turn giving almost 11 g CO_2 on combustion. But one of the commonest CO_2-absorbers used in scintillation counting, phenethylamine, only absorbs 0.6 g CO_2/10 ml, or say 0.9 g CO_2 in 15 ml which is the maximum volume that can be used in a standard liquid scintillation counter. Even allowing for some loss of carbon during processing there is still an order of magnitude difference between the amount of CO_2 that must be absorbed and what is possible.

Procedures vary, but in general a soil sample containing 3 or more g of organic carbon is taken and sieved to remove roots and recognizable pieces of plant material, dried and ground to pass a 60 or 80 mesh sieve. The sample must be kept acid (pH 6.0) to avoid adsorbtion of atmospheric CO_2. The sample is shaken in 0.1 N HCl (1:15 ratio) and centrifuged at 2000 r.p.m. for 15–20 min and the supernatant discarded. This is repeated as necessary to remove undecomposed organic matter, followed by drying. Inorganic carbonate is removed by boiling in 1N HCl for 5 min, and the sample dried by evaporation. Combustion in a furnace yields CO_2 which is absorbed in 200–250 ml of 25% NH_4OH (or NaOH, precipitating $CaCO_3$ by addition of saturated $CaCl_2$). In the case of the former the solution can be poured into 500 ml of 20% hot $SrCl_2$ solution to form $SrCO_3$ which is cooled, filtered, washed and dried *in vacuo* to prevent atmospheric CO_2 contamination.

The carbonate can then be taken for conversion to acetylene for proportional gas counting or further conversion to benzene, or to CO_2 for ratio analysis of $^{12}C/^{13}C$ by the mass spectrometer. No two reaction lines are the same, being dependent on the ingenuity of the worker and the components available. A direct quotation of the procedure due to Scharpenseel *et al.* ([115]) is given here as an example.

> The direct conversion of the fractions' fulvic acid, hymatomelanic acid, brown humic acid, gray humic acid, and humine, as well as the conversion of humus coal via combustion and precipitation as $SrCO_3$, into benzene, were carried out in an inexpensive reaction line, shown in Fig. 11.7. For direct conversion of the humic material into carbide, the silica-gel tower and liquid nitrogen trap (I) were connected to the lithium oven. Li + organic matter were together heated under vacuum to 800°C to produce the lithium carbide. $SrCO_3$ had first to be treated in (II) with $HC10_4$ in order to release the CO_2, which is then channelled into the lithium oven for carbide production.
>
> The carbide is later treated with old, dead water out of a coal mine, the developed acetylene is frozen out after being dried during passage through a $-80°C$ trap and freed from NH_3 by a U-tube with concentrated H_3PO_4-wetted

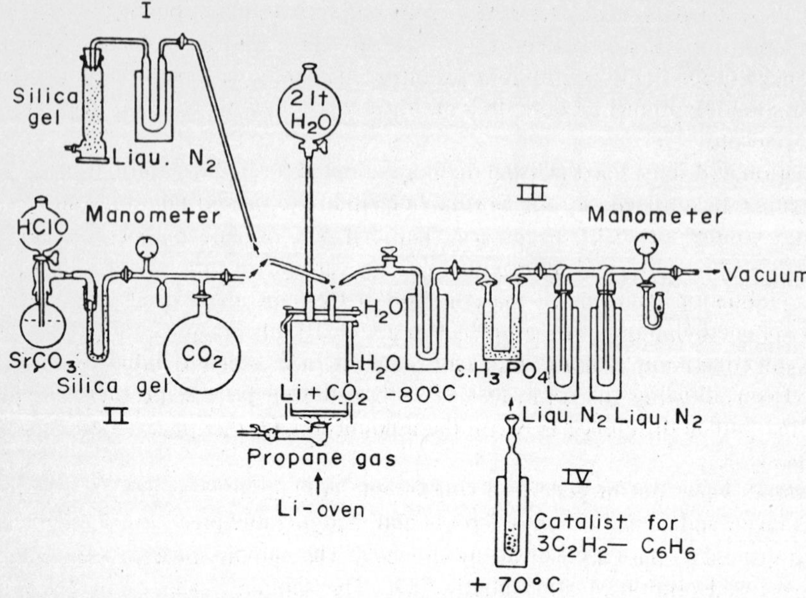

FIG. 11.7 Vacuum line for benzene synthesis from carbon-containing materials (115).

glass pearls or Raschig-rings. The hydrogen gas, also developed from excess lithium reacting with water, is allowed to leave the reaction line by a terminal back-stroke gauge.

After the acetylene has been collected in the liquid nitrogen traps the line is disconnected between the traps and the U-tube. A cylindric tube (IV) with "Perlkatalysator NEU" (Kali-Chemie Hannover) is attached to the ball joint and heated in a larger dewar with water at a temperature of 80°C. The acetylene is allowed to stream into the catalyst tube by removing the nitrogen traps. The developed benzene is subsequently extracted under vacuum by heating the catalyst tube to 200°C. The obtained benzene is gas-chromatographically pure and can be used directly for dating. Three cubic centimetres of benzene are mixed with 1 cm^3 of a standard toluene solution with PPO and di-methyl POPOP scintillator. The sample is measured for its ^{14}C-radioactivity in a liquid-scintillation spectrometer in periods of 100 min for at least 24 h.

Typical data show mean residence times of some organic fractions from 300 to over 5000 years ([72,102,115]) the latter being from 95 cm depth. Residence times of 1000–1500 years seem common for non-hydrolysable humic fractions compared with 10–25 years for hydrolysable humic fractions ([109]).

NITROGEN TRANSFORMATION AND RELATED STUDIES IN SOILS

We know that the carbon and nitrogen components of the active soil fractions are closely linked. It is not surprising therefore that studies with the stable isotope ^{15}N

assumed increasing importance and are contributing to a much better understanding of this relationship. Some studies ([104,122]) have used ^{14}C and ^{15}N techniques together. A full consideration of ^{15}N methods is given in Chapter 7, with an introduction to the extensive soils literature.

REFERENCES FOR FURTHER READING

Soil Chemistry, Nutrient Supply and Movement

1. AHMED, I., Subbiah, B. V. and Dakshinamurti, C. Studies in the migration of sulphate ion through the soil using ^{35}S as a tracer. Pl. Fd. Tokyo (1971).
2. ALEKSIC, Z., Broeshart, H. and Middelboe, V. Shallow depth placement of $(NH_4)_2SO_4$ in submerged rice soils as related to gaseous losses of fertilizer nitrogen and fertilizer efficiency. *Plant Soil* **29**, 338 (1968).
3. BARBER, S. A. A diffusion and mass flow concept of soil nutrient availability. *Soil Sci.* **93**, 39–49 (1962).
4. BARBER, S. A. The role of root interception, mass-flow and diffusion in regulating the uptake of ions by plants from soil. *In: Limiting Steps in Ion Uptake by Plants from Soil.* IAEA Tech. Rept. No. 65, 39–45, IAEA, Vienna (1966).
5. BARBER, S. A., Walker, J. M. and Vasey, E. H. Principles of ion movement through soil to the plant root. Int. Soil Conf. N.Z., 121 (1962).
6. BARBER, S. A., Walker, J. M. and Vasey, E. V. Mechanisms for the movement of plant nutrients from the soil and fertilizer to the plant root. *Agric. Food Chem.* **11**, 204 (1963).
7. BEEK, H. Determination of the cation exchange capacity of a soil. Lecture E 28, IAEA/FAO International Training Course on the Use of Radioisotopes and Radiation in Soil and Plant Nutrition Studies. IAEA, Wageningen (1970).
8. BORLAND, J. W. and Reitemeir, R. F. *Soil Sci.* **69**, 251–59 (1950).
9. DE HAAN, F. A. M. and Bolt, G. H. *Soil Sci. Soc. Amer. Proc.* **27**, 636–40 (1963).
10. DEIST, J. and Talibudeen, O. Ion exchange in soils from the ion-pairs K-Ca, K-Rb and K-NO. *J. Soil Sci.* **18**, 125 (1969).
11. DEIST, J. and Talibudeen, O. Thermodynamics of K-Ca exchange in soils. *J. Soil Sci.* **18**, 138 (1967).
12. DEIST, J. and Retief, W. L. Use of barium-137 for determining the cation exchange capacity of soils. *Agrochemophysica* **1**, 27 (1969).
13. DYER, B. *Trans. Chem. Soc.* **65**, 115–67 (1894) and *Proc. Roy. Soc.* **11** (1901).
14. FRIED, M. and Dean, L. A. A concept concerning the measurement of available soil nutrients. *Soil Sci.* **73**, 263–71 (1952).
15. FRIED, M. *J. Agr. Food Chem.* **2**, 241–44 (1954).
16. FRIED, M. 'E', 'L' and 'A' values. Trans. 8th Int. Congr. Soil Sci. Bucharest, IV 29 (1964).
17. GRAHAM-BRYCE, I. J. Diffusion of cations in soils. *In: Plant Nutrient Supply and Movement.* IAEA Tech. Rept. 48, 42–56 (1965).
18. GRUNES, D. L., Haas, H. J. and Shih, S. H. *Soil Sci.* **80**, 127–38 (1955).
19. GUNNARSSON, O. and Fredriksson, L. A method for determining 'plant available' phosphorus in soil by means of P-32. *Proc. Isotope Tech. Conf. Oxford* **1**, 427–31 (1953).
20. JENNY, H. *In: Growth in Living Systems.* (Zarrow, Ed.) Basic Books Inc. (1961).
21. KEAY, J. and Wild, A. The kinetics of cation exchange in vermiculite. *Soil Sci.* **92**, 54 (1961).
22. KEMPER, W. D. Water and ion movement in thin films as influenced by the electrostatic charge and diffuse layer of cations associated with clay mineral surfaces. *Soil Sci. Soc. Amer. Proc.* **24**, 10 (1960).
23. KLUTE, A. and Letey, J. The dependence of ionic diffusion on the moisture content of non-adsorbing porous media. *Soil Sci. Soc. Amer. Proc.* **22**, 213 (1958).
24. LAGERWERFF, J. V. and Bolt, G. H. Theoretical and experimental analysis of Gapon's equation for ion exchange. *Soil Sci.* **87**, 217 (1959).
25. LAGERWERFF, J. V. The contact-exchange theory amended. *Plant and Soil* **13**, 253 (1960).
26. LAGERWERFF, J. V. Types of limitations to ion uptake by plants. *In: Limiting Steps in Ion Uptake by Plants from Soil.* IAEA Tech. Rept. 65, 71–81, IAEA, Vienna (1966).

27. LAHAV, N. and Bolt, G. H. *Nature* **200**, 1343–44 (1963).
28. LARSEN, S. *Bull. Docum.* (ISMA) **8**, 1.9 (1950).
29. LARSEN, S. The use of P-32 in studies on the uptake of phosphorus by plants. *Plant and Soil* **4**, 1–10 (1952).
30. LARSEN, S. and Cooke, I. J. The influence of radioactive phosphate level on the absorbtion of phosphate by plants and on the determination of labile soil phosphate. *Plant and Soil* **14**, 43–48 (1961).
31. LARSEN, S. and Sutton, C. D. The influence of soil volume on the absorbtion of soil phosphorus by plants and on the determination of labile soil phosphorus. *Plant and Soil* **18**, 77–84 (1963).
32. Lcf4ARSEN, S. Gunary, D. and Sutton, C. D. The rate of immobilization of applied phosphate in relation to soil properties. *J. Soil Sci.* **16**, 141–48 (1965).
33. LAUDELOUT, H. and Van Bladel, R. Interpretation of thermodynamic and thermochemical measurements of ion exchange reaction in clays. *In: Limiting Steps in Ion Uptake by Plants from Soil.* IAEA Tech. Rept. 65, 8–18, IAEA, Vienna (1966).
34. LAVY, L. T. and Barber, S. A. Movement of molybdenum in the soil and its affect on availability to the plant. *Soil Sci. Soc. Amer. Proc.* **28**, 93 (1964).
35. LEGG, J. O. and Stanford, G. Utilization of soil and fertilizer N by oats in relation to the available N status of soils. *Soil Sci. Soc. Amer. Proc.* **31**, 215–19 (1967).
36. LEWIS, D. G. and Quirk, J. P. Diffusion of phosphate to plant soils. *In: Plant Nutrient Supply and Movement.* IAEA Tech. Rept. 48, 71–77 (1965).
37. LOPEZ, P. L. and Graham, E. R. Labile pool and plant uptake of micronutrients. II. Uptake of manganese, iron, and zinc by Ladino clover (*Trifolium refens*) and its relation to soil labile pools. *Soil Sci.* **115**, 380 (1973).
38. MCAULIFFE, C. D., Hall, N. S., Dean, L. A. and Hendricks, S. B. Exchange reactions between phosphates and soils: hydrocyclic surfaces of soil minerals. *Soil Sci. Soc. Amer. Proc.* **12**, 119–23 (1948).
39. MATTINGLY, G. E. G. and Close, B. M. Rept. 1959 Rothamsted Exp. Sta. 50 (1959).
40. MATTINGLY, G. E. G. and Talibudeen, O. Rept. Rothamsted Expt. Sta. 246–64 (1960).
41. NYE, P. H. and Spiers, C. The movement of ions to roots by simultaneous diffusion and mass flow. Trans. 8th Int. Congr. Soil Sci. Bucharest (1964).
42. OLSEN, S. R. *In: Soil and Fertilizer Phosphorus in Crop Nutrition*, 89–122, Academic Press, New York (1953).
43. OLSEN, S. R., Watanabe, F. S., Cooper, H. R., Lardon, W. E. and Nelson, L. B. *Soil Sci.* **78**, 141–51 (1954).
44. OLSEN, S. R., Kemper, W. D. and van Shaik, J. C. Self-diffusion coefficients of phosphorus in soil measured by transient and steady state methods. *Soil Sci. Soc. Amer. Proc.* **29** (1965).
45. OLSEN, S. R. Phosphorus diffusion to plant roots. *In: Plant Nutrient Supply and Movement*, IAEA Tech. Rept. 48, 130–39 (1965).
46. PASSIOURA, J. B. *Plant and Soil* **18**, 225 (1963).
47. REEVE, N. G. and Sumner, M. E. Determination of exchangeable calcium in soils by isotope dilution. *Agrochemophysica* **1**, 13 (1969).
48. REINIGER, P., Lahav, N. and Bolt, G. H. Determination of cation exchange characteristics of calcareous soils. Proc. VIII Int. Cong. Soil Sci. II/51 (1964).
49. REINIGER, P. and Lahav, N. Self-diffusion of calcium-45 into certain carbonates and calcareous soils. *In: Plant Nutrient Supply and Movement*, IAEA Tech. Rept. 48, 85–90, IAEA, Vienna (1965).
50. REISENAUER, H. M. Mineral nutrients in soil solution. *In: Environmental Biology* (Altman & Dittmer, Eds.) 507–8, Federation Amer. Socs. Exp. Biol., Bethesda (1966).
51. RENNIE, D. A. and McKercher, R. B. Adsorption of phosphorus by four Saskatchewan soils. *Can. J. Soil Sci.* **39**, 64–75 (1959).
52. RENNIE, D. A. and Spratt, E. D. The influence of fertilizer placement on A-values. Trans. 7th Int. Congr. Soil Sci. Madison, Wisc. III, 535–43 (1960).
53. RULE, J. H. and Graham, D. R. Labile pools of Mn, Fe and Zn as measured by chemical equilibrium and by plant uptake. *Amer. Soc. Agron. Abst.* **86** (1973).
54. RUSSELL, R. S., Rickson, J. B. and Adams, S. N. Isotopic equilibria between phosphates in soil and their significance in the assessment of fertility by tracer methods. *J. Soil Sci.* **5**, 85–105 (1954).
55. RUSSELL, R. S., Russell, E. W. and Marais, P. G. Factors affecting the ability of plants to absorb phosphate from soils. *J. Soil Sci.* **8**, 248–67 (1957).
56. SHAPIRO, R. E. and Fried, M. Relative release and retentiveness of soil phosphates. *Soil Sci. Soc. Amer. Proc.* **23**, 195–98 (1959).
57. SCHEFFER, F. *et al. Z. PflErnähr. Düng. Bodenk.* 91, 224–32 (1960).

58. SCHOFIELD, R. K. *Soils and Fertilizers* **18**, 373–75 (1955).
59. SCHOFIELD, R. K. and Graham-Bryce, I. J. *Nature* (Lond.) **188**, 1048 (1960).
60. SMITH, D. H. *et al. Agr. and Food Chem.* **1**, 67–70 (1953).
61. STOUT, P. R. and Overstreet, R. Soil Chemistry in relation to inorganic nutrition of plants. *Ann. Rev. Plant Physiol.* **1**, 305–42 (1950).
62. TALIBUDEEN, O. *Proc. 2nd Radioisotope Conf. Oxford*, vol. 1, 405–11 (1954).
63. TALIBUDEEN, O. *J. Soil Sci.* **9**, 120–29 (1958).
64. THOMAS, H. C. Towards a connection between ionic equilibrium and ionic migration in clays. *In: Plant Nutrient Supply and Movement.* IAEA Tech. Rept. 48, 4–19, IAEA, Vienna (1965).
65. ULRICH, B. Die Wechselbeziehungen von Boden und Pflanze. Ferd. Ente Verlag, Stuttgart (1961).
66. WIKLANDER, L. Ann. Royal Agr. Coll. Sweden **17** (1950).

Soil Organic Matter Studies

67. ANDERSON, A., Nielsen, G. and Sorensen, H. Growth chamber for labelling plant material uniformly with radiocarbon. *Plant Physiol.* 14, 378 (1961).
68. BARTHOLOMEW, W. V. and Kirkham, D. Mathematical description and interpretation of culture induced soil nitrogen changes. *Trans. 7th Int. Congr. Soil Sci.* Madison, Wisc. **3**, 471 (1960).
69. BINGEMAN, C. W. *et al.* The use of ^{14}C to determine the effect of added organic matter on the biological decomposition of peat soil. USAEC Rept. TID 7512, 381 (1956).
70. BOVARD, P. *et al.* Effet chelatant de la matière organique et son influence dans la migration des produits de fission dans les sols. *In: Isotopes and Radiation in Soil Organic Matter Studies*, 471, IAEA, Vienna (1968).
71. BROADBENT, F. E. Nitrogen release and carbon loss from soil organic matter during decomposition of added plant residues. *Soil Sci. Soc. Amer. Proc.* **12**, 246 (1947).
72. CAMPBELL, C. A., Paul, E. A., Rennie, D. A. and McKallum, K. J. Factors affecting the accuracy of the carbon dating method in soil humus studies. *Soil Sci.* **104**, 81 (1967).
73. CHAMINADE, R. Effet physiologique des constituants de la matière organique des sols, sur le metabolisme des plantes, la croissance de le rendement. *In*: FAO/IAEA Technical Meeting, 35. Pergamon, Oxford (1966).
74. CLARK, F. E. and Paul, E. A. The microflora of grassland. *Advances in Agronomy*, **27**, 375 (1970).
75. COSTA, F. *et al.* Influencia de la materia organica de origen diferente sobre la movilidad de hierro y manganeso en suelos calizos. *In: Isotopes and Radiation in Soil Organic Matter Studies*, 433, IAEA, Vienna (1968).
76. DATTA, N. P. and Goswami, N. N. Transformation of organic matter in soil in relation to the availability of nutrients to plants. Radioisotopes in Soil-Plant Nutrition Studies. *In: Proc. Symp. Bombay 1962*, 223, IAEA, Vienna (1962).
77. EVEN-HAIM, A. Etude sur l'absorbtion de la matière organique par les plantes supérieures au moyen de leurs racines. *In*: FAO/IAEA Technical Meeting 1963, 49, Pergamon, Oxford (1966).
78. FÜHR, F. and Sauerbeck, D. The uptake of straw decomposition products by plant roots. *In: The Use of Isotopes in Soil Organic Matter Studies.* FAO/IAEA Technical Meeting 1963, 73, Pergamon, Oxford (1966).
79. FauUHR, F. and Sauerbeck, D. Decomposition of wheat straw in the field as influenced by cropping and rotation. *In: Isotopes and Radiation in Soil Organic Matter Studies*, 24, IAEA, Vienna (1968).
80. GEYH, M. A., Benzler, J. H. and Roeschman, G. Problems of dating pleistocene and adocene soils by radiometric methods in paelopedology. *In*: (D. H. Yaalon, Ed., 1971) ref. 46 (1971).
81. GOSWAMI, N. N. and Datta, N. P. Tracer studies on the decomposition of ^{14}C and ^{32}P tagged organic matter and nutrient availability relationship in soils. *J. Ind. Soc. Sci.* **9**, 269 (1961).
82. IAEA. *Tracer Manual on Crops and Soil.* Tech. Rept. 171, IAEA, Vienna (1976).
83. JANSSON, S. L. Balance sheet and residual effects of fertilizer nitrogen in a six year study with ^{15}N. *Soil Sci.* **95**, 31 (1963).
84. JENKINSON, D. S. The production of ryegrass labelled with carbon-14. *Plant and Soil*, **13**, 279 (1960).
85. JENKINSON, D. S. Studies on the decomposition of plant material in soil. I. Losses of carbon from ^{14}C labelled ryegrass incubated with soil in the field. *J. Soil Sci.* **16**, 104 (1965).
86. JENKINSON, D. S. The turnover of organic matter in soil. *In*: FAO/IAEA Technical Meeting 1963, 187. Pergamon Press, Oxford (1966).
87. JENKINSON, D. S. Studies on the decomposition of plant material in soil. II. Partial sterilization of soil and the biomass. *J. Soil Sci.* **17**, 280 (1966).

88. Jenkinson, D. S. Radiocarbon dating of soil organic matter. Rep. Rothamsted Exp. Sta. 73 pp. (1969).
89. Jenkinson, D. S. Studies on the decomposition of ^{14}C labelled organic matter in soil. *Soil Sci.* **111**, 64 (1971).
90. Jenkinson, D. S. and Powlson, D. S. The effects of biocidal treatments on metabolism in soil. V. A method for measuring soil biomass. *Soil Biol. Biochem.* **8**, 209 (1976).
91. Jenkinson, D. S. The effects of biocidal treatments on metabolism in soil. IV. The decomposition of fumigated organisms in soil. *Soil Biol. Biochem.* **8**, 203 (1976).
92. Kononowa, M. M. *Soil Organic Matter: Method of Tizurin, I.V.* Pergamon Press, Oxford (1961).
93. Ladd, J. N., Amato, M. and Parsons, J. W. Studies of nitrogen immobilization and mineralization in calcareous soils. III. Concentration and distribution of nitrogen derived from the biomass. *In: Soil Organic Matter Studies*. Proc. Symp. Braunschweig 1976. I 301–10, IAEA, Vienna (1977).
94. Larsen, S. The influence of soil organic matter on loss of available phosphate in soil by chemical reactions. *In: Isotopes and Radiation in Soil Organic Matter Studies*, 377, IAEA, Vienna (1968).
95. Libby, W. F. *Radiocarbon Dating*. Univ. Chicago Press, Chicago (1965).
96. Libby, W. F. History of Radiocarbon Dating. *In: Radioactive Dating and Methods of Low-Level Counting*, 3, IAEA, Vienna (1967).
97. Martell, Y. A. and Paul, E. A. The use of radiocarbon dating of organic matter in the study of soil genesis. *Soil Sci. Soc. Amer. Proc.* **38**, 501 (1974).
98. Martin, J. K. The chemical nature of the carbon-14 labelled organic matter released into soil from growing wheat roots. *In: Soil Organic Matter Studies*. Proc. Symp. Brauschweig 1976 I 197–203, IAEA, Vienna (1977).
99. Martin, J. P. and Haider, K. Decomposition in soil of specifically ^{14}C labelled DHP and corn stalk lignins, model humic acid type polymers and coniferyl alcohols. *In: Soil Organic Matter Studies*. Proc. Symp. Baunscheig 1976 II 23–32, IAEA, Vienna (1977).
100. Mayaudon, J. Use of radiorespirometry in soil microbiology and biochemistry. *Soil Biochemistry* **2** (A. D. McLaren & J. Skuijns, Eds.), 202, Marcel Dekker, New York (1971).
101. Mayaudon, J. Stabilisation biologique des protéines ^{14}C dans le sol. *In: Isotopes and Radiation in Soil Organic Matter Studies*, 177. IAEA, Vienna (1968).
102. Paul, E. A., Campbell, C. A., Rennie, D. A. and McCallum, K. J. Investigations of the dynamics of soil humus utilizing carbon dating technique. Trans. 8th Int. Congr. Soil Sci. Bucharest, 201 (1964).
103. Paul, E. A. Characterization and turnover rate of soil humic constituents. *Pedology and Quaternary Research*, Proc. Symp. Edmonton, 63 (1969).
104. Paul, E. A. and McGill, W. B. Turnover of microbial biomass, plant residues and soil humic constituents under field conditions. *In: Soil Organic Matter Studies*. Proc. Symp. Braunschweig 1976 I 149–57. IAEA, Vienna (1977).
105. Rosell, R. A. and Babcock, K. L. Precipitated manganese isotopically exchanged with ^{54}Mn and chelated by soil organic matter. *In: Isotopes and Radiation in Soil Organic Matter*, 453. IAEA, Vienna (1968).
106. Ruhe, R. V., Rubin, M. and Scholtes, W. H. Late pleistocene radiocarbon chronology in Iowa. *Amer. J. Sci.* **255**, 671 (1957).
107. Sauerbeck, D. A critical evaluation of incubation experiments on the priming effect of green manure. FAO/IAEA Technical Meeting, 1963, 209. Pergamon, Oxford (1966).
108. Sauerbeck, D. and Führ, F. Experiences in labelling whole plants with ^{14}C. *The Use of Isotopes in Soil Organic Matter Studies*. FAO/IAEA Tech. Meeting, 1963, 391. Pergamon, Oxford (1966).
109. Sauerbeck, D. Stability of recently formed humus compounds in soil. *In: Isotopes and Radiation in Soil Organic Matter Studies*, 57, IAEA, Vienna (1968).
110. Sauerbeck, D. Comparison of plant material and animal manure in relation to their decomposition in soil. *In: Isotopes and Radiation in Soil Organic Matter Studies*, 319, IAEA, Vienna (1968).
111. Sauerbeck, D. and Führ, F. Alkali extraction and fractionation of labelled plant material before and after decomposition: a contribution to the technical problems in humification studies. *In: Isotopes and Radiation in Soil Organic Matter Studies*, 3, IAEA, Vienna (1968).
112. Sauerbeck, D. and Johnen, B. G. Root formation and decomposition during plant growth. *In: Soil Organic Matter Studies*. Proc. Symp. Braunschweig 1976 I 141–48, IAEA, Vienna (1977).
113. Scharpenseel, H. N. Zur Herstellung von allseitig C-14-marketem Pflangen-und Haminsausematerial. *Londw. Forschung* **14**, 42 (1961).
114. Scharpenseel, H. N. Experimental techniques using tritium, including the production of labelled

materials, apparatus required and costs involved. *In: Use of Isotopes in Soil Organic Matter Studies*. FAO/IAEA Tech. Meeting, 1963, 471. Pergamon, Oxford (1966).

115. SCHARPENSEEL, H. N., Ronzani, C. and Pietig, F. Comparative age determination on different humic matter fractions. *In: Isotopes and Radiation in Soil Organic Matter Studies*, 67. IAEA, Vienna (1968).
116. SCHEFFER, F. and Ulrich, B. Lerhsbuch der Agrikultur—Chemie und Bodenkunde, III. Humus und Humus—düngung, I. Morphologie, Biologie, Chemie und Dynamik des Humus. F. Enke, Stuttgart (1960).
117. SHIELDS, J. A. Decomposition of ^{14}C labelled plant material under field conditions. *Can. J. Soil Sci.* **53**, 297 (1973).
118. SMITH, J. H., Allison, F. E. and Mullins, J. F. Design and operation of a carbon-14 biosynthesis chamber. Pub. 911 ARS, U.S.D.A. Washington (1962).
119. SMITH, J. H. Some inter-relationship between decomposition of various plant residues and loss of soil organic matter as measured by ^{14}C labelling. I. FAO/IAEA Technical Meeting, 1963, 223, Pergamon, Oxford (1966).
120. SMITH, R. L. and Shoukry, K. S. M. Changes in the zinc distribution within three soils and zinc uptake by field beans caused by decomposing organic matter. *In: Isotopes and Radiation in Soil Organic Matter Studies*, 397, IAEA, Vienna (1968).
121. Sørensen, L. H. Studies on the decomposition of ^{14}C labelled barley straw in soil. *Soil Sci.* **95**, 45 (1963).
122. SORENSEN, L. H. and Paul, E. A. Transformation of acetate carbon into carbohydrate and amino acid metabolites during decomposition in soil. *Soil Biol. Biochem.* **3**, 173 (1971).
123. SORENSEN, L. H. Factors affecting the biostability of metabolic materials in soil. *In: Soil Organic Matter Studies*. Proc. Symp. Braunschweig 1976 II 3–14, IAEA, Vienna (1977).
124. SUESS, H. E. Natural radiocarbon measurements by acetylene counting. *Science*, **120**, 5 (1954).
125. VAN SLYKE, D. D. and Folch, J. *J. Biol. Chem.*, **136** (2), 509–41 (1940).
126. WAREMBOURG, F. R. and Paul, E. A. The use of $^{14}CO_2$ canopy techniques for measuring carbon transfer through the plant-soil system. *Plant and Soil* **38**, 331–45 (1973).
127. WILZBACH, K. E. *J. Amer. Chem. Soc.* **79**, 1013 (1957).
128. YAALON, D. H. Paleopedology, Origin, Nature and Dating of Paleosols. Int. Soc. Soil Sci. and Israel Univ. Press (1971).
129. ZELLER, A., Oberländer, H. E. and Hiemer, F. A growth chamber for raising ^{14}C labelled plants. *In: Use of Isotopes in Soil Organic Matter Studies*, FAO/IAEA Tech. Meeting, Pergamon, Oxford p. 401 (1963).
130. ZELLER, A., Oberländer, H. E. and Roth, K. A field experiment on the influence of cultivation practices on the transportation of ^{14}C-labelled farmyard manure and ^{14}C-labelled straw into humic substances. *In: Isotopes and Radiation in Soil Organic Matter Studies*, 265, IAEA, Vienna (1968).

Additional

131. DALAL, R. C. and Hallsworth, E. G. Measurement of exchangeable soil phosphorus and interrelationship among parameters of quantity, intensity, and capacity factors. *Soil Sci. Soc. Am. J.* **41**, 81–86 (1977).
132. MEKHAEL, D., Amer, F. and Kadry, L. Comparison of isotope dilution methods for estimation of plant available soil phosphorus. *In: Proc. Symp. Isotope and Radiation in Soil-Plant Nutrition Studies*, 437–48, IAEA, Vienna (1965).

CHAPTER 12

Isotopic Tracers In Field Experimentation

WHEN radioisotopes became generally available, labelled fertilizers were also soon produced for experimental purposes. It was immediately realized that isotopically labelled fertilizer could provide a means of quantitatively determining the fate of specific nutrient elements, and also a precise method for obtaining essential data for efficient fertilizer practice. Much intensive effort was put into placement and availability studies, both in the field and in the glasshouse, mainly using ^{32}P ([5,14,23,25,29,,30,31,43,44,45]). The initial work was primarily in the scientifically developed countries because of the availability of equipment, but such studies are now common in most countries with a reasonably developed agricultural research capability.

The development of the use of ^{15}N in field studies came somewhat later than radioisotopes, because of the cost and supply situation of ^{15}N and the scarcity of mass spectrometers and the back-up required to support them. Two of the early major coordinated research programmes of the FAO/IAEA, those on rice and maize fertilization ([17,18]), not only demonstrated for the first time that the use of ^{15}N in field studies was economically practical and stimulated its widespread use, but had a major influence on extending the use of isotope techniques in field experiments to many developing countries.

PRINCIPLES OF THE USE OF LABELLED FERTILIZERS

It will be appreciated from dilution principles that for a known constant amount of radioactivity, the specific activity is inversely proportional to the total amount of test element present in the system or product. Thus in fertilizer studies, if we label the fertilizer with an isotope such as ^{32}P, apply it to the soil and subsequently grow a crop, then it will be possible to determine what proportion of the phosphorus in the crop came from the fertilizer and what proportion came from the soil.

The proportion of such a nutrient ion e.g. phosphorus present in the plant which has come from the fertilizer, as opposed to soil derived nutrient ion, can be expressed in the following relationship:

$$\begin{array}{l}\text{\% phosphorus in plant}\\ \text{derived from the}\\ \text{fertilizer (\%Pdff)}\end{array} = \frac{\text{Specific activity of plant sample } (S_p)}{\text{Specific activity of fertilizer } (S_f)} \times 100 \qquad (1)$$

With a stable isotope such as ^{15}N, the term "specific activity" is of course inapplicable. However, exactly the same principle applies and the term $\%^{15}N$ excess is used in a corresponding manner:

$$\text{\% nitrogen in plant derived from the fertilizer (\%Ndff)} = \frac{\%^{15}\text{N excess in sample}}{\%^{15}\text{N excess in fertilizer}} \times 100 \qquad (2)$$

The ability to label the fertilizer and thereby determine directly the proportion of the nutrient ion taken up by the crop which has been derived from the fertilizer (%dff), is the unique feature of the isotope technique. It should be noted that the isotope procedure is itself yield-independent. In other words it is possible to determine the proportion of a nutrient ion in the crop which has been derived from the fertilizer without taking any yield data. In practice however this is undesirable, because possessing yield data and assuming that the isotopic analysis is representative of the bulk of the crop, it is also possible to estimate the total uptake of nutrient and hence the amount of applied fertilizer that has actually been used by the crop. Together with yield data in relation to fertilizer level this permits a much more valuable analysis of the experiment.

The broad possibilities of isotope investigations in fertilizer use efficiency can be briefly summarised as follows:

(i) clearly defining the optimum conditions of placement of phosphorus and nitrogen fertilizers

(ii) clearly defining the relative efficiency of the major nitrogen sources

(iii) achieving a better understanding of the effect of time of application on the efficiency of nitrogen and phosphorus utilization

(iv) obtaining by direct measurement, precise information on the proportion of applied fertilizers actually taken up by the crop

(v) Obtaining a clear—numerate—understanding of the penalty paid in terms of fertilizer wastage, if inefficient placement or incorrect nitrogen source is adopted

(vi) for measuring certain residual effects

(vii) determining if better placement would have made less fertilizer necessary

(viii) the determination of 'A' value under field conditions, as a means of evaluating availability of soil nutrients and the effect of management practices.

The literature has now become very extensive and only a relatively few references can be given to illustrate the scope of labelled fertilizer use in the field. General references 5, 15, 23, 25, 36, 39 and 48. For maize ([18,24]), for rice ([17]), for wheat ([19, 34]) and groundnuts ([33]). On availability, utilization, placement and timing ([1,10,13,17, 18, 21,29,30,31,34,37,44]). On fertilizer recovery and residual effects ([8,11,27,35]). Especially on nitrogen references 4, 6, 9, 16, 17, 18, 19, 22, 32, 38 and 41. Especially on phosphorus references 17, 18, 19, 20, 21, 28, 31, 43 and 44.

Fertilizer Use Efficiency

We may define fertilizer use efficiency as: *the relative ability of a fertilizer application technique to ensure the maximum uptake of that fertilizer nutrient by the crop*. It should be noted that the ability or efficiency of the plant in utilizing this fertilizer for the production of economic dry matter is a separate question.

The concept of efficient fertilizer utilization i.e. the efficient uptake of the fertilizer by the crop had not, until the oil crisis, been greatly stressed in those countries with long-established traditions of fertilizer research. However, greatly raised fertilizer prices have much increased awareness of the need for efficiency. Labelled fertilizer methods provide a means not only of estimating the uptake of fertilizer nutrient by the crop, but also the utilization of soil nutrient, and the effect of fertilization and other management practices on the uptake of soil nutrient, whether P or N.

The importance of the need for efficient fertilizer use can be deduced from some of the data of the FAO/IAEA rice fertilization programme, which used labelled fertilizers ([48]). Even under the best conditions of nitrogen fertilizer application i.e. ammonium fertilizer ploughed-in, the amount of fertilizer taken up by the rice crop seldom exceeded 30–35% of that applied and was often less, even where there was a significant yield response to 60 kg/ha of nitrogen. Under the worst possible conditions, nitrate fertilizer applied at planting time, only 3% or less of the fertilizer applied was actually taken up by the crop. In the case of phosphorus about 5–10% was utilized in the case of the most efficient, surface application, treatments and as little as 2% or less for depth or planting point treatments.

Even with such a well-researched crop as maize it required ^{15}N experiments ([18]) to give an adequate estimate of fertilizer use efficiency and to show that utilization of fertilizer nitrogen by the whole plant varied between 20 and 70% with location and management, with a mean value of about 40%. Considering only grain nitrogen, the efficiency of conversion of fertilizer nitrogen into grain protein was found to average not more than 25%.

Many factors may be responsible for the inefficient use of applied fertilizers. Leaching of nitrate and gaseous N losses under continuous or temporary reduced soil conditions may be the cause of a large proportion of the applied nitrogen fertilizer not being taken up by the crop. Phosphate fertilizers on the contrary do not suffer from leaching and gaseous losses. Here the problem is that phosphate from the fertilizer moves continuously into the soil complex where it may be more or less tightly bound, leaving a concentration of phosphorus in the soil solution which is far below the phosphorus concentration required for maximal phosphorus uptake. The continuous export of plant nutrients with the harvests together with "fixation", leaching losses and gaseous losses of nutrients, sooner or later requires regular application of fertilizers, to maintain crop yields.

Practical solutions to avoid soil-fertilizer interactions, or to reduce such interactions to a minimum extent are to apply the fertilizers at such a place where the plant roots will take up the nutrients most rapidly, to apply the fertilizers at such time that uptake is so rapid that the unfavourable soil-fertilizer interaction period is minimized, to apply the fertilizers in the most suitable chemical forms to avoid interaction with soil and at the same time providing for a high nutrient availability to the crop. The critical examination of such practices is what is implied by fertilizer utilization efficiency studies.

Fertilizer use efficiency can be studied by (i) direct measurement of how much of the fertilizer is taken up, by using isotopically labelled fertilizers, (ii) indirect measurement of the amounts of fertilizer taken up by calculating the difference between the yield of nutrient from a treated plot and a non-fertilized control plot, (iii) indirectly by observing yield responses to the various methods and times of fertilizer application, (iv) indirectly by a more sophisticated extension of ii, by applying linear regression to estimate crop recovery of applied nutrient, assuming that several rates of fertilizer were used. In this case $Y = a + bX$, if Y is the total nutrient in the plants, a is the intercept representing total nutrient in plants from the control plots (no applied nutrient), b is the regression coefficient, 100b is the percent applied nutrient recovered, and X is the total amount of applied nutrient.

In the isotope technique, labelled fertilizers are applied to sub-plots, involving different treatments of placement/time/chemical form of the nutrient/etc. according to the specific objectives of the experiment. Plant samples may then be taken once or on a number of occasions for isotopic analysis and subsequent calculation of the per cent nutrient in the plant derived from the fertilizer. Where ^{32}P is being used as a label (half-life = 14 days), usually the last practicable sampling date is about 60 days from the time of application, though ^{15}N samples may be taken at final harvest. Normally, yield sub-plots are also grown, both to enable a calculation of the amount of applied fertilizer actually taken up by the crop in the particular experiment, and also to establish a point on the "yield curve" at which the experiment was conducted. It is not necessary to have yield sub-plots to determine the per cent nutrient in the plant derived from the fertilizer, however their inclusion clearly permits a much more complete evaluation of the experiment than is otherwise possible.

"Direct" Isotope Method Complementary to "Indirect" Methods in Fertilizer Studies

One of the problems regularly encountered in experiments where yield response or the content of a nutrient are used to compare placement, source, and time of fertilizer application, is that the results lack consistency from year to year, and from location to location. This is easily understood, because as the yields of the various treatments of placement trials are usually approaching maximum and the differences are small, the first requirement of such an experiment is that the levels of application of fertilizer should be adjusted in such a way that the magnitude of the treatment responses will be still within the steep part of the yield response curve (Fig. 12.1). In practice it is difficult to select such levels of application prior to carrying out the experiment. Moreover, it is obvious that high applications of fertilizer would bring the yield level of all treatments to the flat part of the yield response curve and no difference in yield between two different methods of placement or two times of application could be observed.

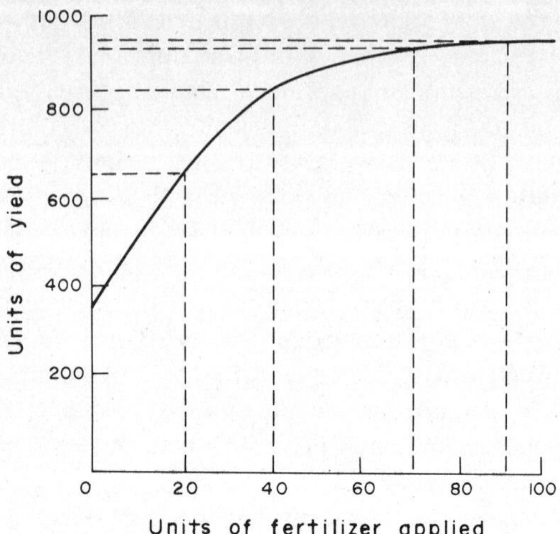

FIG. 12.1 The normal yield curve. As yield approaches the maximum the same increment of fertilizer produces an ever-decreasing yield increment.

Under certain conditions that do not necessarily correspond to those present on farmers or experiment station fields, yield techniques can certainly be used to determine indirectly the efficiency of fertilizer uptake. This is particularly so on soils of extremely low fertility where yield response to fertilizer is very high. Under such conditions it would be expected that conclusions derived from "direct" isotope methods and "indirect" yield techniques would be the same. Unfortunately on such soils the results tend to be confounded due to very marked increases in root growth from fertilization exploiting a larger volume of soil, hence acquiring more nutrients from the soil, changing the whole environment of the root and confusing the interpretation of fertilizer utilization. The results are then applicable only to this one soil condition and any conclusions on the effect of a particular treatment would only be valid for crops growing under deficiency conditions. Thus, extrapolation of the findings to normal conditions where fertilizers are applied to maintain good yields might not be justified. Furthermore, such yield experiments are necessarily large to include several levels of fertilizer application and to provide adequate statistical replication and require repeating in a number of successive years.

In practice, finding suitable experimental fields for critical yield experiments is also difficult, because the fields of experimental farms are almost universally of high fertility, while obtaining suitable sites of farmers' land may be near-impossible in countries where subsistence farming is the rule. The question of residual effects must also be considered in some cases. Residual effects upwards of ten years have been reported for phosphate applications to rice soils ([40]), while on very poor rice soils instances have been reported where the addition of a single major nutrient, e.g. N, P or K actually caused a depression in yield, due to intensification of the deficiency of other nutrients ([3]).

The indirect measurement of the amount of fertilizer taken up by calculation from the difference between the yield of nutrient of a non-fertilized control plot and a treated plot is subject to similar errors as in the case of straight yield determination experiments. Moreover, two errors are involved in any calculation by subtraction, while another serious error is that the yield of nutrient from the control plot is usually significantly lower (due to poorer root growth) than the yield of soil-derived nutrient in the treated plot, making calculation by subtraction inaccurate.

Indirect methods are therefore comparatively inefficient and can be used successfully only under selected conditions. Additionally, these experiments give little indication if similar responses could have been obtained with smaller quantities of fertilizer. The use of labelled fertilizers will normally make it possible to critically determine actual fertilizer uptake in relation either to placement, time of application or fertilizer source, with results which are essentially independent of vegetative development or grain yield. For example, if selection of too high a nitrogen fertilizer level results in the crop "lodging" and a loss of yield, it will still be possible to determine from the isotope data which of a number of treatments resulted in the most efficient uptake of the fertilizer. The isotope procedure is thus valid for virtually every experiment regardless of soil or location and the ability of the technique to give in the vast majority of cases a positive answer to the stated experimental objectives, is one of its most valuable practical features.

However, it has been pointed out ([88]) that when using ^{15}N labelled fertilizer the fertilizer uptake by the crop could be underestimated through immobilization and exchange of the fertilizer-N. This could happen if some of the labelled fertilizer was first immobilized and then subsequently re-mineralized. This re-mineralized nitrogen is then likely to have a much lower ^{15}N concentration than had the fertilizer before immobilization, and yet the calculation to determine the per cent plant nitrogen derived from the fertilizer is based on the original ^{15}N content of the fertilizer. In practice this does not seem to be a serious limitation.

Limitations of Labelled Fertilizers

The expense of isotopically labelled fertilizers has been put forward as an objection to their use in field experiments. If a comparison is made merely of the cost of labelled and unlabelled fertilizers this view has some validity. However, if the true cost of obtaining data of the same reliability from yield techniques is computed, i.e. the increased labour costs for carrying out the necessarily larger experiments and their need for repetition in successive years, then isotope techniques compare very favourably. It will be appreciated that the cost of putting an isotope experiment out in the field and its subsequent sampling, does not differ significantly from a corresponding type of experiment using unlabelled fertilizers.

In fact, with current costs for ^{15}N and ^{32}P labelled materials there is no need for a properly designed field experiment to cost more than \$1000–\$1500 for labelled fertilizer. This of course assumes that labelled fertilizer is used in a sub-plot and unlabelled fertilizers are applied to the yield sub-plots.

A major limitation of the isotope technique in the field is its present restriction to nitrogen and phosphorus studies, because of the lack of adequate isotopes of potassium and calcium. ^{42}K has too short a half life, while the expense of the stable ^{41}K is at present a limiting factor. It may be possible to use ^{86}Rb as a label in placement studies, but its use is precluded from any experiment in which fertilizer interaction is studied. This is because although Rb chemistry is almost completely analogous to potassium, rates of reaction are different, and thus while rubidium may be used as a qualitative indicator e.g. which is the best placement for most effective uptake of potassium fertilizer by the crop, it cannot be used for quantitative studies. A practical disadvantage of ^{86}Rb for large scale experiments is the radiation protection necessary when handling such a relatively energetic gamma emitter. In comparison this problem scarcely arises with ^{32}P where the glass wall of the bottle is an adequate shield.

Although ^{45}Ca has an adequate half-life it is at present too expensive for any but very small experiments, and it also has very poor counting characteristics. ^{89}Sr ($t_{1/2}$ = 50 days) may be used as a qualitative tracer for calcium in the same way that Rb can be used for K. There is however distinct biological discrimination between Ca and Sr, ruling out exact quantitative studies. ^{33}P ($t_{1/2}$ = 25 days) would offer valuable extra time in phosphorus experiments but is at present too expensive.

It should be understood that isotope techniques are neither necessary nor suitable for indicating the amount of fertilizer to apply for optimum yield. Their role is to indicate the most efficient placement or time of application. In the economics of fertilizer use there are really two basic questions involved:

(1) What is the best method (placement/time) of getting the fertilizer into the crop?

(2) What level of fertilizer application will give the best, or most economic increase in yield if applied in the optimum manner?

For the most scientific approach to fertilizer studies the first question should ideally be answered first, and for this isotopes are the best method available to us. The second question does not require labelled fertilizers and is a matter for determination according to local conditions of soil and climate, taking into account the generalized answers to placement and timing problems previously obtained with isotopes.

The isotope method is especially valuable for studies of the type under consideration because labelling or "tagging" the fertilizer offers a direct approach to determining the amount of fertilizer taken up by the crop. The actual fertilizer taken up can be detected and measured, and enables a straightforward quantitative comparison of the amount of fertilizer element taken up for each experimental treatment.

Field-determined A-values

The concept of A-value as a measure of the availability of a soil nutrient in terms of a standard has been considered in Chapter 11 which should be read in conjunction. In general the A-value has tended to be regarded as a laboratory technique to be carried out in pots. However, considerable evidence exists which suggests that the concept has much wider applicability ([36b,37a,37b]) to include field fertilizer studies.

For field experiments it is most convenient to modify equation (6) of Chapter 11 as,

$$\text{A-value} = \frac{\%\ \text{plant-P derived from the soil}}{\%\ \text{plant-P derived from the fertilizer}} \times \text{rate of application of fertilizer P (kg/ha or lb/ac)} \quad (3)$$

An exactly similar expression may be used for nitrogen, or indeed any appropriate element.

It has already been noted that the A-value is a quantitative value and as such it enables the plant-available soil nutrients or fertilizers to be expressed in quantitative values rather than in relative terms. However, as the A-value is a measure of available soil nutrients in terms of a fertilizer standard, if the standard is changed in some manner the size of the A-value will change also. Change in the standard can be caused by changes in the fertilizer (e.g. fixation), or by the placement of the fertilizer in relation to the plant roots. Extremely high or low rates of fertilizer application may also give anomalous values and should be avoided.

It is now accepted that under some circumstances where ^{32}P-labelled fertilizer is used there will be significant irreversible isotope exchange and/or the P-fertilizer standard reverts to a less available form. Clearly if this occurs the A-values will not stay constant with increasing rate of fertilizer application. At first sight this might appear to invalidate the A-value concept, at least as far as field measured values are concerned. In fact such irregularities can be utilized to provide an index of isotope exchange or fertilizer fixation.

This can be done by comparing A-values derived from mixed placement of fertilizer with those deriving from band placement, in which isotopic exchange is minimized due to the reduced soil-fertilizer contact ([36a]). This condition is of course a pre-requisite for the ideal A-value experiment. Therefore, in general, field determined A-values must be based on band placement of fertilizer, because this usually gives an uptake pattern of fertilizer and soil phosphorus resulting in the A-value remaining relatively constant for the specific experimental conditions. Such field techniques have been used ([36b,37b]) to determine the effect of organic residues and soil moisture on plant and soil phosphorus.

Over a period of time it is clear that in experiments with banded fertilizer there will be a tendency for the A-value to alter due to the continued growth of the plant roots. It is therefore necessary to establish for a given crop the most appropriate time of sampling for the most meaningful A-value determination. In the case of P-fertilizer this should be at a relatively early stage because it is the availability of P to the young plant that is especially important for eventual yield. The change in A-value due to root growth with banded placement of fertilizer can be used to evaluate environmental factors affecting root distribution and the volume of soil exploited by the roots. Comparative determinations of root growth between varieties of the same crop species could also be made.

Although the A-value was conceived as a means of measuring the availability of a soil nutrient, it can as a corollary also be used to determine the relative effectiveness of different fertilizers when applied to the same soil, as the availability of the soil nutrient is taken to be a constant. In effect, any change in the ratio of $\frac{\text{soil-nutrient}}{\text{fertilizer-nutrient}}$ results in a changed A-value, lower fertilizer nutrient uptake giving a higher A-value and vice versa.

In order to avoid problems in comparing A-values calculated on the basis of different fertilizer standards it is necessary to take one of the fertilizers as a comparative standard. This makes it possible to calculate an "effective rate of application" as follows ([10]):

$$\text{ERA} = \frac{\%\ \text{nutrient derived from fertilizer}}{\%\ \text{nutrient derived from soil}} \times \begin{array}{l}\text{A-value from the}\\ \text{fertilizer standard}\end{array} \qquad (4)$$

When a number of fertilizers are compared they may then be placed in order of relative efficiency, taking the one with the highest availability as 100%.

Determination of Residual Effects

In a similar manner, labelled fertilizer may also be used to estimate residual effects of previous fertilizer treatments. The greater the residual effect and the more residual nutrient is available to the crop, the less will be the uptake of applied labelled fertilizer. An indirect estimation procedure has to be employed when using radioisotopes because of the relatively short half-life, e.g. 14 days for ^{15}P, in relation to the residual period.

In the case of nitrogen, as there is the stable isotope ^{15}N available, it is not only possible to measure residual effects directly, but to do so while applying the same level of nitrogen fertilizer during the current growing period. Moreover, it has been pointed out by Fried ([89]) that it is possible to calculate the quantity of available residual nitrogen remaining in the soil, in the year after initial application of fertilizer. A direct estimation can be made on the basis that residual or carryover nitrogen from the first year can be considered to be equivalent to fertilizer-N in the second year. ^{15}N-labelled fertilizer is applied in the first year to part of the experimental area while the other part receives the same level of unlabelled fertilizer. In the second year the part which initially received labelled fertilizer now receives unlabelled fertilizer, while ^{15}N-labelled material is given at the same level to the area which initially received unlabelled fertilizer.

The plant will have three sources of N in the second year so that, following Fried's notation,

Soil $N = A_1$ (kg N/ha)
Residual N in the soil $= A_2$ (kg N/ha)
Fertilizer $N = F$ (kg N/ha)

and if it is assumed that the uptake from each source is directly proportional to the amount available from each, then

$$\frac{a_1}{A_1} = \frac{a_2}{A_2} = \frac{f}{F} \tag{5}$$

where a_1, a_2 and f are the amounts *taken up by the plant* from the sources A_1, A_2 and F. Now, a_2 can be determined directly from the area which received ^{15}N in the first year but not in the second, while f the fertilizer-N in the crop in the second year is determined from the area receiving ^{15}N in the second year. Calculation of a_2 and f can be made following equations (2) and (13) of this chapter. F, the fertilizer application, is known.

Therefore, as A_2 is the only unknown, this can be calculated,

$$A_2 = a_2 \times \frac{F}{f} \tag{6}$$

which gives the carryover of fertilizer nitrogen from the first year to the second.

PRACTICAL ASPECTS OF THE DESIGN OF FIELD EXPERIMENTS WITH LABELLED FERTILIZERS

Have a limited objective and plan the experiment as simply as possible. The amount of labelled fertilizer required should be kept to a minimum in order to keep down costs. In the case of ^{32}P or other radioisotope this also reduces the amount ofradioactive material it is necessary to handle. This is done by careful design of the experiment and also by keeping the labelled sub-plots as small as practicable.

Labelled fertilizer should be used for the role in which it is uniquely valuable i.e. measuring the actual uptake of fertilizer nutrient as affected by methods of placement, time of application, interaction between fertilizers, chemical nature of the fertilizer, residual effects studies, etc. It is not necessary to use it for yield-only plots, except inasmuch as yield data may also be taken secondarily from labelled fertilizer sub-plots.

Proper precautions should be taken in handling radioactive ^{32}P-labelled fertilizer. In practice this is not too difficult and once the phosphorus is applied to the soil it is immediately adsorbed and overall radioactivity is rather low.

Choice of Experimental Site

The soil type selected will normally be representative of a large major soil area. The experiment site should be uniform and should not be on recent previous experimental sites, to avoid residual treatment effects.

The fertility level is not of major importance, but it is desirable that with appropriate fertilization the site will produce yields approaching the maximum for the area.

Labelled Fertilizers

It should be clearly understood that ^{32}P-labelled superphosphate or other phosphates, such as "ammophos", must be labelled with the ^{32}P at the time of manufacture. This is to ensure uniform labelling of the fertilizer. It is not possible to "label" phosphate fertilizers by, for instance, spraying and mixing the ^{32}P into the already-prepared fertilizer. It is also not practicable to activate phosphorus fertilizers such as rock phosphate, in the reactor. Neither of these methods ensures uniform labelling and experimental results are quite invalid. It is normal in the preparation of ^{32}P-superphosphate to add the ^{32}P at the acid-digestion stage of manufacture, when ground rock phosphate is treated with 70% sulphuric acid containing phosphoric acid-^{32}P.

If preparation of ^{32}P-labelled fertilizer is to be undertaken in the laboratory then ammonium phosphate will be found the most convenient fertilizer to prepare. Technical grade $NH_4H_2PO_4$ is dissolved in the minimum of water and the appropriate amount of carrier-free ^{32}P added in the form of H_3PO_4. The resulting product is evaporated and crystallized out. Laboratories should think seriously before undertaking the preparation of ^{32}P-superphosphate in any great quantity, because there are substantial health and safety problems to be overcome due to the amount of radioactivity involved and the dry dusty nature of the product.

Two major suppliers of ^{32}P-labelled superphosphate are the Tennessee Valley Authority, Muscle Shoals, Alabama, U.S.A. and the radiochemical Centre, Amersham, England, the former making other ^{32}P fertilizers such as nitric phosphates, ammophos, etc. Charges are normally based on price per mCi plus preparation charge.

It is possible to do an elaborate calculation as to the activity required for the ^{32}P-fertilizer based on the potential isotope dilution, the anticipated P-uptake by the crop, the half-life and the proposed counting system. Such calculations are a total waste of time, as sufficient field experiments have now been carried out for it to be appreciated that if ^{32}P radioactivity determinations are to be made at 60–70 days (effectively after 5 half-lives) the highest specific activity is needed commensurate with safety in preparation and handling. In practice the minimum activity required is 0.2 mCi ^{32}P per g P_2O_5 (about 0.5 mCi ^{32}P per g ^{31}P) but it is usually more satisfactory to order material with either 0.4 or 0.5 mCi per g P_2O_5, this latter specific activity being effectively 100 mCi ^{32}P per kilo of superphosphate. It is quite easy to lose one whole half life between the time of shipping the fertilizer and the time it is applied in the field, due to bad weather or to customs clearance if it is shipped between countries.

Weighing-out ^{32}P-labelled fertilizers may be a problem for a small laboratory. Normally the amount required for each row of each plot of each treatment should be carefully worked out. Then the amount necessary for each row is weighed out into a glass bottle, ^{32}P being a β-emitter, most of the radiation is shielded by the glass; the fertilizer is then sprinkled down the row direct from the bottle.

^{32}P fertilizer should be weighed out in a perspex-shielded glove-box. It will usually be necessary to leave the balance in the glove-box until natural decay of the ^{32}P activity has occurred, as it is almost inevitable that it will become contaminated. Care must be taken that the outside of the glass bottle is not contaminated. To avoid this possibility a special glove-box can be constructed by means of which the weighed fertilizer is placed in the bottles by means of a funnel which passes outside the box. Thus the bottles are held up close to the aperture of the funnel for filling, but never actually go inside the glove-box and remain free from external contamination.

^{15}N-labelled fertilizer materials can be obtained from a number of major sources, as listed in Chapter 7. In addition ^{15}N can also be obtained from Japan and is also manufactured in the U.S.S.R. but is difficult for individuals to obtain.

^{15}N is still comparatively expensive though prices have got more favourable in recent years. The cheapest compounds are ammonium and nitrate salts, urea usually costing quite a lot more. Low enrichment materials are usually quite a lot cheaper than high enrichment compounds, and above 85% enrichment the price usually doubles, or even more. However, such enrichments are not needed for fertilizer experiments. Most suppliers will prepare any enrichment if given sufficient notice. A discussion on the use of ^{15}N-depleted materials is given in Chapter 7.

TABLE 12.1

Experimental treatments of a typical labelled fertilizer field experiment ([17]) (for details see text)

Treatment	Placement: ^{15}N-fertilizer	Placement: ^{32}P-fertilizer	Fertilizers mixed or separate	Plot numbers
A	Applied in rows on the surface	Applied in rows on the surface	Fertilizers mixed before application	6, 14, 24, 27, 38, 41
B	Applied in rows on the surface	Applied in rows on the surface	Fertilizers are applied separately in rows without previous mixing	4, 15, 23, 28, 36, 47
C	Applied in rows at 5 cm depth	Applied in rows on the surface	Fertilizers are applied separately in rows without previous mixing	3, 9, 19, 32, 33, 44
D	Applied in rows at 5 cm depth	Applied in rows at 5 cm depth	Fertilizers mixed before application	8, 10, 22, 29, 40, 43
E	Applied in rows at 5 cm depth	Applied in rows at 5 cm depth	Fertilizers are applied separately in rows without previous mixing	1, 11, 17, 30, 37, 45
F	The nitrogen and phosphorus fertilizers are mixed, broadcast on the surface and mixed into the top 5 cm of soil (puddled)			5, 12, 20, 31, 39, 42
G	This is the yield response plot and N and P are applied mixed in the row at 5 cm depth and at a rate of 120 kg N/ha and 60 kg P_2O_5/ha			7, 13, 18, 25, 34, 48
H	This is the zero fertilizer control plot (yield check) and no fertilizer is applied except for the basic potassium application + 60 kg P_2O_5/ha in a row at 5 cm depth			2, 16, 21, 26, 35, 46

For field work with fertilizers, in which the ^{15}N in the plant is subsequently determined, N-fertilizer with about 1% atom excess enrichment is required. Probably as little as 0.7% ^{15}N atom excess is quite practicable for many experiments with experience and refined instrumentation, but the higher figure is suggested for initial use. A 1% enrichment means of course that the fertilizer N has a total ^{15}N abundance of 1.37%, as the natural abundance of ^{15}N is 0.37%. The use of ^{15}N and the enrichment required has been discussed extensively in Chapter 7. The use of ^{15}N-depleted material is likely to come into more general use though relatively few experiments have been carried out hitherto.

Treatment and Field Layout

Treatments and replications will obviously vary according to the specific requirements of the experiment, but in general it will be found that six treatments involving labelled fertilizer, plus a further two treatments for Zero Fertilizer Control and a high-fertilizer-level Yield Reponse plots will prove to be as large as most investigators wish to handle. Intermediate harvests may be superimposed on the yield sub-plots. Randomized block designs with four to six replications have been found satisfactory.

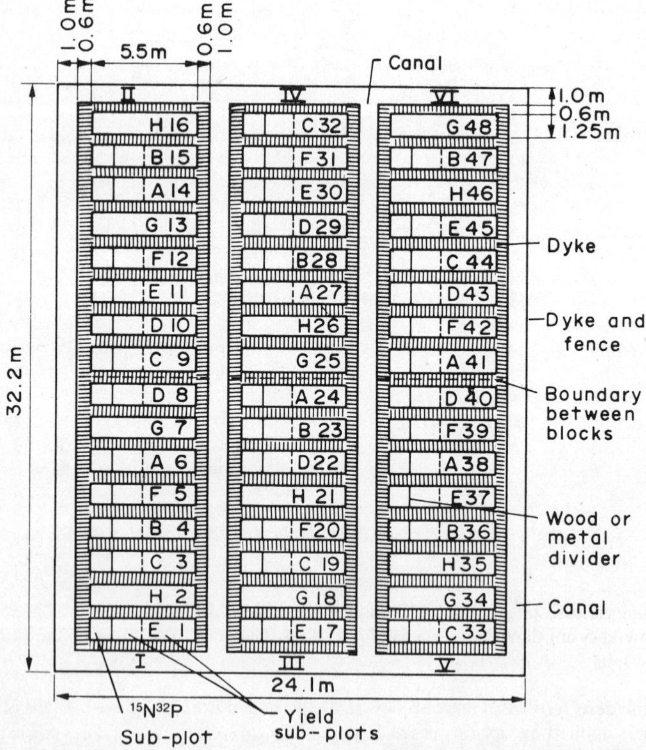

FIG. 12.2 Layout of typical labelled fertilizer field experiment for rice ([17]).

For small grains such as rice, wheat etc., rectangular plots have been used as in normal fertilizer experiments. In the case of row crops such as maize or sorghum, then "plots" must be based on widely spaced rows. This discussion is primarily concerned with the former; that is small grains grown in plots consisting of several rows, as the practical problems are greater.

In a typical experiment each main plot is divided into three sub-plots. The smallest sub-plot is the labelled fertilizer sub-plot of 5 or 6 rows of plants, from which samples are taken for the determination of ^{15}N and/or ^{32}P at a sixty-day intermediate harvest and for the determination of ^{15}N in the grain. The largest sub-plot is of 10 or 12 rows, equivalent to half the main plot, and is for determining yield at the final harvest, while adjacent to it is a sub-plot of 5 or 6 rows of plants from which the dry matter yield at sixty days is determined.

The practice of only applying the labelled fertilizer to small sub-plots enables the cost of each experiment to be kept as low as possible. It should be noted that the unlabelled superphosphate used for the yield sub-plots should be as close as possible in type to the labelled material, the same particle size being especially important. Ideally it should be made by the same process at the same time (to have the same ageing effects), but in practice using ordinary commercial fertilizer of the same particle size appears to give satisfactory results. Nitrogen fertilizers are so soluble that this problem does not arise.

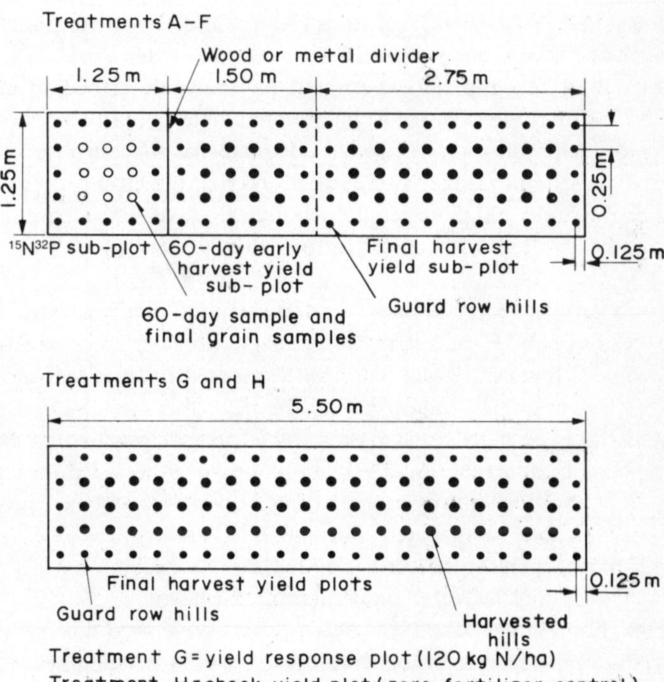

FIG. 12.3 Diagram of single plots, showing division into isotope labelled sub-plots, harvest plots and yield check plots ([17]).

As a typical experiment it is proposed to describe one of the experiments of the FAO/IAEA Rice Fertilization Programme ([17]) because it proved successful in practice in a number of widely separated countries, and because experiments involving irrigated, or permanently flooded soils as for rice, represent the most difficult situations. Therefore, other experimental conditions are comparatively easy. The treatments are shown in Table 12.1, while Fig. 12.2 shows the layout of the whole experimental field. Enlargements of single plots are shown in Fig. 12.3.

The size of each individual plot was 1.25 m × 5.5 m giving an area of 6.87 m^2 of which the $^{15}N^{32}P$ sub-plot as 1.56 m^2, the sixty day yield sub-plot 1.87 m^2, and the final yield sub-plot 3.44 m^2. The yield response plots were 1.25 m × 5.5 m, giving an area of 6.87 m^2. The arrangement of the harvested hills and the guard-row hills will be clear from Fig. 12.3. In the case of rice the plants are transplanted into hills of three plants. The spacing between the hills was 25 cm and between the guard rows and dykes 12.5 cm. The spacing between the hills of the $^{15}N^{32}P$ plot and the yield sub-plot was also 25 cm. These spacings gave 5 × 22 = 110 hills per main plot, in which 5 × 5 = 25 hills were in the $^{15}N^{32}P$ sub-plot, 5 × 11 = 55 hills in the final yield sub-plot and 5 × 6 = 30 hills in the 60-day yield sub-plot.

One outside row adjacent to the dykes and one row on each side of the boundary between sub-plots were guard rows. These were found to be adequate. Consequently there were 9 centre hills for harvesting from the $^{15}N^{32}P$ sub-plot, 27 centre hills for the final yield sub-plot and 12 centre hills for the 60-day yield sub-plot, excluding guard rows. For Zero Fertilizer Control and Yield Response plots of full size, then the number of harvested centre hills was 3 × 20 = 60.

The layout of this typical experiment has been described in considerable detail because it was found in practice to be of convenient overall size, not too complicated, to have plots and sub-plots of a size adequate for the purpose but not so big that unnecessary labelled fertilizer is required, and that the arrangement of guard rows and harvested rows and yield plots proved wholly effective. It is clear that the basic general design can be applied with modifications to a variety of crops grown in relatively narrow rows.

As rice is grown under submerged conditions the plot layout is complicated by the necessity to have dykes and irrigation canals, which are not necessary with dryland grown crops. However, when intermittent irrigation is necessary, even if not for continuously submerged conditions as for rice, then provision has to be made for feeder and drainage canals and dykes, as is normal practice in experimentation on irrigated soils. In general, flood irrigation should be avoided, because of the risk of significantly "moving" nitrogenous fertilizers in placement experiments. Plots should as far as practicable be irrigated individually. However, in one or two cases where rice experiments were completely flooded due to exceptional rainfall, there was in fact no evidence that fertilizer movement took place.

In the experiment described, irrigation canals were used to separate pairs of blocks, while earth dykes separated plots and blocks, with a wooden or metal divider making the boundary between the $^{15}N^{32}P$ sub-plots. No divider was used between the yield sub-plots. The dykes were prepared from soil adjacent to the experimental field but not from the top soil of the experimental plots. Polyvinyl plastic sheeting may also be used successfully to separate the plots. The sheeting is held upright by light wooden posts, and the base of the sheet buried in the earth. The dyke width was 60 cm and the width of the irrigation canal around each pair of blocks was 100 cm. For an experiment with 48 plots (e.g. 8 treatments and 6 replications) the total field area including the irrigation canal surrounding the plots was 24.1 m × 32.2 m = 776 m^2, not including the main enclosing dyke with the fence.

Fertilizers

The fertilizer levels chosen should be at a level normally used for the production of high yields in the area. The levels used for the experiment described above were comparatively low with 60 kg N/ha and 60 kg P_2O_5/ha, but these levels were chosen for various specific reasons, and clearly individual experimenters must choose levels suited to their own conditions.

Where N and P are the experimental fertilizers any possible deficiency of potassium and/or minor elements should be made good by applications to the experimental area during tillage operations. Where any doubt exists, an overall dressing of say 60 80 kg K/ha should be given.

In order to obtain information as to what point on the yield curve the experiment is being conducted, it is useful to have Zero Fertilizer control plots to which no fertilizer at all is applied except for the basic potassium application. At the same time the Yield Response plots provide an indicator of the upper limit of the yield curve. To these plots P and K applications are given at the same level as for the rest of the experiment but the nitrogen level applied is substantially greater, at least 50–100% higher. Although this is obviously a very incomplete means of determining the yield curve, in practice it serves adequately enough for the purpose of a simple check to determine if the level of fertilizer chosen for the overall experiment was correct.

Fertilizer Placement

Where fertilizers are applied broadcast there are few application problems, but where depth of placement or mixing effects are being investigated it is necessary to apply the fertilizer in rows. Where a combined seed and fertilizer drill is available this is no problem otherwise rows have to be marked out with string, and furrows for the seed and fertilizer opened up either with a wooden "skid" marker or else with hoes. In rice experiments where the plants are transplanted, the fertilizer rows must be carefully marked out with strings, and the strings left in position until the transplanting is completed. This ensures accurate planting.

Achieving accurate depth placement of fertilizers applied to wet paddy soils can be extremely difficult. One way used to overcome some of the problems is to make a V-shaped "bottomless" metal trough, which is pressed into the mud to the required depth, the fertilizer applied through the open base and the trough then removed.

Before starting to lay out the experiment, every lot of fertilizer for each single individual row, whether labelled or unlabelled should be carefully weighed out in advance, in either glass bottles—for active material—or polythene bags for inactive fertilizer. In this way the experiment can be put out as quickly as possible.

Caution: always remember to keep a sample of the isotopically labelled fertilizer for later determination of specific activity or, in the case of ^{15}N, atom % ^{15}N excess.

Sampling

For intermediate yield harvests separate sub-plots are required. However, for periodic sampling for $^{15}N^{32}P$ this is not necessary, as enough leaves may be taken for analysis from the plants of the central harvest hills. Not more than one leaf is taken from each plant and a standard leaf, say the one next to the flag leaf, is chosen. Although the yield of the $^{15}N^{32}P$ sub-plot is actually not required, because of the separate yield sub-plot, it is of interest to note that the yield of the plants sampled in this manner is not significantly affected.

The time chosen for the $^{15}N^{32}P$ sampling is at the choice of the experimenter, but normally 60 days is the last practicable date that ^{32}P samples can be usefully taken. Normally, ^{15}N samples will be taken at the final harvest and it is often useful to determine the ^{15}N abundance of both straw and grain.

Data

As in most such fertilizer experiments, adequate soil sampling can usually give useful supplementary information.

The records taken are entirely up to the investigator, but should include all appropriate dates of sowing or transplanting, applying fertilizers, sampling, intermediate and final harvests etc. Data on the plants may include number of tillers per plant, average plant height, number of ears per hill, panicle weight, 1000 grain weight, total dry matter and grain yields etc. Such data although not critical to the isotope determinations or interpretation of the results, are frequently useful for comparison with existing data, and in drawing conclusions for practical application.

Isotopic Analysis

In the determination of ^{32}P a fairly large sample is usually required to get an adequate count to reduce the natural standard deviation to 1%. 2–10 g of dry ground plant material is either dry ashed in the muffle furnace at 500°C and the ash taken up in dilute acid, or it may be wet digested. The wet digestion may be carried out with a ternary mixture of nitric, perchloric, and sulphuric acids (12:3:5). If nitrogen is to be determined on aliquots of the same digest then a sulphuric acid/hydrogen peroxide digest should be used.

^{31}P may be determined colorimetrically on a suitable aliquot of the digest using the vanado-molybdate method.

^{32}P is usually most conveniently determined directly on the digest, either in a halogen quenched liquid counter such as the Mullard MX124 or 20th Century Electronics M6H counter; or alternatively Cerenkov counting may be used.

To determine the fertilizer standard weigh out a portion of fertilizer estimated to contain 20 mg P (about 0.23 g of superphosphate) into a 200 ml volumetric flask, add 20 ml of dilute (1:2) nitric acid, dissolve and make up to volume. Take 10 ml of fertilizer solution and determine ^{31}P content as for the plant digest. Determine ^{32}P activity on a suitable aliquot of fertilizer.

Calculate the specific activity for both plant material and fertilizer on the basis of c.p.m./mg P. The percent nutrient derived from the fertilizer is next calculated.

Then
$$\%\ \text{P dff} = \frac{\text{c.p.m. per mg P plant material}}{\text{c.p.m. per mg P fertilizer}} \times 100 \tag{7}$$

and
$$\%\ \text{P df soil} = 100 - \%\ \text{P dff} \tag{8}$$

Determinations of ^{14}N and ^{15}N may be carried out as described in Chapter 7. ^{15}N abundance is calculated from the formula

$^{15}N(\text{atom}\%) = \frac{100}{2R+1}$, and $^{15}N\%$ atom excess is calculated by deducting 0.37 from the total $^{15}N\%$.

Then
$$\%\ \text{N dff} = \frac{\%^{15}\text{N excess in plant material}}{\%^{15}\text{N excess in fertilizer}} \times 100 \tag{9}$$

and
$$\%\ \text{N df soil} = 100 - \%\ \text{N dff} \tag{10}$$

Secondary Data

If desired the *A-value* may now be calculated:

$$\text{A-value} = \frac{\%\text{P (or N) df soil}}{\%\text{P (or N) dff}} \times \text{rate of P (or N) fertilizer application, kgP (or N)/ha} \tag{11}$$

The *total amount of P (or N) in the crop* is of course given by:

$$\text{yield (kg/ha)} \times \%\text{P or N} = \text{kg P (or N)/ha} \tag{12}$$

and the *total amount of P or N in the crop derived from the fertilizer* is then:

$$\text{total P or N (kg/ha)} \times \%\text{dff} = \text{kgP (or N) dff/ha} \tag{13}$$

The *per cent utilization of applied fertilizer* is then calculated from:

$$\frac{\text{total crop P or Ndff (kg/ha)}}{\text{applied fertilizer P or N (kg/ha)}} \times 100 = \% \tag{14}$$

or alternatively combine (10) and (13), so that

$$\begin{matrix}\text{\% utilization of} \\ \text{fertilizer nutrient}\end{matrix} = \frac{\text{yield (kg/ha)} \times \%\text{P or N} \times \%\text{P or N dff}}{\text{applied fertilizer P or N (kg/ha)}} \tag{15}$$

ISOTOPE TECHNIQUES FOR ROOT STUDIES

Root systems are of particular importance as they are responsible for the uptake of water and nutrients by the plant. A knowledge of where the active roots are located in the soil can be valuable for practical studies of fertilizer placement and timing, and possibly irrigation. However, in the past the study of root systems has been somewhat neglected because of the labour required. The classical techniques for root investigation involved elaborate and painstaking excavations of the root systems, characterized by the work of Weaver ([86]) and his school. However, these methods are only suitable for investigating a few plants because of the labour involved in digging profiles, tracing and drawing roots etc. Furthermore, methods of excavating root systems only indicate the presence of roots, not whether they are active roots contributing to the nutrition of the plant.

Comparatively little work has been carried out on the structure and distribution of root systems and yet as plant breeding provides us with new crop varieties it is becoming clear that differences in root habit exist between varieties, just as much as do differences between the above ground parts. Root distribution investigations will therefore be a continuing study. Isotope technique enables the investigator to handle large numbers of plants, either annuals or trees, without destroying them and with greatly reduced labour.

Using tracer techniques has made it possible to determine the effects of reduced cultivation on root growth, and suggested that the most economical method of land preparation is one which ensures root establishment to the minimum depth required for an adequate supply of water and nutrients ([79]). Wheat has been shown to have considerable varietal differences in root depth and form, which may have significance for cultivation and fertilization ([74,82,85]). In India a deep-rooted wheat variety in rotation with a deep-rooted cotton crop aggravates zinc deficiency, but is greatly reduced if a wheat variety is selected with spreading lateral roots ([83]). In Japan, the study of rooting patterns of onion and rice crops with incompatible nitrogen fertilization requirements, lead to the development of fertilizer placement practice using less nitrogen to enable growing onions and rice in succession and at reduced cost ([32]). The method has been applied to the study of root activity and nutrient uptake by mixed crops from different depths ([68]).

Isotope techniques now make it comparatively easy to determine the distribution of active roots, both in terms of area around individual plants and also in relation to rooting depths in the soil. Reliable information on these questions is not available for many crops, and especially perennial crops such as orchard and plantation trees. The significance of such information lies ultimately in determining the best method of fertilizer placement, the most suitable form of fertilizer to use and the optimum time and frequency of fertilizer application. Fertilizer use efficiency can thus be maximized; increasing yields and reducing costs of production.

Such studies are of particular value in the case of tree crops, because normal fertilizer trials are notoriously difficult with these crops. This is because, due to the reservoir of nutrients in the trunk, the results of fertilizer trials on trees often do not become clear for a period of years. In fact, fertilizing may be as significant for increasing the useful life of the plantation as for increasing yields.

Principle

The principle of methods for determining *root activity* consists of injecting soution containing a suitable isotope at various distances and depths from the base of the plant into the soil and taking the amount of the isotope found in the above ground parts as a measure of root activity.

Data obtained from root activity measurements may be used as a guide to plan subsequent fertilizer placement experiments with labelled fertilizer. Since the cost involved in injection experiments is in general much less than in experiments with labelled fertilizers, particularly with trees, the former is a useful preliminary approach to improving fertilizer placement.

Suitable isotopes are those that are readily translocated and equilibrate with the pool of nutrient in the plant, for instance ^{32}P. The use of γ emitters, e.g. ^{86}Rb, may offer a small advantage in respects of counting, but this advantage is usually offset by the health hazard involved. ^{131}I has also been used. The technique for injection and sampling is usually different for annual and perennial tree crops.

Somewhat different to root activity studies are *root distribution* studies. In this case the comparative distribution of living roots can be determined in relation to depth, lateral distance, or in response to growing conditions, but the method does not give information as to whether the roots are actually absorbing nutrients from the soil. The presence of a living root is no guarantee that it is contributing effectively to the plant; it may in effect be dormant.

For root distribution studies the plant is sprayed or injected with a solution containing a radioactive isotope which is easily translocated to the roots. By sampling the soil in a regular pattern as regards distance and depth, the activity of the roots contained in the soil sample is a measure of the root mass in the location from where the sample was taken. Suitable isotopes in this respect are again ^{32}P and ^{86}Rb as both elements are translocated readily.

In the case of ^{32}P, the roots may have to be separated from the soil samples since absorption of the emitted electrons (β rays) by the soil tends to make the counting inefficient. This however depends on the relative amounts of soil and roots in the sample. ^{86}Rb which emits γ rays (1.08 MeV) has the advantage that no separation of soil and roots has to be made and the γ can easily be counted by a scintillation detector in the entire soil sample.

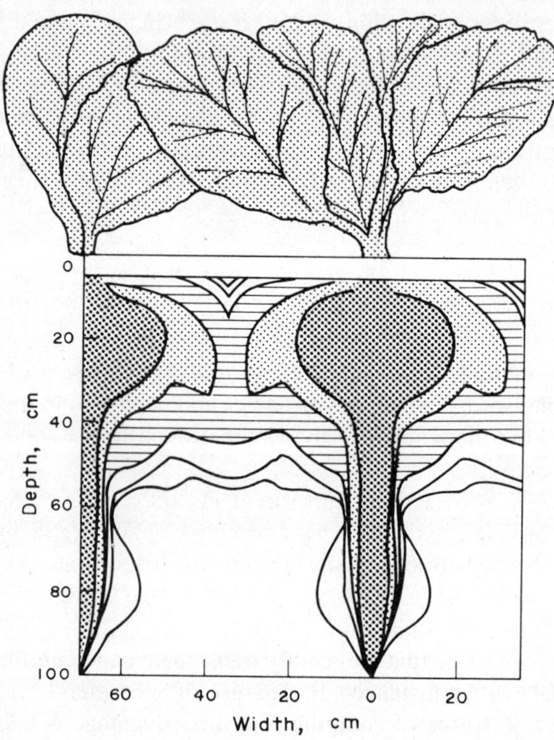

FIG. 12.4 Root activity distribution map of the gobo plant (burdock, *Arctium lappa*), Nishigaki ([75]).

Techniques for Annual Crops

The soil injection technique for studying root systems was developed by Hall *et al.*, ([60]) using ^{32}P in a study of maize, tobacco, cotton and peanut root systems. The technique was later used for such different applications as the study of root systems of grasses and alfalfa ([57]), ([69]) and for fertilizer placement for vegetable crops ([61]). ^{32}P injection techniques have also been used for the study of root systems in grassland communities ([52,53,54]), cotton ([55]), sorghum ([67]) and tomatoes ([66]).

Suitable injection systems using syringes have been developed, usually using ^{32}P, contained in solution with a carrier concentration of the order of 100–2000 p.p.m. P. Small aliquots of this solution, about 1–2 ml, are injected in rings around the base of the plants at the required distance and depth. Carrier free solutions should not be used because of rapid fixation in the soil and the possible difference in degree of dilution with labile soil P depending on depth and/or distance. On the other hand, a too high carrier concentration i.e. too low a specific activity will reduce the amount of ^{32}P taken up by plants. An optimal range of specific activity is generally of the order of 10 to 20 μCi ^{32}P/g P.

As an alternative to injecting the ^{32}P in a solution of ^{31}P carrier, it is also possible to adsorb ^{32}P labelled PO_4 to very finely pulverized anion exchange resin, which may be similarly injected ([75]). This may have the advantage of placement of the radioisotope.

After a suitable time, usually of the order of about two weeks, the plants are harvested, dried, and ground. A representative sub-sample is assayed for ^{32}P activity. The activity in the sample is a measure of the root activity at the point of injection. A 1–5 g dry plant sample is ashed and taken up in 2 N HNO_3 or HCl for liquid G-M or Cerenkov counting. Alternatively, the activity in the ash can be counted, using an end window β counter.

With the large number of injections that it is possible to make with annual plants e.g. a point every 10 cm, both horizontally and vertically, so as to cover a cubic metre, it is possible to plot a root activity distribution map, as in Fig. 12.4. From 300–500 plants are required for such a map. Much work has been done on these lines in Japan, on onion, rice, wheat, soybeans, sugar beet and gobo (burdock, *Arctium lappa*) ([75]).

An interesting technique for determining the distribution and extent—but not the nutrient uptake capacity—of living roots of annual crops has been developed ([77,78]). In essentials, an isotope usually ^{32}P, is injected into the stem of a plant such as wheat and after a period of time has elapsed its presence is determined in soil cores removed from the rooting zone. ^{86}Rb may also be used, being easier to count.

About 0.05 ml of a carrier-free ^{32}P in dilute HCl with an activity of 8 mCi of ^{32}P per ml is injected by means of a micro syringe into two or more tillers of a cereal plant just above the first node. After the needle is withdrawn the hole is sealed with collodion. At least six hours, overnight is convenient, are allowed to elapse before soil cores are taken, preferably with a hydraulic soil corer, the above ground parts of the plant and the "crown" being previously removed. Cores are taken at varying radii from the plant, each core being subsequently sectioned into 10–15 cm lengths.

The cores are air dried, a representative sample, say 50–60 g, is ashed at 500°C, mixed, and counted as an "infinitely thick" sample using an end-window counter. The "percentage distribution of active roots" as a function of soil depth is expressed by taking the count rate for each section as a percentage of the sum of the count rates for the total depth (i.e. number of sections) sampled. In order to obtain a quantitative measure of active plant roots in each core, the roots present in 2 or 3 cores must be physically separated and actual root weights related to activity levels. Cores from the surface layer should be taken for this purpose, due to the large number of dead roots present in the lower layers.

The method is apparently not very suitable for solid stem plants ([58]), and it is also destructive of the plant and disruptive of the experimental plots. Coupled with classical root washing and separation techniques the method is capable of determining the actual percentage of living, as opposed to dead and decomposing roots at any given point. Nevertheless, it primarily shows that a root is living and not that it is participating in nutrient uptake processes at the time of sampling. Thus the method is mainly appropriate for the determination of gross root growth patterns and structure, as influenced by species, variety, soil type and cultivation.

Techniques for Perennial Tree Crops

Radioisotope techniques were used quite early for fertilizer placement studies in the vine ([84]) and in the oil palm ([55]) using ^{32}P. The lateral and vertical spread of longleaf pine roots was determined by soil injections of ^{131}I ([59,70]). Coffee was also studied ([80]).

^{32}P is now the most commonly used isotope, and because of the large quantities of labelled solutions that have to be injected around trees, the syringe technique is not really adequate. Usually several hundreds of mCi of ^{32}P are involved in tree studies, and it is not advisable to carry around stock solutions containing such large amounts of activity. In the case of an accident, a health hazard might be involved, and this is one of the reasons why relatively little injection work with trees was carried out in the past. In addition, the dilution of a few ml ^{32}P solution containing several hundred mCi, require appropriate shielding and handling facilities which are not often available in tree plantations.

Over a period of 5 years the IAEA organized ([64,65,73]) the longest and most extensive coordinated programme on root activity studies in tropical and subtropical fruit trees, such as olive, orange, apples, coconut, coffee, cocoa, and oil palm hitherto carried out. A suitable point placement technique was developed which proved very satisfactory. Glass ampoules are filled with 4 ml of a solution containing 300 μCi ^{32}P and a carrier concentration of 1000 p.p.m. P. The ampoule filling is carried out using a normal commercially available ampoule filling machine with suitable shielding.

The ampoules containing the labelled solution should be properly packed in polystyrene for transport to the experimental fields. Up to 5 mCi are given to each tree, i.e. 16 injections around each tree. The ampoules are placed at the required distance and depth in rings around the tree by means of a simple device consisting of a steel tube with a central steel rod. The device is driven into the soil at the appropriate position and to the required depth. The central rod is then withdrawn, the ampoule dropped down and then crushed with the steel rod. It is convenient to prepare the holes, by means of a soil auger, a day or two in advance, plugging the holes with rods to prevent in-filling.

Although the ampoule technique was developed partly to overcome problems of safe transportation and handling, in practice it also proves to be a very precise and relatively easy means of placement.

At regular intervals after the application, the trees are sampled for ^{32}P assay. It is essential to sample leaves of the same age and morphological position and to choose trees of the same age and physiological cycle. The treatments resulting in the highest activity in the leaf samples correspond with the location of highest root activity.

The standard leaves chosen for sampling will naturally vary with the crop, and the following are given as examples:

Banana: third fully opened leaf.
Citrus: young outer leaves of the centre part of the tree.
Cocoa: recently mature leaves in upper part of tree.
Coffee: young leaves from lower part of tree, specifically shoot tips, up to 3rd leaf.
Coconut: leaflets from mid portion of 6th frond.
Oil palm: leaflets from 17th frond.

Single trees are usually used as experimental units in a randomized block design.

It is advisable to include a sufficient number of replicates in the experiments, at least four or five, because the variation in ^{32}P counts for each treatment may be relatively high. The standard deviation for the activity in a single sample may be as high as 50 or even 100%, depending on the tree species. Despite the high variation in activity, the technique has proved very useful, because the differences in root activity as a function of distance and depth are often of the order of 200 or 300% and can be easily detected.

Choice of Isotope

^{32}P has been the main isotope used, because of its relatively low price and safe handling characteristics. However, it has a relatively short half life and in some soils is powerfully adsorbed. ^{86}Rb may be used as it has a slightly longer half life and better counting characteristics. Nevertheless, experiment has shown ([62]) that the advantages, though real, are not sufficient to outweigh the increased health hazard and difficulty of handling. ^{15}N has no health hazard and no half life restrictions. It can be used satisfactorily, but it is largely ruled out due to the expense of the 95% ^{15}N atom excess material required to overcome the great dilution factor. Moreover, inorganic nitrogen is highly mobile in the soil which makes it unsuitable for general use.

Analytical Technique

Methods as described earlier may be used for ashing or digesting leaf samples for ^{32}P determination. As the activity is normally rather low it may be necessary to take as much as 10 g dry wt of plant material. Correction for naturally occurring ^{40}K will almost certainly be necessary: this may be done either by determining the ^{40}K content in leaves from untreated trees or else by carrying out a second count after allowing several days to elapse for the ^{32}P to decay further.

Source and Reduction of Errors

It has been ([65,73]) pointed out that the main sources of error are:

- Unequal probability of roots of individual trees contacting the applied ^{32}P due to the number of injection points being too small.
- Unequal probability of the ^{32}P coming into contact with roots at the different distances tested, because of the number of injection points per unit length of circumference for distances close to the tree being greater than that for circumferences further out from the trunk.
- Soil heterogeneity.
- Natural variation between trees.
- Leaf sampling, due to non-uniform distribution of ^{32}P in leaves.

A number of experiments were analyzed to examine these sources of error including an experiment in which twice the normal number of ^{32}P injection points were used. The conclusion was that there was little advantage in increasing the number of injection points. Further, that the main contributory factor to the high variability between individual trees is leaf sampling error, arising out of non-uniform distribution of ^{32}P, even between leaves of similar position and age. All other sources of error are comparatively small and subordinate to leaf sampling error. Reduction of this source of error must be through taking a greater number of sub-samples ([73]).

For the basic technique to be valid it is necessary that root activity ratios as determined by isotope assay of the leaves are independent of the morphological position of the sampled leaf. This can be tested by injecting a mixture of ^{32}P and ^{33}P isotopes into two distinct areas of the rooting zone. For the same time of sampling the $^{32}P/^{33}P$ of leaf samples from different positions should be the same.

If translocation and redistribution are slow then the isotope may accumulate principally in leaves on the side of the tree where the injection was done. In practice this would have to be overcome by sampling leaves on the side of the tree where the injections were made, or alternatively by means of a frequently spaced ring of injections around the tree at the appropriate distance and depth.

The Interpretation of Data from Root Activity Studies

Root activity studies may be a good preliminary approach in the case of tree crops for the assessment of depths and distances to be tested in subsequent fertilizer placement studies. The quantities of labelled fertilizers that have to be used for fertilizer placement studies in tree crops, are generally very high. Results of root activity studies enable one to make an intelligent choice of treatments of interest for fertilizer placement experiments.

However, great care must be taken in making quantitative comparisons from root activity studies with tree crops, particularly when there is a considerable time interval between injection and sampling of a particular leaf. The amount of nutrient taken up by the root is not exactly proportional to the activity found in the leaves since translocation from leaves to other organs of the tree takes place continuously. Thus the results of root activity studies are of a semi-quantitative nature, i.e. one can indicate for instance that in a particular experiment the highest activity was found at 10 cm depth or at 3 m distance. The method can be used to study the effects of different soils and of soil heterogeneity, the difference in root activity between wet and dry seasons, the differences due to age of tree, competition between trees and the influences of shade and irrigation.

Typical Applications and Results

This technique is capable of giving significant useful information in a relatively short time and a few results of IAEA contractors' work illustrate this ([56,65]). A notable feature was that work carried out on cocoa, coffee, coconuts, citrus and oil palm showed that highest root activity is found in the upper soil layers, especially from 0–10 cm depth. However, significant differences were found in the horizontal distribution of active roots.

Coffee

In Kenya, the roots of coffee appear to be virtually inactive during the January/February growth period. During the April/May period the majority of active roots were shown to be concentrated near the stem of the coffee plant. In the past, fertilizer had traditionally been applied at a considerable distance from the trunk ([63,76]).

Orange trees

8 year old orange trees in Taiwan had their most active system within a radius of 1–2 m from the trunk at a depth of 10 cm. In contrast, much older trees of navel orange, grown in Spain, showed that the most favourable depth of placement was 30 cm and the optimum distance from the trunk 2–3 m ([65,71,87]).

Oil palm

Highest root activity was found in the surface layer of the soil. There was little or no root activity below 25 cm. There was little difference in root activity between 1 and 4 m from the trunk ([65]).

Coconuts

10 year old trees grown in the Philippines showed the greatest root activity at a distance of 1–2 m from the trunk and a depth of 15–30 cm. Ceylon experiments have indicated that a marked improvement in fertilizer uptake can be expected if the fertilizer

is applied close to the stem. With 50 year old palms greatest uptake was observed at 10 cm depth and 0.5 m from the trunk ([50,65,81]).

Cocoa

In Ghana it has been found that root activity is about ten times greater in the top few cm of soil than at depths of 30 cm or more. There was a marked difference between varieties, the more recently developed variety WAE 11 having by far the most active root system. The pure Amelonado type had the poorest root activity, though it had relatively more activity at the greater depths ([49]).

Although quite a lot of work has now been carried out on the root systems of perennial tree crops, very much more work needs to be done, particularly to compare the differences between different varieties. Forest trees and trees in natural habitats have scarcely been touched. Reference 65 provides the best source of general information and technique.

REFERENCES FOR FURTHER READING

Fertilizer Studies

1. AHMED, S., Azmi, A. R. and Malik, M. B. Uptake of fertilizer phosphorus and nitrogen by lowland rice in W. Pakistan under different methods and time of application. *In: Proc. Symp. Isotopes and Radiation in Soil-Plant Nutrition Studies*, Ankara, 1965, 449, IAEA, Vienna (1965).
2. ALEKSIC, Z., Broeshart, H. and Middelboe, V. Shallow depth placement of $(NH_4)_2SO_4$ in submerged rice soils as related to gaseous losses of fertilizer nitrogen and fertilizer efficiency. *Plant Soil* **29**, 338 (1968).
3. ALLEN, E. F. Paddy manurial trials in the 1950–51 season. *Malaya Agric. J.* **15**, 3 (1952).
4. ATANSIU, N., Bakhati, H. and Hamdi, H. Evaluation of the utilization of ammonium and nitrate nitrogen using tracer technique. *U.A.R. J. Soil Sci.* **11**, 171 (1971).
5. BARBER, S. A., Mitchell, J. and Spinks, J. W. T. Soil studies using radioactive phosphorus. *Can. Chem. Process. Ind.* **31**, 757–61 (1947).
6. BARTHOLOMEW, W. V. ^{15}N in research on the availability and crop use of nitrogen. *In: Nitrogen-15 in Soil Plant Studies* (Proc. Panel, Sofia, 1969) 1–20, IAEA, Vienna (1971).
7. BRIDGENS, A. P. and Pretorius, T. P. The effect of placement of ^{32}P-labelled superphosphate and/or radiation and nitrogen on the nutrient uptake of *Eragostis curvula*. *Proc. Natn. Conf. Nucl. Energy*, Pretoria 403 (1963).
8. BROADBENT, F. E. and Nakashima, T. Plant uptake and residual value of tagged fertilizer. *Soil Sci. Soc. Amer. Proc.* **32**, 388 (1968).
9. BROADBENT, F. E. Field measurements of N-utilization efficiency and nitrate movement in soils using ^{15}N-depleted fertilizer. *In: Proc. Symp. Isotope Ratios as Pollutant Source and Behaviour Indicators*, Vienna, 1974, 373, IAEA, Vienna (1975).
10. CALDWELL, A. C. and Blanchar, R. W. Phosphorus uptake by plants and ion concentration in fertilizer zones as affected by ammonium, potassium and dilution. *In: Proc. Symp. Ankara*, 1965, 413, IAEA, Vienna (1965).
11. CARTER, J. N., Bennett, O. L. and Pearson, R. W. Recovery of fertilizer nitrogen under field conditions using ^{15}N. *Soil Sci. Soc. Amer. Proc.* **31**, 50 (1967).
12. DATTA, N. P. and Venkateswarlu, J. Uptake of fertilizer phosphorus and nitrogen from different methods of application of lowland rice growing on major Indian soils. *9th Int. Congr. Soil Sci. Trans.* **4**, 9 (1968).

13. DATTA, N. P., Banerjee, N. K. and Prasada Rao, D. M. V. A new technique for study of nitrogen balance sheet and an evaluation of nitro-phosphate using ^{15}N under submerged conditions of growing paddy. *In: Proc. Int. Symp. Soil Fert. Eval.* Indian Soc. Soil Sci., New Delhi **1**, 631 (1971).
14. ENSMINGER, L. E. and Pearson, R. W. Use of macronutrient isotopes in soil fertility. Atomic Energy and Agriculture, *Am. Assoc. Adv. Sci.* **49**, 19–47 (1957).
15. HERA, C. Response and current fertilization practice of grain legume crops in Romania. *In: Use of Isotopes for Study of Fertilizer Utilization by Legume Crops.* Proc. Panel Meet., Vienna 1971, 157–182, IAEA-149, Vienna (1972).
16. HERA, C. Field experiments for increasing soil productivity with the aid of nitrogen-15. *In*: Proc. Symp. Soil Organic Matter Studies, *Braunschweig* 1976, vol. I, 315–30, IAEA, Vienna (1977).
17. IAEA *Rice Fertilization.* Tech. Rept. Series no. 108, pp. 177, IAEA, Vienna (1970).
18. IAEA *Fertilizer Management Practices for Maize*: Results of Experiments with Isotopes. Tech. Rept. Series no. 121, pp. 78, IAEA, Vienna (1970).
19. IAEA *Isotope Studies on Wheat Fertilization.* Tech. Rept. Series no. 157, pp. 99, IAEA, Vienna (1974).
20. KAMATH, M. B. *Phosphate Utilization by Crops Under Mixed Cropping Conditions.* Ann. Rept. Chemistry Div., Indian Agriculture Res. Inst. (1970).
21. KHIN WIN, U. Placement of phosphate fertilizer in rice cultivation. Union of Burma, *J. Life Sci.* **1**, 109 (1968).
22. LEGG, J. O. and Stanford, G. Utilization of soil and fertilizer N by oats in relation to the available N status of soils. *Soil Sci. Soc. Amer. Proc.* **31**, 215 (1967).
23. LOW, A. J. The use of isotopes in agricultural research II. *Chem. and Ind.* 1124–28 (1951).
24. LUGO, C. J. Eficiencia del Isótopo Estable ^{15}N en la Determinación del Nitrógeno en el Maíz. Estación Experimental Agrícola de la Molina, Boletin 18 (1967).
25. MATTINGLY, G. E. G. *Soils and Fertilizers* **20**, 59–68 (1957).
26. MENZEL, R. G., Eck, H. V. and Champion, D. F. Effect of placement depth and root-inhibiting chemicals on uptake of strontium-85 by field crops. *Agron. J.* **54**, 70–72 (1967).
27. MISTRY, K. B. Quantitative evaluation of the residual values of phosphatic fertilizers in soils. *In: Proc. Symp. Radioisotopes in Soil-Plant Nutrition Studies.* Bombay, 1962, 363, IAEA, Vienna (1962).
28. MISTRY, K. B. Relative efficiencies of phosphatic fertilizers for wheat through radioactive tracer technique. *In: Proc. Symp. Radioisotopes in Soil-Plant Nutrition Studies*, Bombay, 1962, 427, IAEA, Vienna (1962).
29. MITCHELL, J., Dehm, J. E. and Dion, H. G. Availability of fertilizer and soil phosphorus to grain crops and the effect of placement and rate of application on phosphorus uptake. *Sci. Agric.* **32**, 511–25 (1952).
30. NELSON, W. L. *et al.* Application of radioactive tracer techniques to studies of phosphate fertilizer utilization by crop: II. Field experiments. *Soil Sci. Soc. Amer. Proc.* **12**, 113 (1947).
31. NELSON, W. L., Krantz, B. A., Welch, C. D. and Hall, N. S. Utilization of phosphorus as affected by placement: II. Cotton and corn in North Carolina. *Soil Sci.* **68**, 137 (1944).
32. NISHIGAKI, S. The use of ^{15}N as a tracer in fertilizer efficiency study in Japan. *In: Isotopes and Radiation in Investigations of Fertilizer and Water Use Efficiency.* Bangkok 1969, IAEA-20, 161, IAEA, Vienna (1970).
33. OFORI, C. S. The effect of nitrogen and sulphur treatments on phosphorus absorbtion and yield of groundnuts on soils of sandstone origin. *In: Use of Isotopes for Utilization by Legume Crops.* Proc. Panel Meeting, Vienna 1971, 183–94, IAEA-149, Vienna (1972).
34. OOMEN, P. K. and Oza, A. M. Utilization of applied phosphorus by wheat varieties. Sonora 64 as revealed by field experiments using radiotracer technique. *Indian J. Agric. Sci.* **38**, 986 (1968).
35. OZA, A. M. and Subbiah, B. V. A study of the residual fertilizer nitrogen in soil under multiple cropping conditions using ^{15}N. *Indian Soc. Nucl. Tech. Agric. Biol. Newsl.* **2**, 55 (1973).
36a. RENNIE, D. A. and Spratt, E. D. The influence of fertilizer placement on 'A' values. 7th Int. Congr. Soil Sci. Madison, Wisc. Trans. III 535–43 (1960).
36b. SPRATT, E. D. and Rennie, D. A. Factors affecting and the significance of 'A' values using band placement. *In: Proc. Symp. Radioisotopes in Soil-Plant Nutrition Studies.* Bombay 1962, 329–42, IAEA, Vienna (1962).
37a. RENNIE, D. A. The significance of the 'A' value concept in field fertilizer studies. Proc. Study Group Isotopes and Radiation on Investigations of Fertilizer and Water Use Efficiency, *Tech. Rept.* **120**, 132 IAEA (1970).
37b. RENNIE, D. A. and Clayton, J. S. An evaluation of techniques used to characterize the comparative

productivity of soil profile types in Saskatchewan. Trans. Comm. II and IV Int. Soc. Soil Sci. Aberdeen, 365–76 (1966).

38. RENNIE, D. A. and Fried, M. An interpretative analysis of the significance in soil fertility and fertilizer evaluation of N-15 labelled fertilizer experiments conducted under field conditions. *Proc. Int. Symp. Soil Fert. Eval.*, Indian Soc. Soil Sci., New Delhi **1**, 639 (1971).
39. RENNIE, R. J. and Rennie, D. A. Standard isotope versus nitrogen balance for assessing the efficiency of nitrogen sources for barley. *Can. J. Soil Sci.* **53**, 73 (1973).
40. RHIND, D. and Tin, U. Residual effects from superphosphate, basic slag and bone meal on some paddy soils in lower Burma. *Trop. Agric.* **108**, 102–7 (1952).
41. SACHDEV, M. S., Oza, A. M. and Subbiah, B. V. Possibilities of improving the efficiency of nitrogenous fertilizers. *In: Proc. Symp. Soil Organic Matter Studies, Braunschweig* 1976, vol. I, 371–82, IAEA, Vienna (1977).
42. SINHA, M. N. Note on the response of cotton to placement of phosphorus using labelled superphosphate. *Indian J. Agron.* **15**, 197 (1970).
43. SPINKS, J. W. T. and Barber, S. A. Study of fertilizer uptake using radioactive phosphorus. *Sci. Agr.* **27**, 145–56 (1947).
44. STANFORD, G. and Nelson, L. B. Utilization of phosphorus as affected by placement: I. Corn in Iowa, *Soil Sci.* **68**, 129 (1949).
45. TALIBUDDEEN, O. *Soils and Fertilizers* **20**, no. 4 (1957).
46. VASEY, E. H. and Barber, S. A. Effect of placement on the absorbtion of ^{86}Rb and ^{32}P from soil by corn roots. *Soil Sci. Soc. Amer. Proc.* **27**, 193 (1963).
47. VENKATACHALAM, S., Premnathan, S., Arunchalam, G. and Vivekanandam, S. N. Soil fertility studies in Madras State using radiotracer technique. 1. Placement of phosphate to hybrid sorghum. *Madras Agric. J.* **66**, 104 (1969).
48. VOSE, P. B. A review of the coordinated research programme on the application of isotopes to rice fertilization studies, 1962–68. *In: Isotopes and Radiation in Investigations of Fertilizer and Water Use Efficiency*. Bangkok 1964, IAEA-20, 6, IAEA, Vienna (1970).

Root Studies

49. AHENKORAH, Y. A study of the distribution of the root activity of mature cocoa (*Theobroma cacao* L.) using ^{32}P and injection technique. *Ghana J. Agric. Sci.* **2**, 97 (1969).
50. BALAKRISHNAMURTIE, T. Isotope studies on efficiency studies of fertilizer utilization by coconut palms. *Ceylon Coconut Quarterly* **20**, 111 (1969).
51. BASSETT, D. M., Stockton, J. R. and Dickens, W. L. Root growth of cotton as measured by ^{32}P uptake. *Agron. J.* **62**, 200 (1970).
52. BOGGIE, R. Hunter, R. F. and Knight, A. H. Studies of the root development of vegetable crops as measured by radioactive phosphorus injection technique. *Agron. J.* **55**(4), 329–33 (1963).
53. BOGGIE, R. and Knight, A. H. Studies of root development in a grass sward growing in deep peat using radioactive tracers. *J. Br. Grassld. Soc.* **15**(2), 133–36 (1960).
54. BOGGIE, R. and Knight, A. H. An improved method for the placement of radioactive isotopes in the study of root systems growing in deep peat. *J. Ecol.* **50**(2), 461–62 (1962).
55. BROESHART, H. The application of radioisotope techniques to fertilizer placement studies in oilpalm cultivation. *Neth. J. Agr. Sci.* **7**, 95–109 (1959).
56. BROESHART, H. and Nethsinghe, D. A. Studies on the pattern of root activity of tree crops using isotope techniques. *Isotopes and Radiation in Soil-Plant Relationships including Forestry* (Proc. Symp. Vienna 1971) p. 453, IAEA, Vienna (1972).
57. BURTON, G. W., DeVane, E. H. and Carter, R. R. Root penetration, distribution and activity in southern grass measured by yields, drought symptoms and ^{32}P. *Agron. J.* **46**(5), 229–33 (1954).
58. CHAN, K. Y., Halm, B. J. and Stewart, J. W. B. Phosphorus uptake by rapeseed from different depths in the soil. *Proc. 1973 Soil Fertility Workshop*, Univ. Saskatchewan, Sask. 148–67 (1973).
59. FERRILL, M. D. *Root Extension in a Plantation of Longleaf Pine: Investigation of a technique Using I-131*. Ph.D. dissertation, Duke University, Durham, N.C., 129 pp (1963).
60. HALL, N. S., Chandler, W. F., van Bavel, C. H. M., Reid, P. H. and Anderson, J. H. A tracer technique to measure growth and activity of plant root systems. *North Carolina Agric. Exp. Sta. Bull.* 101 (1953).
61. HAMMES, J. K. and Bartz, J. F. Root distribution and development of vegetable crops as measured by radioactive phosphorus injection technique. *Agron. J.* **55**(4), 329–33 (1963).
62. HUXLEY, P. A., Patel, R. Z. and Kabaara, A. M. ^{86}Rb as a tracer for root activity studies in Kikuyu

red loam. *E. Afric. Agric. For. J.* **35**, 340 (1971).

63. HUXLEY, P. A., Patel, R. Z. and Mitchell, H. W. Tracer studies with ^{32}P on the distribution of functional roots of *Arabica* coffee in Kenya. *Ann. Appl. Biol.* (1974).
64. *IAEA Annual Report of the Laboratory*. Tech. Reports nos. 98 and 103 (1969 and 1970).
65. IAEA. *Root Activity Patterns of Some Tree Crops*. Tech. Rept. no. 170. IAEA, Vienna (1975).
66. DEJONG, E., Otinkorang, E. S. Measurement of the root distribution of irrigated tomatoes with ^{32}P injection technique. *Can. J. Plant Sci.* **49**, 69 (1969).
67. KAFKAFI, U., Karhi, Z., Albral, N. and Roodick, J. Root activity of dryland sorghum as measured by radiophosphorus uptake and water consumption. *Isotopes and Radiation in Soil-Plant Nutrition Studies* (Proc. Symp. Ankara 1965) 481, IAEA, Vienna (1965).
68. KAMATH, M. B. and Subbiah, B. V. Phosphorus uptake pattern by crops from different soil depths. Proc. Int. Symp. Soil Fert. Eval., *Indian Soc. Soil Sci.* New Delhi, **1** (1971).
69. LIPPS, R. C., Fox, R. L. and Koehler, F. E. Characterizing root activity of alfalfa by radioactive tracer techniques. *Soil Sci.* **84**, 195 (1957).
70. MCCORMACK, M. L. *A Study of Techniques for Determining Root Extension Using Radioactive Tracers*. Ph.D. dissertation. Duke University, Durham, N.C. 193 pp. (1963).
71. MELLADO, L. and Caballero, F. Estudio de la distribuición de raíces activas en el naranjo utilizando ^{32}P. An. Inst. Nacl. Invest. Agr., Serie: *Producción Vegetal* **4**, 7 (1974).
72. NETHSINGHE, D. A., Rennie, D. A. and Vose, P. B. Radiotracers in Tree Culture. *Agricultural Chemicals*, November, 18–20 (1968).
73. NETHSINGHE, D. A. The validity of assumptions and control of errors in using ^{32}P soil injection techniques for studying the root activity of trees. FAO/IAEA Research Coordination Meeting on the Use of Isotopes to Study the Efficient Use of Fertilizers in Tree Culture. Vienna, January 1970. Mimeo. PL-283-2, 1–8 (1970).
74. NEWBOULD, P., Ellis, F. B., Barnes, B. T., Howse, K. R. and Lupton, F. G. H. *Intervarietal Comparisons of the Root Systems of Winter Wheat*. Agricultural Research Council Letcombe Laboratory, Ann. Rept. 1969, 38 (1970).
75. NISHIGAKI, S. Study on the distribution of root activity in the soil. FAO/IAEA Research Coordination Meeting on the Use of Isotopes to Study the Efficient Use of Fertilizers in Tree Culture. Vienna, January 1968. Mimeo. 1–9 (1968).
76. PATEL, R. Z. and Kabaara, A. M. Isotope studies on the efficient use of P-fertilizers by coffee *Arabica* in Kenya. 1. Uptake and distribution of ^{32}P labelled KH_2PO_4. *Exp. Agric.* **11**, 1 (1975).
77. RACZ, G. J., Rennie, D. A. and Hutcheon, W. L. The ^{32}P injection method for studying the root system of wheat. *Can. J. Soil Sci.* **44**(1), 100–8 (1964).
78. RENNIE, D. A. and Halstead, E. H. A ^{32}P injection method for quantitative estimation of the distribution and extent of cereal grain roots. *Isotopes and Radiation in Soil-Plant Nutrition Studies*. (Proc. Symp. Ankara, 1965) 489, IAEA, Vienna (1965).
79. RUSSELL, R. S., Clarkson, D. T. and Newbould, P. Tracer studies of the root systems of crop plants. *Int. Conf. Peaceful Uses of Atomic Energy* (Proc. Conf. Geneva, 1971) **12**, 215, IAEA, Vienna (1972).
80. SAIZ DEL RIO, J. F., Fernandez, C. E. and Bellavita, O. Distribution of absorbing capacity of coffee roots determined by radioactive tracers. *Proc. Amer. Soc. Hort. Sci.* **77**, 240–44 (1961).
81. SANTOS, I. S. Progress in applying nuclear methods to increase production of rice, soybean and coconut in the Philippines. *4th Int. Conf. Peaceful Uses of Atomic Energy* (Geneva, 1971) **12**, 179, IAEA, Vienna (1972).
82. SUBBIAH, B. V., Katyal, J. C., Narasimham, R. L. and Dakshinamurti, C. Preliminary investigations on root distribution of high yielding wheat varieties. *Int. J. Appl. Radiat. Isotopes* **19**, 385 (1968).
83. SWAMINATHAN, M. S. The role of nuclear techniques in agricultural research in developing countries. *In: Proc. Fourth Int. Conf. Peaceful Uses of Atomic Energy*, Geneva 1971, vol. 12, 3–32, IAEA, Vienna (1972).
84. ULRICH, A., Jacobson, L. and Overstreet, R. Use of radioactive phosphorus in a study of the availability of phosphorus to grape vines under field conditions. *Soil Sci.* **64**, 17–28 (1947).
85. VIJAYALAKSHMI, K. *Root Growth Studies in Wheat and Pulse Crops using Radioisotope Injection Technique*. Ph.D. thesis. Indian Agricultural Research Inst. (1971).
86. WEAVER, J. E. *Root Development of Field Crops*. McGraw-Hill, New York (1926).
87. YEH, Y. Estimation of the functional root distribution of citrus trees by radiotracer method. *Mem. Coll. Agric. Natl. Taiwan Univ.* **11**, 2, 79 (1969).
88. HAUCK, R. D. Critique: field trials with isotopically labelled fertilizer. *In: Nitrogen in the Environment*, Vol. I (D. R. Nielsen & J. G. MacDonald, Eds.) 63–77, Academic Press, New York (1978).
89. FRIED, M. Critique: field trials with isotopically labelled fertilizer. *ibid.*, 43–62 (1978).

CHAPTER 13

Nuclear Techniques In Plant Science

THIS chapter contains examples of nuclear techniques specific to the study of plant function. There can be little doubt that the present development of plant physiology and biochemistry would have been quite impossible without isotopes, and they are now used in all manner of studies, including photosynthesis, translocation of assimilates, ion uptake by cells and whole plants, the movement of herbicides and pesticides, the biochemistry of carbon assimilation, the synthesis of nitrogen compounds, enzyme studies, intermediary metabolism etc. There are also some limited uses of radiation apart from mutation induction (Chapter 15), for determination of water content and biomass.

METABOLISM STUDIES

Methods were quite early developed and established for biochemistry ([3,8]). They use the basic principle of isotope dilution and tracer kinetics that have already been discussed in Chapter 6, to establish processes of turnover, exchange, accumulation and removal. Similarly, product-precursor relationship can be studied and the turnover rate of metabolites established.

In studies of intermediary metabolism the isotopically labelled (usually ^{14}C with ^{3}H and ^{35}S) precursor is added to a system of tissue homogenates or tissue slices, immediately followed by a much larger quantity of the substance considered to be an intermediary. Now, the biological system metabolizes the labelled precursor and converts it into labelled intermediate, but as such a large amount of non-labelled intermediate has been added to the system the labelled intermediate undergoes little further metabolism, as the unlabelled intermediate has greatly diluted it and also because the total pool of intermediary is too large for the system to convert more than a small amount to the following intermediary. The pool of labelled and unlabelled material is then isolated and purified, and if following re-purification the specific activity remains constant the postulated intermediary is considered to be confirmed. Successive repetition of the technique using other postulated intermediates of later metabolic steps, combined with the use of specific inhibitors, permits the construction of the complete metabolic pathway.

^{14}C coupled with paper chromatography and other separative methods has proved to be an exceptionally powerful technique for investigating all manner of plant metabolic pathways.

In enzyme studies ^{14}C labelled compounds can be used both for routine enzyme assays and for investigating kinetic and control functions of enzymes. As radioisotope enzyme assays are extremely sensitive and highly specific, they permit the use of very small samples and enable assays to be made on even crude plant extracts where less specific methods can be affected by other enzymes or inhibitors. Separation of labelled product from labelled precursor is carried out by ion-exchange chromatography or solvent extraction.

At the cellular or sub-cellular level, localization of metabolite synthesis or of enzymes in the cell may be carried out by appropriate radioisotope labelling followed by microautoradiography, as described in Chapter 10.

Isotope techniques are now so widely used in all the varied aspects of metabolism studies that giving here a meaningful list of references is no longer practicable, but current work can be found in such journals as *J. Biol. Chem.*, *Arch. Biochem. Biophysics*, *Biochem. J.*, *Phytochemistry*, *Plant Cell Physiol. (Tokyo)*, *Plant Physiol.*, *J. Expt. Bot.*, etc.

Some biochemical methods have developed beyond basic research. For example, the need to develop screening methods for the essential amino acids in plants has lead to isotopic analytical procedures which could be put on a routine basis. They provide an instructive insight into principles and techniques. One method involves the radioisotope assay of carbon dioxide derived from the action of decarboxylase enzymes on specific amino acids ([19]). The CO_2 evolved is measured as insoluble $^{133}BaCO_3$ following reaction with $^{133}BaOH$. No chemical separation of amino acids is needed. Another technique ([6]) for amino acids depends on double isotope labelling after hydrolysis of the protein. ^{14}C-labelled DNP derivatives of the specific amino acid are added to the hydrolysate, followed by reaction with 1-fluoro-2, 4-dinitrobenzene-^{3}H to give ^{3}H-labelled dinitrophenyl derivatives. DNP-amino acids are then separated by thin layer chromatography and the ''spots'' counted by liquid scintillation. The ^{3}H activity in the derivative corresponds to the amount of amino acid in the hydrolysate while ^{14}C activity gives a recovery factor for the ^{14}C amino acid added, and therefore permits quantification.

Of recent years a whole area of applied studies has developed relative to the uptake, transport, metabolism, persistence and degradation of pesticides and herbicides in plants, their ecosystems and food products. Isotopes have played a noticeable role in many of these studies, and reference 31 will provide a useful introduction to the literature.

PHOTOSYNTHESIS STUDIES

Isotopes have been critical in establishing knowledge of the mechanisms of photosynthesis, since the first use of ^{14}C to follow the pathway of carbon ([49]) and the use of ^{18}O to show that the oxygen evolved in photosynthesis comes from water and not from oxygen in the CO_2 molecule ([50]). Later the work of Calvin and his co-workers on further elucidating the pathways of carbon fixation was largely dependent on ^{14}C and lead to the definitive work on the determination of isotopic carbon before liquid scintillation counting became available ([9]).

Subsequently there have been innumerable publications concerned with ^{14}C in photosynthesis studies, and an increasing number of these have been on determining and comparing photosynthetic rates. The dilution of fixed ^{14}C carbon and isotopic discrimination during long term experiments ([63]) are potential practical problems, but the Shimshi ([53]) rapid method which uses short pulse $^{14}CO_2$ seems to overcome these difficulties. A number of modifications of this method have since been developed ([32,39,54,58,62,67]) and successfully applied, the technique being usable in the field if desired.

Determination of Photosynthetic Rate

Principle

In essentials the technique involves the construction of a small plexiglass flow chamber in two sections, which can be clamped either side of a leaf by means of surgical scissors or a similar device. With the leaf maintained under saturation light conditions, excess $^{14}CO_2$ is then allowed to flow through the chamber for the required short time. Leaf discs are then cut out of the treated portion of the leaf, frozen in liquid nitrogen and the radioactivity subsequently determined. The method described here is that of Sondahl *et al.* ([54]).

Apparatus

The apparatus is essentially as in Fig. 13.1. The leaf chamber is made in two halves from 7–8 mm thick plexiglass, each half fitted to the arms of a pair of surgical scissors to form a chamber about 6 mm in diameter and about 4 mm high. The size and shape of the chamber can be altered to suit any particular purpose. The gas flow is split by means of Y tubes so that $^{14}CO_2$ gas mixture is applied to each side of the leaf, the excess being absorbed in sodium hydroxide. A stock cylinder is filled with $^{14}CO_2$ gas mixture either by means of a compressor and CO_2 absorbing tower or from a pressurized cylinder of CO_2-free gas. The "working" gas is then obtained by filling the steel sub-stock cylinder at a pressure of 1 atm (10^5 Pa), this being connected through a valve to a reinforced glass decompression cylinder. Both cylinders have a volume of about 265 cm^3. Except for the stock cylinder the apparatus may be placed in a portable case for field use.

Procedure

Preliminary work should define the optimum time of day for $^{14}CO_2$ incorporation experiments, by following daily changes in stomatal aperture and transpiration resistance by means of porometer studies. Measurement is carried out by means of releasing $^{14}CO_2$ gas mixture into the decompression cylinder until a pressure of 1.6

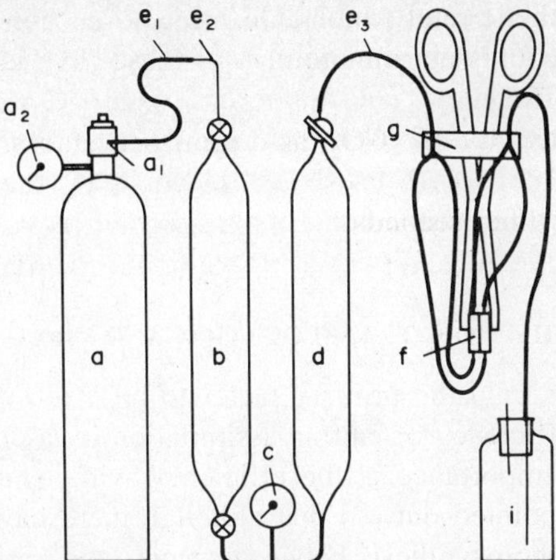

FIG. 13.1 Modified Shimshi apparatus: (a) stock cylinder, (b) substock cylinder, (c) sphingomanometer, (d) decompression cylinder, (e) tubing, e_1 is a 1.0 cm diam. copper pipe, e_2 is a 1.0 cm diam. plastic tube, and e_3 is a 0.1 cm diam. and 2.0 m long plastic capillary, (f) plexiglass incorporation chamber, (g) surgery scissors, (h) plexiglass flux divider, (i) flask of NaOH absorber ([54]).

$\times$ 10 Pa is recorded. The leaf chamber is clamped with light pressure to the mid part of a typical leaf, avoiding primary and large secondary veins. The valve in Fig. 13.1 is opened to allow the pressure to drop from 1.6×10^4 to 0.8×10^4 Pa as indicated by the sphingomanometer. For the particular system of Sondahl *et al.* ([54]) this pressure drop occurred in 20 sec, corresponding to a gas volume of about 26 ml flowing at a velocity of 280 cm min^{-1}, equivalent to 755 leaf chamber volumes per minute. The advantage of the short exposure time is that with a short period of incorporation there is less likelihood of evolution of recently fixed $^{14}CO_2$, as there are reports of $^{14}CO_2$ being evolved after times of between 30 and 120 sec ([26,34]).

A cork borer is used to cut 1.0 cm diameter leaf discs from the treated area, frozen in liquid nitrogen and transferred to test tubes kept in solid CO_2 + ethanol until combusted. Combustion may be carried out in any of the biological oxidizers available for liquid scintillation sample preparation, the $^{14}CO_2$ absorbed in appropriate reagent plus scintillation mixture as discussed in Chapter 5. Two replicate leaf discs should be taken from each leaf.

A quite different and simplified technique of sample preparation has been described by Antozewski ([67]) and reported to be effective. After 1 min exposure, leaf discs are cut out, kept in a test tube in the light until arrival in the laboratory when they are placed between blotting paper and dried on a photographic drier. The dry uncreased discs are then counted by GM or semi-conductor counter, the data being corrected for self-absorbtion.

Whatever the method of sample counting the data are calculated to express photosynthetic rate as μmol CO_2 $dm^{-2}h^{-1}$, as follows:

$$PR = \frac{\text{c.p.m.} \times 3600}{\text{SpA}^* \times \varepsilon \times a \times t}$$

where PR = photosynthetic rate, μmol CO_2 dm^{-2} h^{-1}
c.p.m. = count in counts per minute
ε = counter efficiency
SpA^* = specific activity of $^{14}CO_2$, as d.p.m. per μmol CO_2
a = area, cm^2
t = exposure time, seconds

CARBON ASSIMILATION, TRANSLOCATION AND YIELD COMPONENTS

Photosynthetic rate is only the primary factor to be taken into account in consideration of the overall efficiency of carbon assimilation by a crop plant. Of equal, or probably even greater importance is the efficiency with which the assimilates are translocated. Stoy has pointed out ([55]) that although there have been many cases of differences in rates of photosynthesis between species and varieties ([2,14,25,56]) there is almost no clear-cut relationship found between photosynthetic activity and yield performance.

The use of ^{14}C to study the translocation of carbohydrates in relation to yield in cereals was pioneered by Birecka and Dakic-Wlodkowska ([7]), Stoy ([56]) and Lupton ([35]). Later work ([36,37,55]) simultaneously evaluated photosynthesis and translocation and made it possible to estimate the contribution of different plant parts to total photosynthesis.

Thus Lupton ([36,37]) sealed ears or upper internodes of wheat, still attached to the intact plant, into Perspex containers forming part of a closed circuit containing $^{14}CO_2$, with a circulation pump and gas scintillation cell connected to a ratemeter and recorder. The rate of photosynthesis could then be measured by monitoring the rate of fall of $^{14}CO_2$ in the circuit and the translocation pattern by measuring the proportion of total ^{14}C assimilated and later transported to the grain or other plant part. Differences in the pattern of translocation were found and the relative importance in determining grain yield was estimated by integrating over the ''grain filling period'' the product of the photosynthetic rate for each organ with the proportion of ^{14}C assimilated that was translocated to the grain.

A rather different pulse labelling approach was used by Stoy ([55,56,57]) to measure source and sink relationship and capacity. The procedure was to place plants or parts of plants in a $^{14}CO_2$ atmosphere for a relatively short period of time, e.g. 10 min, then some plants were immediately harvested while remaining ones were allowed to grow under the original conditions so that they were able to translocate the labelled photosynthetic products. These treatments were made at weekly intervals. At harvest the individual plant parts were separated and the radioactivity determined by scintillation counting after combustion. Thus the pattern and amount of $^{14}CO_2$ absorbtion (i.e. source capacity) could be followed. Similarly, $^{14}CO_2$ was given to individual upper leaves of wheat at different stages of plant development after ear emergence, and the distribution of ^{14}C determined one week after treatment. This type of approach was able to demonstrate that individual leaves (sources) supply different sinks with varying proportions of the total exported assimilates, and moreover, that the sink capacity of an individual plant part can vary substantially at different development stages.

The contribution of different parts of a plant canopy to total assimilation was studied by Austin *et al.*, ([4,5]) using a combination of CO_2 infra-red gas analysis (IRGA) and $^{14}CO_2$. A 0.4 m^2 area of crop was enclosed in a transparent container and the CO_2 content of the air was monitored by IRGA before and after passing though the container, to provide an estimate of the nett CO_2 exchange. The contribution of different parts of the canopy was determined periodically by introducing $^{14}CO_2$ into the circuit and estimating the ^{14}C content of the different plant parts immediately following treatment. It was possible to demonstrate that the nett assimilation rate of a wheat variety with erect leaves was greater than one with prostrate leaves, due to increased carbon assimilation by the lower leaves of the more erect variety. The prostrate leaved variety compensated by greater translocation from stems and leaf sheaths.

Following earlier typical investigations on e.g. soybean ([11,60]) and sugarcane ([24,46]), a feature of more recent ^{14}C investigations of assimilation and translocation patterns has been relating them to yield, environment, variety and growth stage. Typical studies have covered a wide spectrum of crop species, including contrasting yield varieties in wheat ([51]), the strawberry ([1]), orange ([33]), pecan ([15]), rice ([12]), grasses ([38]), and sunflower ([10]).

Such yield-oriented assimilate distribution studies are now firmly based on radiotracer confirmation of the classical work on translocation of Mason and his co-workers in the 20s and 30s ([40,41,48]). Moreover, the basic source-sink relationship has been fully established. Webb and Gorham ([61,65]) demonstrated that in the squash plant *Curcubita* sp. there is a movement of carbohydrate in the phloem which is governed by the demands of growth. They followed ^{14}C through the plant and found a multi-directional pattern of movement, which is switched from leaf to leaf as each matures and no longer required carbohydrates assimilated by other leaves. Thus, lower leaves translocated downwards, upper leaves upwards but nodes in the mid-stem region could be involved in upward or downward movement according to growth stage. It has indeed been shown ([44]) by autoradiography that the direction of phloem transport even within a single leaf can be altered according to the relative position of source and sink.

Hale and Weaver had earlier found ([22]) in the vine, *Vitis vinifera*, that if they took an autoradiograph six hours following exposure of a young leaf to $^{14}CO_2$ then the ^{14}C remained in the treated leaf. If on the other hand an older leaf was treated then the autoradiograph showed ^{14}C both in the treated leaf and in the younger leaves above it. This showed that young leaves initially need all self-produced photosynthate for their own development, but as they mature they are able to export carbohydrates to even younger leaves.

The lateral movement of assimilates from a leaf into and across the stem was studied ([46,47,64]) by exposing the leaf to $^{14}CO_2$ and then detecting the radioactivity of honeydew from an aphis colony feeding on the opposite side of the stem. As aphis feed from the phloem cells which transport the assimilates from the leaf, it was thus possible to obtain a fairly precise indication of the cells in which labelled photosynthetic products were present.

Translocation studies were an essential and prominent part of the early work on herbicides. Crafts and his co-workers carried out a classical series of autoradiographic studies involving ^{14}C-labelled herbicides ([13,28]).

Experimental

Experimental set-ups for photosynthesis studies are of course individual to the experiment and the investigation but there are certain general features. The $^{14}CO_2$ is prepared by the action of dilute acid on solid $Ba^{14}CO_3$ or $Na_2{}^{14}CO_3$ aqueous solution, the former usually being preferred. Activities are normally 20–50 mCi/mmole, but specific activities greater than 50 mCi/mmole can be obtained and may be useful for larger gas flow systems. $^{14}CO_2$ may also be purchased directly in gaseous form in 1 mCi quantities in "break-seal" ampoules with a specific activity greater than 40 mCi/mmole, but most workers prefer the flexibility of preparation at the time of use.

Individual leaves or fruits may be contained, while on the plant, in small perspex chambers or in polyethylene bags for short term experiments. The $^{14}CO_2$ may be generated *in situ* by using a hypodermic syringe to place acid on the carbonate, or the small chamber may be part of a larger continuous flow system, the latter being preferable for ensuring repeatability of conditions. For larger perspex chambers containing whole plants a continuous flow system is almost essential to ensure adequate mixing of the gas and to avoid "stratification". A closed system is usually used and circulation may be done by means of a simple aquarium pump, and provisions should be made for monitoring the gas flow, either by a flow meter or by a simple bubbler device. A flask fitted with serum stopper, or other device to permit introduction of acid to the ^{14}C-carbonate, will be part of the flow system.

Whatever the system it should be designed and used so that neither $^{14}CO_2$ or light will be limiting factors. In the case of larger chambers and longer experiments it is almost essential to have provision for cooling, if plant conditions are to be both repeatable and approximating to free-growing conditions. Ensuring repeatability of conditions and treatment is particularly important in studies involving possible varietal differences.

The activity required for short pulse-labelling experiments is very substantially higher than that appropriate for long term uniform ^{14}C labelling of plants for organic matter studies (Chapter 11), where the possibility of radiation damage must be taken into account. For example, 250 μc of ^{14}C might be an appropriate level of activity for containers up to 5 liters with a CO_2 concentration of 0.03–0.1% (approx. 0.1 mc/gC for 0.1% CO_2) for short experiments of several to 24 hours, and it could be necessary to have an activity as high as 1 mc ^{14}C in the same volume of gas mixture for a short pulse labelling period of 10–15 minutes, if subsequent biochemical analysis was to be carried out, or a long post-treatment growth period was anticipated. Frequently, higher concentrations of CO_2 are used than the atmospheric 0.03% by volume

(320 p.p.m.), because photosynthesis by C_3 plants is often doubled when CO_2 concentration is increased from 320 to 600 or 1000 p.p.m. ([18]).

Known amounts of $^{14}CO_2$ may be prepared by taking appropriate calculated amounts of reagents using the data: density of CO_2 is 1.9768 g/liter, the conversion factors for $BaCO_3$ to CO_2 is 0.2229, for CO_2 to $BaCO_3$ is 4.4847, and for CO_2 to C is 0.2728.

Carbon-11, with a half-life of only 20.4 min has had comparatively little use. However, it was used by Moorby in translocation studies directly following preparation in the cyclotron ([43]). The advantage of ^{11}C is that after the activity of the $^{11}CO_2$ treated plants has been allowed to decay it is possible to use the same plants again under different experimental conditions, thus avoiding inherent plant variability.

More recent work ([157]) has used ^{11}C to study the effects of short term freezing, chilling, shading and water stress on the dynamics of translocation in the moonflower, *Ipomoea alba*, results being consistent with mass flow theory. ^{11}C was supplied in 10 minute pulses, and the activity in the stem followed by a linear array of scintillation detectors. 50-second counts were followed by a 10-second print-out, with the printer connected to an on-line computer which permitted either storing of the data for future processing, or displayed in real time on a video terminal to assist experiment control.

PLANT PATHOLOGY

There has been limited use of tracer techniques in plant pathology and host parasite relations. IAEA Technical Report no. 66 ([30]) and the review by Mendgen ([42]) provide some insight into the possibilities. Probably a majority of studies have been on the effect of the parasite on transport and distribution of solutes in the infected plant ([20,21,27,52]), but a growing number of reports have been concerned with the infection process and the morphological development of the parasite in the host, primarily using ^{14}C, ^{3}H or ^{35}S in metabolites such as ^{3}H uridine or ^{14}C orotic acid ([16,17,42,45,59]). Both light and electron microscope autoradiography have proved to be useful techniques.

MEASUREMENT OF LEAF WATER STATUS BY β-RAY GAUGING

Mederski ([81]) first used the attenuation of β-rays as a means of determining the moisture content of plant leaves, and the method has since been investigated or used by a number of workers ([66,77,78,83,85,86,87,89]). The principle is basically simple: a low activity source of β-rays is placed on one side of a leaf and a detector on the other. Now, radiation is attenuated by the leaf, but as the basic composition of the leaf is constant in the short term, any variation in attenuation should be due to variation in water content. The preparation of a calibration curve, essentially by plotting attenuation against relative leaf water content as determined by drying and weighing detached leaves, should make it possible to determine non-destructively water content of leaves actually growing on the plant. In practice there are a number of complications.

Theory

The theory of γ-ray attenuation has been considered in Chapter 14 and that for β-radiation through a leaf is not significantly different. Taking equation (1) (Chapter 14) we have

$$I = I_o e^{-\mu' T} \tag{1}$$

Where I is the intensity (count rate) of the radiation beam I_o after transmission through a thickness T, and μ' is the linear absorption coefficient (cm^{-1}) of the matter.

Therefore, for radiation passing through a leaf equation (1) can be re-stated as:

$$I = I_o \exp[-T(\mu_o \rho_o + \mu_w \rho_w)] \tag{2}$$

where, μ_o = mass absorption coefficient (cm^2g^{-1}) of the leaf organic matter
μ_w = mass absorption coefficient (cm^2g^{-1}) of leaf water
ρ_o = density ($g\ cm^{-3}$) of leaf organic matter
ρ_w = density ($g\ cm^{-3}$) of water
T = thickness of leaf (cm)

Unfortunately, leaf thickness is not constant as it is altered by variation in turgidity, therefore equation (2) must be developed to account for this:

$$I = I_o \exp[-(\mu_o x_o + \mu_w x_w)] \tag{3}$$

where, x_o = mass/unit leaf area of organic matter ($mg\ cm^{-2}$)
x_w = mass/unit leaf area of water ($mg\ cm^{-2}$)

Now, the relative water content of a leaf is

$$\text{RWC} = \frac{\text{fresh weight leaf} - \text{dry weight leaf}}{\text{turgid weight leaf} - \text{dry weight leaf}}$$

Clearly the greater portion of the weight of a leaf is due to water, the dry matter content being relatively small and constant. So, if x'_w *is the mass per unit leaf area at turgidity, then*

$$\text{RWC} = 1 - \frac{x_w}{x'_w} \tag{4}$$

but I, the β-radiation passing through the leaf is inversely proportional to the fresh weight per unit leaf area, therefore, I must depend mainly on the value of x_w as x_o is a virtual constant.

It follows that for leaves of a defined species, a plot of RWC against $\ln\left(\frac{I}{I_0}\right)$ should give a calibration curve for estimating leaf water status (Fig. 13.2).

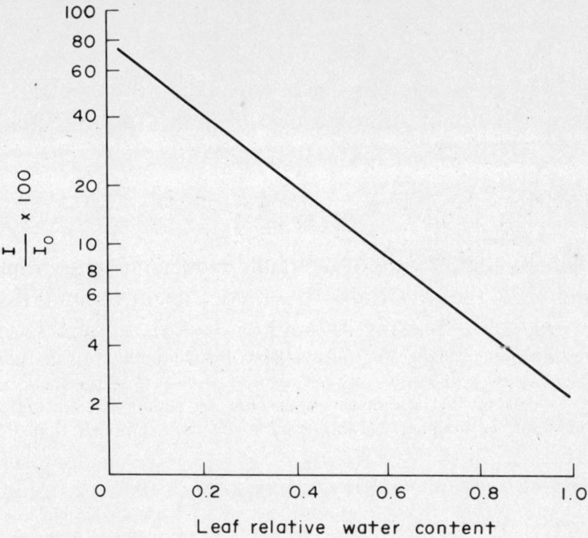

FIG. 13.2 Calibration curve for estimating leaf water status, where I is the intensity (count rate) after transmission through the leaf of the radiation beam I_o.

Experimental

A device has to be constructed and arranged whereby the radioactive source is held in constant geometry to the detector, one on top of the other about 0.3–0.4 cm apart, with the leaf in between them, and so that the leaf is not damaged. One answer ([67]), particularly for field or glasshouse *in situ* determinations, is to attach the detector and source to opposite arms of laboratory clamps or surgical scissors (Fig. 13.3).

Mostly thin window G-M detectors have been used, but with the greater availability of semi-conductor detectors this would now be the detector of choice because of its compactness, low background count, and sensitivity to soft β-radiation from e.g. ^{14}C and ^{45}Ca. A β source of about 5–10 μCi is required, and often ^{14}C has been used ([67, 81, 87]), but ^{99}Tc ([71, 82]), ^{147}Pm ([68, 77, 83, 86, 87]), ^{85}Kr ([70]), ^{204}Tl, and ^{32}P ([67]) have also been employed. ^{55}Fe, an emitter of soft X-radiation, has been tried experimentally ([66]). De Jong ([78]) made a study of the characteristics of β-radiation from suitable sources. In general, higher activity and a stronger β-emitter is more suitable for thicker leaves. Thus while ^{14}C is appropriate for the leaves of 5–20 cm^{-2}, ^{99}Tc and ^{147}Pm are better for thicker leaves from 15–40 mg cm^{-2}. Above this thickness ^{204}Tl or ^{32}P, the latter being somewhat unsatisfactory due to its comparatively short half-life. For any source the isotope should be mounted on plexiglass, covering a circular area about 2 cm in diameter. It should be covered with a thin mica cover.

The device must then be calibrated. First determine I_0 by counting several times without any leaf between source and detector. Repeat this after leaf measurements have been made and take the mean value of all counts to establish the value for I_0.

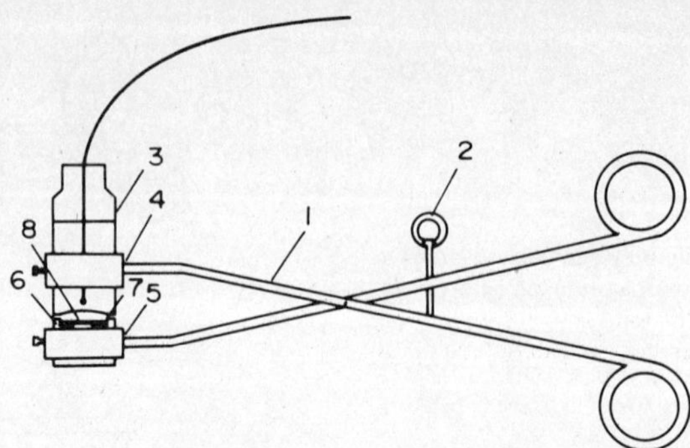

FIG. 13.3 Device for estimating the changes in water content of leaves on intact plants under field conditions. (1) laboratory clamp, (2) screw regulating constant position of G-M tube and beta-ray source, (3) G-M counter or semi-conductor detector, (4) holding ring with screw, (5) support for a metaplex disc, (6) metaplex disc on which the beta-source is mounted, (7) thin mica cover, (8) dry preparation of barium carbonate - ^{14}C, (67).

Prepare a test plant beforehand by watering copiously. Later, remove leaves and place them in a shallow dish of distilled water for 4–6 hours, under sufficiently high light intensity to ensure full opening of the stomata. Then remove replicated leaves from the water and carefully blot the excess water from the leaves with tissue. In turn, place the leaves in the radiation device and count for an appropriate period, followed by weighing each leaf. At this initial weighing RWC = 1, and the fresh weight equals the turgid weight.

Allow the leaves to dry on the bench, periodically weighing each leaf and also taking counts of I with the leaf in the apparatus. Continue weighing and counting until leaf RWC is estimated to be about 0.2–0.25, then place the leaves between blotting paper and dry flat in the oven at 100°C for 60 min. Take the dry weight, and determine I for the dry leaves. Calculate RWC for all weighings and plot against $\ln \frac{I}{I_0}$. This calibration curve can then be used for estimation of RWC of leaves on intact plants.

Limitations

Unfortunately, although the method can be very sensitive to water content change it has severe limitations and many workers have found it disappointing. The main problem is that the thickness of leaves varies with species, age, and also position on the plant (76,77,82,89). Therefore, the calibration curve is only effective for leaves of corresponding maturity. Another more fundamental problem is that not only does the thickness of leaf vary with turgidity but the area of leaf exposed beneath the source also varies with drying and age (77,82,84), and there seems no adequate way of correcting for this. However, some workers have achieved good correlations as for example with citrus (68).

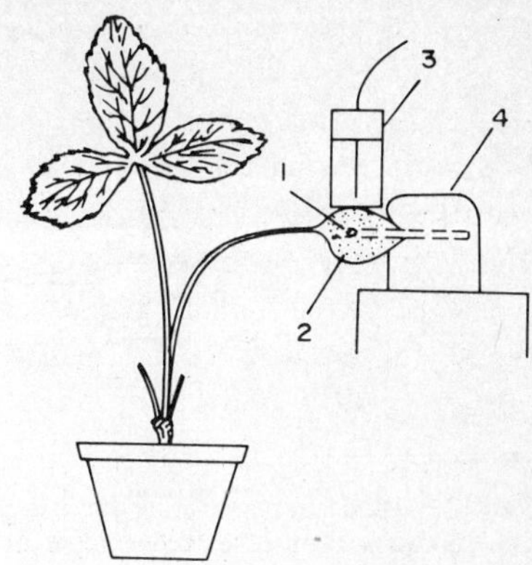

FIG. 13.4 Set-up for studying water changes in fruit. (1) Capillary tube with preparation of beta-emitter $Na_2H^{32}PO_4$, (2) fruit with radioactive capillary inserted, (3) G-M-counter, (4) plasticine block to maintain constant geometry ([67]).

The β-gauging technique of determining leaf water status seems best suited to use in rather carefully defined conditions, for which leaves have been specifically calibrated, as for examples in studies to determine the effect of changes in the plant environment. It seems unsuited to general field use, though making determinations in leaves of the same physiological age, as defined by leaf emergence number, would clearly help to reduce variability. The big advantage of the method is that it is rapid, and also permits a series of determinations to be carried out on a single attached leaf.

An attempt ([66]) was made to overcome some of these problems by using two radiation beams, one of ^{14}C and the other ^{55}Fe, to determine the mass of dry matter and the mass of water simultaneously. The method as used was less precise than determining relative water content of leaves by weighing, though theoretically it reduces the need for such frequent recalibration and destructive sampling.

The beta attenuation method has been used on plant organs other than leaves. Thus Antoszewski ([67]) used it to study water changes in the developing strawberry fruit, with a capillary tube loaded with $Na_2H^{32}PO_4$ being inserted in the centre of the fruit as a radiation source (Fig. 13.4).

ROOT DEVELOPMENT AND ACTIVITY AS A FUNCTION OF SOIL MOISTURE DEPLETION

Of recent years there has been increasing interest in obtaining information on the rooting patterns of different crop varieties, with a view to possibly selecting deep rooting genotypes for potential drought resistance, or for other reasons. Isotopic tracer methods for determining active rooting patterns are discussed at length in Chapter 12 and, while effective, are extremely time consuming for examining more than a limited range of material. Moreover, destructive sampling is involved.

To overcome the difficulties a number of workers have used soil moisture depletion as measured by the neutron moisture meter (Chapter 14) to determine the active rooting depth and water uptake ([69,73,74,75,80]). This makes use of the fact that as water is withdrawn by the plants a distinct moisture depletion zone is created as illustrated in

Fig. 14.12. To apply this method it is necessary to install access tubes for the neutron probe in the centre of each experimental plot, preferably at seeding, to avoid later disturbance to the plants.

Lupton ([80]) was able to show that certain wheat varieties e.g. Pitic 62 exhaust the reserves of soil water more rapidly than others, and to relate this to the known poor yield of these varieties in years of low early summer rainfall. Similarly, Haahr ([74]) was able to use the method to show differences in soil water uptake between some drought-tolerant and drought-susceptible varieties of barley, oats and wheat. He also compared a ^{32}P method for determining root activity with the neutron moisture method and found the latter was the most promising. Carlsberg II barley was shown to take up a larger amount of water from deeper soil layers than a mutant which apparently had a much shallower root system.

DETERMINATION OF PLANT DENSITY OR BIOMASS

The Hannover group have developed methods for the determination by beta or gamma absorbtion of the fresh weight of a single plant ([72,90]) and also of the plant density or biomass within experimental plots. Thus it was found possible to follow by continuous measurement the diurnal fluctuation in weight of a single cabbage plant ([90]).

For determining biomass an apparatus has been developed (Fig. 13.5(a)) which comprises a scanning/counting device recording the absorbtion of a gamma source in a strongly collimated beam ([79]), followed by appropriate computation. The source is present on one side of the experimental plot and the counter opposite. Tests with model plants indicated a 4% maximum error for mass determination and that discrimination between culm and spike mass was possible. Work progressed to measuring the biomass of wheat growing in actual field plots ([158]). The scanning device is mounted on rails and driven automatically by synchronous motors, with the scintillation detector on one side of the plot and the 0.2 mCi ^{241}Am source on the other. Different scanning programmes are possible, and during scanning the counts are recorded by a multichannel analyser, the data output being then transferred to punched tape. It has been found that the accuracy of mass determination is better than 3% (Fig. 13.5(b)), and that weighed crop yields agree with measured values determined directly before harvest.

The method could also be used to determine water uptake by the biomass of a plot during a single day, because although one cannot distinguish between dry matter and water, the daily fluctuation in water content may be separated from the real normal growth on which it is superimposed.

Other radiation applications, such as the β-absorption Dew Meter and the application of radiography to the detection of disease in trees have recently been reviewed (Kühn) ([159]).

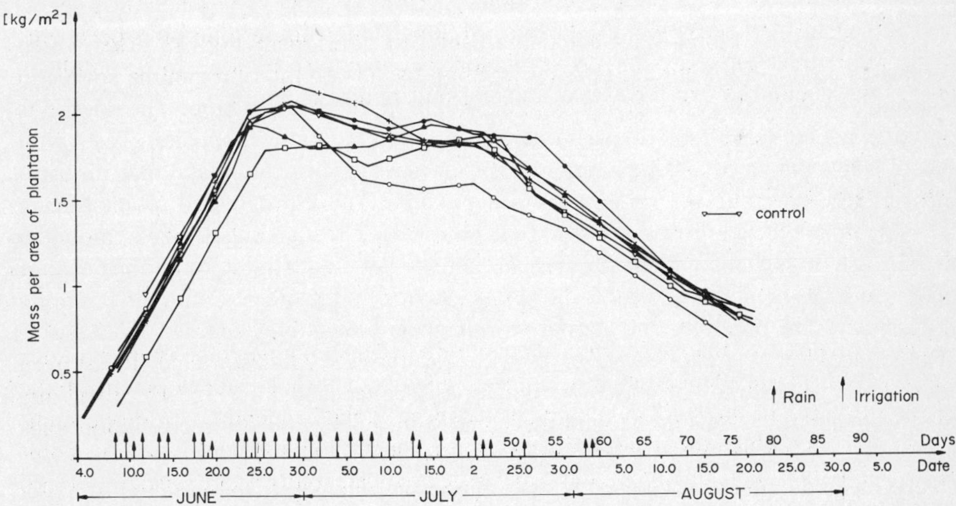

(b)

FIG. 13.5(a) Determination of biomass of experimental field plots of wheat. The total mass of the field plot is measured by gamma-ray absorbtion in a scanning device. Distribution between spikes and culms is possible.

FIG. 13.5(b) shows growth curves of seven field plots, with accuracy better than 3% ([158]).

ION UPTAKE, DISTRIBUTION AND PLANT NUTRITION STUDIES

Isotope techniques have found little place in what might be called the long term aspects of plant nutrition studies, such as the investigation of the essentiality of elements, the determination of specific requirements and levels of elements for optimum growth, and the determination of optimum plant nutrient content. But tracer methods have proved indispensable in research on ion uptake by whole plants and plant parts, and in studies on nutrient ion translocation and redistribution, and in the determination of turnover and labile ion pools.

Ion uptake studies of the macroelements have extensively used ^{32}P, ^{42}K, ^{86}Rb, ^{28}Mg, ^{22}Na, ^{45}Ca, ^{89}Sr, ^{90}Sr, ^{15}N and ^{35}S. Of the microelements, ^{55}Fe, ^{59}Fe, ^{54}Mn, ^{60}Co, ^{99}Mo and ^{65}Zn are the most feasible. Unfortunately ^{26}Al is only available in very restricted amounts and ^{64}Cu has a short $t_{1/2}$ of only 12.9 hr.

With most ion uptake experiments a fairly high level of activity is usually used, to permit reasonable count rates from material exposed to short uptake times. For short term experiments with excised roots or small plants about 10 μCi per litre is an appropriate level. Much higher levels may be necessary, thus 5 μCi ^{54}Mn per 25 ml of uptake solution were used with leaf discs (144) and 50 μCi ^{55}Fe per litre in uptake experiments with excised rice roots over periods of only 5–20 minutes ([134]). However, if maintained over a long period such levels of activity would imply the risk of radiation damage to the plants. Moreover, the total amount of activity that would be present in a solution culture experiment of even quite modest size would represent too great a hazard to personnel involved unless exceptional safety measures were undertaken. Accordingly for longer term nutrition experiments about 1–2 μCi per litre seems an adequate level of activity in most cases. Although long culture solution experiments involving radioisotopes are not very common they may also be used for example in producing labelled forage for animal mineral nutrition research.

Ion Uptake Studies

Theory

The great progress that has been made in our knowledge of ion uptake characteristics and mechanisms of plant roots has largely come about through development of the carrier concept. In turn this development has been greatly indebted to isotope techniques, without which the measurement of ion uptake at low ion concentrations combined with short uptake times would have been substantially more difficult if not impractical.

Advanced initially by van den Honert ([143]) the concept lay largely dormant until developed by Epstein and Hagen ([104]), Hagen and Hopkins ([109]), Leggett and Epstein ([119]), Fried and Noggle ([106]) and others.

It was demonstrated ([104]) that the kinetics of a model visualizing active absorbtion of ions as the transport of ions across barriers by means of ion-binding compounds, are directly analogous to the Michaelis-Menten ([121]) theory of enzyme action, as in the equation:

$$E + S \underset{k_2}{\overset{k_1}{\rightleftharpoons}} C \underset{k_3}{\rightarrow} E + P \qquad (5)$$

where *E* is free enzyme, *S* is substrate, *C* the intermediate enzyme-substrate complex, *k* the rate constant for each reaction, and *P* the reaction product. For ion transport models, *C* represents the ion-carrier complex which can pass a barrier impermeable to free ions but, following transport, the ions are released from the carriers.

This may be written

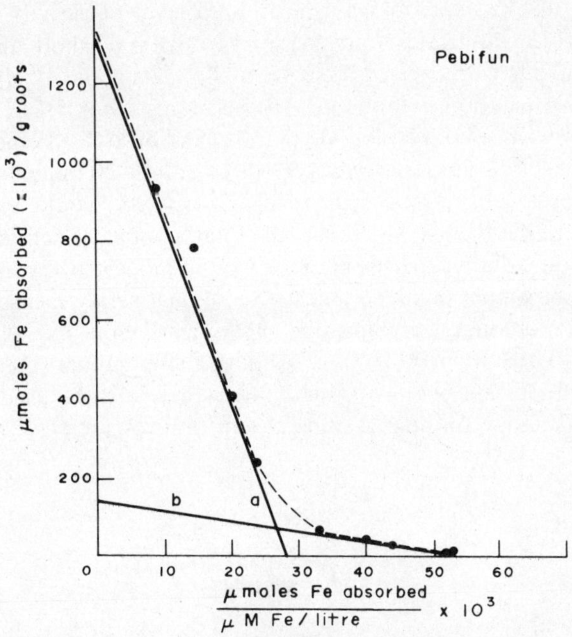

FIG. 13.6 Effect of solution concentration on uptake of Fe^{+++} by excised rice roots in a 20-minute period (broken line). The solid line resolves the uptake graphically into two linear components 'a' and 'b' typical of the Hofstee Plot III ([134]).

$$R + M_{\text{outside}} \underset{k_2}{\overset{k_1}{\rightleftharpoons}} MR \underset{k_3}{\rightarrow} R + M_{\text{inside}} \qquad (6)$$

where *R* is a metabolic carrier site, *M* the ion, *MR* the intermediate product in the root and k_3 is the rate constant for the final reaction.

Extensive references to such kinetic treatment are available ([105]), but we may note for example that it has been applied to all the macrocations ([104,106]) to phosphate ([109, 128]) using ^{32}P, to sulphate ([119]) to iron ([134]) using ^{55}Fe, and to nitrogen ([107]) with ^{15}N.

In a supposedly carrier-mediated uptake process, the specificity of the carrier, the rate of turnover of the carrier, the concentration of the carrier and the relative proportions of the carriers can be most readily described in a kinetic system.

If steady-state conditions are assumed, the observed absorbtion of an ion by the root, v, can be derived from equation (6) as

$$V_{\max} = v + \frac{v}{[M]} - K_m \tag{7}$$

$$v = -K_m \frac{v}{[M]} + V_{\max} \tag{8}$$

$$V_{\max} = k_3[\Sigma R]t \tag{9}$$

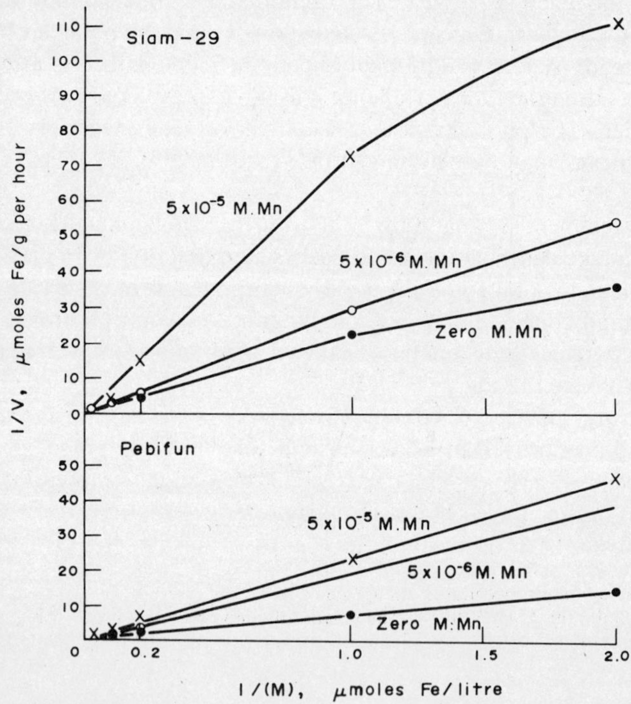

FIG. 13.7 Competitive inhibition of manganese on iron uptake by excised rice roots of varieties Pebifun and Siam-29. Illustrative of the Lineweaver-Burk double reciprocal plot ([134]).

where $V_{\max}$ represents maximum adsorbtion at infinite concentration and K_m is the apparent dissociation constant of the intermediate MR.

Typically, results may be plotted as a ''Hofstee plot III'' ([110]) when if v is plotted against $\frac{v}{M}$ a straight line will result if a first-order reaction is represented, the slope of this line being $-K_m$. If more than one reaction is involved then a curve results, which may be graphically resolved into linear components as shown in Fig. 13.6. Thus the observed absorbtion is the resultant of two first-order reactions acting simultaneously and independently, as shown by lines 'a' and 'b', while the absorbtion maxima, respectively $V_{\max a}$ and $V_{\max b}$, are given by the intercepts with the ordinate.

The Lineweaver and Burk ([120]) double reciprocal plot is often used for deducing specificity or competition for a carrier site, with rate of absorbtion v, plotted against solution concentration, $[M]$. Competitive inhibition by a second ion in solution is indicated by the slope of the line increasing but leaving the intercept unchanged (Fig. 13.7).

The rate of carrier turnover should be determined by the apparent dissociation constant, K_m, of the intermediate, and by k_3, the rate constant for the second reaction, provided that the breakdown of MR is irreversible. K_{ma} and K_{mb} may be calculated by substituting in equation (8) or by calculating the slope of lines 'a' and 'b'. k_3 can be determined from a time-course curve exhibiting steady-state uptake at a concentration appropriate to the reaction, so that each reaction at that concentration accounts for the major part of the uptake. A general statement of k_3 may be obtained by dividing the slope of the steady-state curve by the ordinate intercept, which is MR. If a high ion concentration is used then the calculation must take into account the presence of more than one carrier. The concentration, ΣR, of the respective carriers and hence their relative proportions may be determined by substituting k_3, $V_{\max}$, and t into equation (9).

The above treatment is what may be termed the conventional kinetic approach to handling ion uptake data, but other approaches are possible ([92]); in particular the curve typical of the Hofstee plot III ([110]) is usually interpreted as representing two reactions, one active at high concentrations and the other at low concentrations, but it may not only represent two reactions, but may be the resultant of a number of reactions specific at high or low concentrations.

Although it may not be yet possible to reconcile all aspects of the carrier concept, kinetic analysis has proved to be very useful for developing models of ion uptake, providing an overall statement of ion uptake, and enabling the expression of differences in uptake both between species and varieties.

If such analysis is to be meaningful it is essential that the experimental isotope procedures developed for ion uptake experiments with excised roots and seedlings are precise, standardized, repeatable, and carried out in identical manner each time. Moreover, the method used must do the least damage possible to the experimental material.

Experimental

Growth of material for excised root studies: The seeds for example of barley or rice are surface sterilized with 10% hypochlorite solution for 5 minutes or 0.2 per cent mercuric chloride for 2 minutes, and washed thoroughly with several changes of deionized water. The seeds are distributed on nylon-mesh or stainless steel mesh screen in frames about 30 cm square, supported at the surface of the aerated nutrient solution. Usually "low-salt" material is required, so after germination has commenced in deionized water, the water is replaced by dilute nutrient solution such as 1/20 Hoagland or a simple calcium sulphate solution of 2×10^{-4} M $CaSO_4$. In general the latter is preferable. Adequate calcium nutrition is now regarded as essential for good ion uptake. The seedings should be grown until there is sufficient root growth, i.e. about 10 days, under reasonably controlled and repeatable conditions of light and temperature (about 20°C). When required the roots are then carefully excised with scissors and placed in a large volume of aerated deionized water overnight.

Uptake procedure: It is essential to develop a routine procedure that allows good timing with adequate washing of roots following the uptake period. Any method should also employ a very large volume of uptake solution with quite a small amount of roots, say 0.5–1.0 g roots in 500 ml, so that for practical purposes the concentration of the solution changes very little during the course of the experiment.

Every experiment will naturally vary in objective and detail, but as an illustration of the technique the following experiment is described to determine the effect of complementary cations on the uptake of rubidium by excised barley roots. This has proven to be an instructive and popular class experiment ([147]). Samples of excised barley roots are placed in a labelled solution of 10^{-3} M rubidium, which also contains another cation (10^{-3} M Rb, K, Na, NH_4, Ca). The roots are removed from solution at different time intervals of 2, 5, 15, 30, 60 and 90 minutes. The roots are washed, dried and counted, and the amount of rubidium taken up is initially expressed as μ moles/g dry weight roots.

Material required for each replicate of each time course run: 6 beakers of 600 ml fitted with aerators; 6 Erlenmeyer flasks of 500 ml marked with the volume; 1 large beaker to contain rinsing water; 1 large container for water; 6 glass scintillation counting vials; 8 (about 0.5–1.0 g) lots of excised roots; appropriate stock solutions. The fresh roots are handled and damaged as little as possible. For this reason only take approximate fresh weights by estimation, and determine the dry weight after the experiment. Do not blot the roots as this damages the root hairs, but only drain them dry.

Uptake "run"

1. Place an aliquot of roots in about 200 ml of deionized water in each of the six beakers. Adjust the aeration.
2. Using a glass rod to hold back the roots, carefully pour off the water from the roots. Leave the roots sticking to the sides of the beaker, and allow the beaker to remain upside down on the work tray while preparations are completed. In this way the roots drain but do not dry out.
3. Put approximately 250 ml of deionized water into each 500 ml Erlenmeyer flask.
4. Add 5 ml of 0.1 M $RbCl_2$ stock solution to each flask.
5. Add 5 ml of 5 μCi/ml ^{86}Rb stock solution to each flask.
6. Add 5 ml of 0.1 M complementary ion (either Rb, K, Na, NH_4 or Ca).

7. Make up all the flasks to the 500 ml mark. Do not shake because of possible spilling. Adding the additional water plus pouring out the solution ensures adequate mixing.
8. Turn over the beakers containing the drained roots and insert the aerators.
9. Carefully noting the time or starting a stop-clock, pour the radioactive solutions in turn into each beaker.
10. *Standards*: Place two aliquots of root samples in separate vials, drain off the water and press the roots to the bottom. Pipette 1 ml of labelled solution, taken from the beakers for times 5 minutes and 15 minutes, into the middle of the roots in each vial. Avoid running the solution down the side of the glass. Each 1 ml contains 1 μmole of rubidium. Out of 500 ml of solution, this withdrawal is negligible. Place the vials on one side till the end of the experiment.
11. At the end of 2 minutes the first roots are harvested. Using a glass rod to hold back the roots quickly decant the active solution into the waste container. Then rapidly add about 500 ml water, swirl the roots and again quickly decant.
12. Repeat 3 times, that is a total of 4 washes of water.
13. Carefully take the roots from the beaker (rubber gloves!) and place in the bottom of a weighed counting vial.
14. Place in the oven to dry.
15. Repeat the process at times of 5, 15, 30, 60 and 90 minutes. At the end of the experiment dry the standards too.
16. Weigh counting vial + roots and determine weight of roots.
17. Count all vials by Cerenkov counting.
18. Calculate from the standards the μmoles of rubidium in each sample. Express as μmoles per gram dry or fresh weight roots.

This primary experiment may be repeated for the other complementary ions, or these can be carried out by different groups in a class experiment. Subsequent experiments may include a range of concentrations, say 2×10^{-3} to 10^{-6} M, of both basic and complementary ions. Temperature and inhibitors may also be used as additional variables, and replication must also be taken into account. Experiments such as this provide a basic technique which may be applied to a wide variety of ion uptake situations with seedlings, excised roots, leaf and storage tissue. The results are analysed kinetically in the manner that seems most appropriate for the specific objective.

Note, that although Cerenkov counting is very suitable for many radioisotopes employed in ion uptake studies it is not suitable for ^{45}Ca as the β-energy is too low and the self-absorbtion effect too great. In this case place the roots in metal planchets, ash in the furnace at about 450°C, and when cool add a drop of dilute nitric acid to dissolve and spread the ash. Standards must be treated in the same way. Count using a thin-window G-M tube. Other isotopes may also be counted in this way if liquid scintillation counting equipment is not available. Alternatively, with an energetic radioisotope, root samples may often be counted merely after pressing the roots into a planchet without the need for ashing.

When using a nearly carrier-free radioisotope it is usually not necessary to take into account the actual element content, but with more dilute solutions such as 10^{-6} and 10^{-7} M, this will be necessary. In experiments in which there is a comparison of different ion concentrations then it is usual practice to keep the radioisotope label at the same level of activity for each concentration employed. To retain a constant specific activity is not practicable with such a wide concentration range, as at the highest concentrations the activity would be much too high.

Ion Absorbtion by Leaves, Transport, Redistribution and Metabolism

Foliar fertilization, particularly of nitrogen (urea), and microelements such as iron and copper has been of increasing practical relevance. Theoretically it has been shown that the pattern of ion uptake by leaf cells is not very different from roots, but with the added complication in intact leaves of leaf epidermis and cuticle, the stomatal apparatus, and photosynthesis. Nearly all the developments in the area have been connected with the use of radioisotopes.

Study of foliar application of nutrients has been particularly associated with Wittwer and Tukey and co-workers ([115,139,140,154,155,156]) in Michigan. Early ^{32}P studies showed that phosphorus was actively absorbed by leaves and translocated ([116,138]). It was also found, comparing ^{32}P uptake from the soil or from leaf application, that the utilization from leaf application was twenty times as efficient ([155]).

Ions can be expected to obtain passive access to the intercellular spaces of the leaf through the stomata and by slow penetration of the cuticle. Thereafter the ions can be absorbed actively by the mesophyll cells. The various factors affecting this process have been studied using whole leaves ([91,115]) but there has been a growing tendency to use parts of leaves to avoid the complication of the epidermal barrier. Either leaf discs cut with a cork borer ([144,146]) or leaf slices ([123,131,136]) have been used with or without vacuum infiltration in experiments covering a wide range of radioisotopes e.g. ^{54}Mn ([144]), ^{90}Sr and ^{32}P ([122,147]), ^{36}Cl ([122,123]), $^{86}Rb/K$ ([131,137]). Leaf discs provide the least disturbed material but the uptake is dependent on the diameter of the disc ([136]), though "margin effects" can largely be overcome by vacuum infiltration ([122,123,
Work with isotopes has shown that uptake is greatly influenced by light and carbohydrate status, but there are still many unknown factors.

Closely allied in function to leaf absorbtion studies has been the earlier work concerned with the relative mobility of ions in the plant, phloem transport from the leaf and ion redistribution. Much of this work was carried out by means of leaf application of radioisotope followed by autoradiography at different time intervals. Bukovac and Wittwer ([100]) were able to show in the bean plant that K, Rb, Na, Mg, P, S and Cl are very mobile, while Li, Ca, Sr, Ba and B were virtually phloem immobile and Fe, Mn, Zn, Cu and Mo were of an intermediate nature. Other isotope work has indicated that the heavy metals are in fact reasonably phloem mobile e.g. Zn ([150,153]), Mn ([133,142,145]), Fe ([98,103,150]), Pb ([150]) in most cases.

Ion mobility and redistribution studies are certainly not absolute and, for example, the use of ^{45}Ca has shown that under certain circumstances Ca may be redistributed, as in subterraneum clover *Trifolium subterraneum* ([124]) and oats *Avena sativa* ([132]) although translocation of calcium from leaves is basically poor ([96]). In the case of manganese-54, studies with oats showed ([145]) that substantial manganese might redistribute from old leaves to new, if the stress of actual deficiency in the new leaves was high enough.

Autoradiographic studies ([94]) using ^{35}S and ^{32}P showed that phosphorus and sulphur move out of the leaf via the phloem, and later work by Biddulph *et al.* ([95]) demonstrated that once phosphorus was in the leaf it was redistributed in a similar pattern to the movement of sugars. Leaf phosphorus moved upwards towards young developing leaves, or downward to the roots depending on the relative proximity of the sinks.

The study of manganese movement in cereals provides a nice illustration of the use and development of isotope methods and associated thinking. Single ([135]) used classical and very laborious analytical techniques with wheat, *Triticum* to show that there was little redistribution within the plant. Vose ([145]) later used ^{54}Mn, largely with autoradiographic techniques, to demonstrate that in oats leaf-applied manganese was preferentially distributed to developing leaves or grain, and that Mn was redistributed under conditions of deficiency. Munns *et al.* ([125,126,127]) were able by means of kinetic analysis of ion uptake with ^{54}Mn to show three pools of manganese in the roots, which they defined as a replaceable fraction (probably representing "Apparent Free Space" manganese), a non-labile fraction (conceivably vacuolar) and a third labile fraction (which might represent manganese in the cytoplasm).

Of the Mn-pools, the labile pool was considered to be the means of supplying manganese to the shoot, and this pool was also connected with all other manganese pathways including the substract. In contrast, the non-labile pool was considered to be in contact only with the replaceable and labile pools. It was found that varieties showed considerable variation in the size of the pools, particularly of the non-labile pool.

As a means of determining turnover of manganese in plants Dokiya *et al.* ([102]) used a double tracer technique with ^{54}Mn ($t_{1/2}$ = 300 days) and ^{56}Mn ($t_{1/2}$ = 2.6 hr). Seedlings were first grown in 0.1 p.p.m. ^{2}Mn solution labelled with ^{54}Mn for several days. Absorbtion experiments were then carried out using ^{56}Mn, and afterwards the plants were separated into root and shoot and counted by scintillation counter immediately (A), and also after the elapse of two days (B), following the complete decay of ^{56}Mn.

Comparative specific activity (CSA) was calculated as

$$\mathrm{CSA} = \frac{^{56}\mathrm{Mn}}{^{54}\mathrm{Mn}} = \frac{(A - B)^{e-\lambda t}}{B}$$

where t = time for correction of decay of ^{56}Mn (min)
λ = 0.0045

The CSA was calculated for absorbtion periods of one to eight hours and plotted as in Fig. 13.8. Then the transport of manganese was defined as

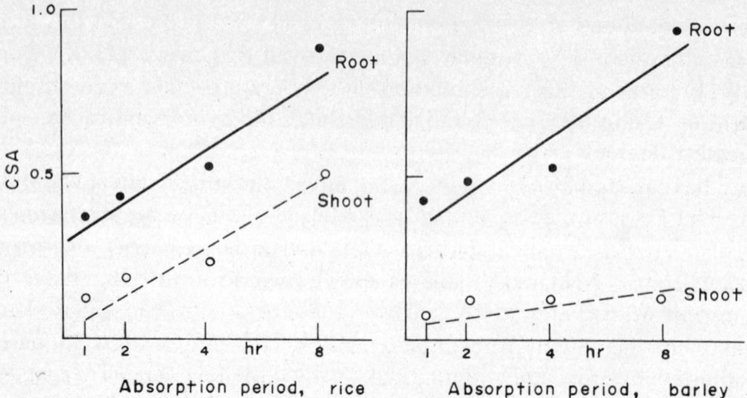

FIG. 13.8 Comparative Specific Activity (CSA) for absorption of ^{56}Mn by plants previously treated with ^{54}Mn. See text for details of the concept ([102]).

$$\text{transport rate} = \frac{d\,(\text{CSA of shoot})}{dt}$$

where *t* is the absorbtion time, and d is determined from the slope of the curve (shoot). It was thus possible to show that rice transports manganese to the shoots 5 times faster than barley.

Such a technique might be applied to the turnover of manganese in chloroplasts, mitochondria etc., or the principle could be used in other cases where isotopes of suitable half lives are available.

In a typical transport study Wallace *et al.* ([148]) were able to show that if beans *Phaseolus vulgaris* were kept in ^{22}Na labelled 1×10^{-4}M NaCl for 24 hours, although there had accumulated in the roots 2.28 μmoles Na/g dry wt, the concentration in the leaves amounted to only 0.01 μmoles Na/g dry wt. This clearly showed that sodium is poorly transported and is substantially retained in the root.

Isotopes have been widely used in nutrient assimilation and metabolism studies. Knowledge of ion incorporation and movement has been greatly advanced, particularly in the comparative metabolic studies involving Hawkeye (HA) Fe-efficient and PI-54619-5-1 (PI) Fe-inefficient varieties of soybean. A classical range of studies have been carried out by Brown and his co-workers, which are summarized in ([99]).

^{65}Zn and ^{59}Fe have been used to demonstrate that chelating agents such as EDTA or DTPA can greatly increase the uptake of these micronutrients from both solution and soil cultures ([151]).

^{32}P was given at intervals to vines grown in sand culture, and it was found that there were two seasonal uptake peaks. These were initially at bud-burst and again after the grapes had been harvested, with a period of reduced uptake in between ([130]). A comparison, using ^{32}P, of phosphorus uptake by ryegrass *Lolium temulentum*, showed that vegetative material had a greater rate of P uptake which declined more slowly with time than was the case with vernalized (reproductive) plants ([117]).

It was found that hybrid vigour in maize seedlings could be expressed in terms of the amount of $^{14}CO_2$ incorporated in a given time per unit weight of leaf. Characteristic differences were found from the point of view of ^{32}P uptake and the formation of organic P compounds ([108]). Similar studies showed that maize genotypes responded differently, in terms of $^{14}CO_2$ assimilation, to temperature, light intensity and nitrogen (^{15}N) nutrition. Using ^{32}P it was found that the heterotic hybrids assimilated phosphorus to the greatest degree ([112]).

Isotopes have played an essential role in many studies on genetical variation in plant nutrition. The Fe-assimilation studies in soybeans ([99]) have been referred to above. ^{55}Fe was also used to demonstrate basic variation in the efficiency of iron uptake by rice varieties ([134]). $^{15}NH_4$ uptake studies showed variation in NH_4 uptake by excised roots of *durum* wheat varieties. A radiation induced mutant had lesser efficiency for NH_4 uptake than the parent Cappelli, and this was reflected in straw N content. Bread wheat had a much lower ammonium uptake efficiency than *durum* wheats ([129]). ^{54}Mn uptake by oat leaf discs was related to variety ([144]).

The mechanisms of zinc tolerance was studied in a zinc-tolerant population of *Agnostis tenuis*, using ^{65}Zn. Autoradiographs showed that differential uptake of Zn was not the mechanism of tolerance. Biochemical analysis and ultracentrifugation indicated that Zn-tolerance is associated with the cell wall fraction ([141]).

^{15}N has made it possible to confirm that the initial phase of nitrogen uptake by plant roots conforms to the general ion uptake pattern of a dual carrier system ([107]). Moreover, other works demonstrated ([97,101,114]) that the initial incorporation of N into single amino acids over relatively short times (less than 60 minutes), confirming that glutamic acid is formed by direct incorporation of ammonia into α-ketoglutaric acid, with other amino acids formed from glutamic acid by transamination. Information on the general factors affecting ammonium and nitrate uptake and utilization by higher plants has also been greatly extended by the use of ^{15}N, e.g. ([93,114]). Reference 111 remains a good literature source.

As Ivankov ([113]) has pointed out, our knowledge of the metabolic pathways of nitrogen assimilation in plants has largely depended on either classical techniques or indirect working with ^{14}C or ^{35}S. The greater availability of the means for determining ^{15}N and the small amounts that can now be determined by emission spectrometer enables the confirmation of existing work by direct studies and opens up new possibilities.

REFERENCES FOR FURTHER READING

Carbon Assimilation, Yield Components, Translocation, Plant Pathology

1. Antoszewski, R. and Dzięciol, V. Translocation and accumulation of ^{14}C-photosynthates in the strawberry plant. *Hortic. Res.* **13**, 75 (1973).
2. Apel, P. Potentielle Photosynthesintensität von Gerstensorten des Gaterslebener Sortiments. *Kulturpflanze* **15**, 161 (1967).
3. Aronoff, S. *Techniques of Radiobiochemistry*. Iowa State College Press (1956).

4. AUSTIN, R. B. and Longden, P C. A rapid method for the measurement of rates of photosynthesis using $^{14}CO_2$. *Ann. Bot.* **31**, 245 (1969).
5. AUSTIN, R. B., Ford, M. A. and Edrich, J. A. *Photosynthesis, Translocation and Grain Filling in Wheat*. Rep. Pl. Breed. Inst. Cambridge, 1973, 159 (1974).
6. BEALE, D. and Whitehead, J. K. The determination of sub-microgram quantities of amino acids with ^{3}H and ^{14}C-labelled 1-fluoro-2, 4-dinitrobenzene. *In: Tritium in the Physical and Biological*. Proc. Symp. Vienna, 1961, 179, IAEA, Vienna (1962).
7. BIRECKA, H. and Dakic-Woldkowska, L. Photosynthesis, translocation and accumulation of assimilates in cereals during grain development III Spring wheat-photosynthesis and the daily accumulation of photosynthates in the grain. *Acta Soc. Bot. Pol.* **32**, 631 (1963).
8. BRODA, E. *Radioactive Isotopes in Biochemistry*. Van Nostrand (1960).
9. CALVIN, M., Heidelberger, C., Reid, J. C., Tolbert, B. M. and Yankwich, P. F. *Isotope Carbon, Techniques in its Measurement and Chemical Manipulation*. Wiley, New York (1949).
10. CHOPOWICK, R. E. and Forward, D. F. Translocation of radioactive carbon after the application of ^{14}C-alanine and $^{14}CO_2$ to sunflower leaves. *Plant Physiol.* **53**, 21 (1974).
11. CLAUSS, H. *et al. Plant Physiol.* **39**, 269–73 (1964).
12. COCK, J. H. and Yoshida, S. Accumulation of ^{14}C-labelled carbohydrate before flowering and its subsequent redistribution and respiration in the rice plant. *Nippon Sakumotsu Gakkai Kiji* **41**, 226 (1972).
13. CRAFTS, A. S. Translocation of Herbicides. *Hilgardia* **26**, 287–415 (1956).
14. CURTIS, P. E., Ogren, W. L. and Hageman, R. H. Varietal effects in soybean photosynthesis and photorespiration. *Crop Sci.* **9**, 323, (1969).
15. DAVIS, J. T. and Sparks, D. Assimilation and translocation patterns of carbon-14 in the shoot of fruiting pecan trees, *Canja illinoensis*. *J. Amer. Soc. Hort. Sci.* **99**, 468 (1974).
16. EHRLICH, M. A. and Ehrlich, H. G. Electron microscope radioautography of ^{14}C transfer from rust uredospores to wheat rust host cells. *Phytopath. Z.* **60**, 1850 (1970).
17. FAVALI, M. A. and Marte, M. Electron microscope autoradiography of rust-affected bean leaves labelled with tritiated glycine. *Phytopathology* **76**, 343 (1973).
18. GAASTRA, P. *Meded Landbouwhogesch.* Wageningen **59**, 1 (1959).
19. GALE, W. *Methods of Biochemical Analysis* **4**, 285 (1957).
20. GARRAWAY, M. O. and Pelletier, R. L. Distribution of ^{14}C in the potato plant in relation to leaf infection by *Phytophthora infestans*. *Phytopathology* **56**, 1184 (1966).
21. GAUNT, R. E. and Manners, J. C. Host-parasite relations in loose smut of wheat. II The distribution of ^{14}C labelled assimilates. *Ann. Bot.* **35**, 1141 (1971).
22. HALE, C. R. and Weaver, R. J. The effect of developmental stage on direction of translocation of photosynthate in *Vitis vinifera*. *Hilgardia* **33**, 89–131 (1962).
23. HARTT, Constance E. *et al. Plant Physiol.* **38**, 305–18 (1963).
24. HATCH, M. D. and Glasziou, K. T. *Plant Physiol.* **39**, 180–84 (1964).
25. HESKETH, J. D. and Moss, D. N. Variation in response of photosynthesis to light. *Crop Sci.* **3**, 107 (1963).
26. HEW, C. S. and Canvin, D. T. *Plant Physiol.* **44**, 671–77 (1969).
27. HOLUKOWICZ, T. and Fisher, W. *Translocation of ^{32}P in Raspberry Cores Infected with Spur Blight Didymella applanata*. Trans. 3rd Symp. on Accumulation and Translocation of Nutrients and Regulators, Warsaw, 14–18 May 1973, 437 (1973).
28. IAEA. *Isotopes in Weed Research*. Proc. Symp. 1965, IAEA, Vienna (1966).
29. IAEA. Fourth International Conference on the Peaceful Uses of Atomic Energy. Vol. 12, *Nuclear Methods in Food Production*. IAEA, Vienna (1972).
30. IAEA. *Isotopes and Radiation in Plant Pathology*. Tech. Rept. no. 66, IAEA, Vienna (1966).
31. IAEA. *Origin and Fate of Chemical Residues in Food, Agriculture and Fisheries*. Proc. Coordination Meeting, Vienna 1974. IAEA, Vienna (1975).
32. INCOLL, L. D. and Wright, W. H. A field technique for measuring photosynthesis using $^{14}CO_2$. *Spec. Bull. Soils Conn. Agric. Exp. Stn. XXX/100*, 1 (1969).
33. KADOYA, K. and Tanaka, H. Studies on the translocation of photosynthates in Satsuma orange. 1. Effect of summer cycle shoot and bearing fruit on the translocation and distribution of ^{14}C. *Engei Gakkai Zasshi* **41**, 23 (1972).
34. LUDWIG, L. J. and Canvin, D. T. *Can. J. Bot.* **49**, 1299–313 (1971).
35. LUPTON, F. G. H. Translocation of photosynthetic assimilates in wheat. *Ann. Appl. Biol.* **57**, 355 (1966).
36. LUPTON, F. G. H. The analysis of grain yield of wheat from measurements of photosynthesis and translocation in the field. *Ann. Appl. Biol.* **64**, 363 (1969).

37. LUPTON, F. G. H. Further experiments on photosynthesis and translocation in wheat. *Ann. Appl. Biol.* **71**, 69 (1972).
38. LUSH, W. M. and Evans, L. T. Translocation of photosynthetic assimilates from grass leaves, as influenced by environment and species. *Aust. J. Plant Physiol.* **1**, 433 (1974).
39. MCWILLIAM, J. R., Phillips, P. J. and Parkes, R. R. Measurement of photosynthesis using labelled carbon dioxide. *Tech. Pap. Div. Pl. Ind. CSIRO Aust.* **31**, 1 (1973).
40. MASON, T. G. and Maskell, E. J. Studies on the transport of carbohydrates in the cotton plant. II The factors determining the rate and the direction of movement of sugars. *Ann. Bot.* **42**, 571–636 (1928).
41. MASON, T. G., Maskell, E. J. and Phillis, E. Further studies on transport in the cotton plant. III Concerning the independence of solute movement in the phloem. *Ann. Bot.* **50**, 23–58 (1936).
42. MENDGEN, K. Tracer techniques for the study of host-parasite relations. *In: Tracer techniques for Plant Breeding*. IAEA, Vienna (1975).
43. MOORBY, J. *et al. J. Exp. Bot.* **14**, 210–20 (1963).
44. NAKATA, S. and Leopold, A. C. Radioautographic study of translocation in bean leaves. *Plant Physiol.* **54**, 769–72 (1967).
45. ONOE, T., Tani, T. and Naito, N. The uptake of labelled nucleosides by *Puccinia coronata* grown in susceptible oat leaves. *Rep. Tottori Mycol. Inst.* **10**, 303 (1973).
46. PEEL, A. J. and Weatherley, P. E. *Ann. Bot.* **26**, 633–46 (1962).
47. PEEL, A. J. *J. Expt. Bot.* **15**, 104–13 (1964).
48. PHILLIS, E. and Mason, T. G. Further studies on transport in the cotton plant. IV On the simultaneous movement of solutes in opposite directions through the phloem. *Ann. Bot.* **50**, 161–74 (1936).
49. RUBEN, S., Hassid, W. Z. and Kamen, M. D. *J. Amer. Chem. Soc.* **61**, 661 (1939) and **62**, 3443, 3450, 3451 (1940).
50. RUBEN, S., Randall, M., Kamen, M. D. and Hyde, J. L. *J. Amer. Chem. Soc.* **63**, 877 (1941).
51. RUCKENBAUER, P. *Yielding Ability and Translocation Pattern of $^{14}CO_2$ Labelled Assimilates in Contrasting Wheat Varieties*. Trans. 3rd Symp. on Accumulation and Translocation of Nutrients and Regulators, Warsaw, 14–18 May 1973, 129 (1973).
52. SHAW, M. and Samborski, D. J. The physiology of host parasite relations. 1. The accumulation of radioactive substance at infections of facultative and obligate parasites including tobacco mosaic virus. *Can. J. Bot.* **34**, 389 (1956).
53. SHIMSHI, D. A rapid method for measuring photosynthesis with labelled carbon dioxide. *J. Expt. Bot.* **20**, 381 (1969).
54. SONDAHL, M. R., Crocomo, O. J. and Sodek, L. Measurements of ^{14}C incorporation by illuminated intact leaves of coffee plants from gas mixtures containing $^{14}CO_2$. *J. Expt. Bot.* **27**, 1189–95 (1976).
55. STOY, V. Use of tracer techniques to study yield components in seed crops. *In: Tracer Techniques for Plant Breeding*. Proc. Panel, Vienna 1974, 43–55, IAEA, Vienna (1975).
56. STOY, V. Photosynthesis, respiration, and carbohydrate accumulation in spring wheat in relation to yield. *Physiol. Plant Suppl.* IV, 1 (1965).
57. STOY, V. The translocation of ^{14}C-labelled photosynthetic products from the leaf to the ear in wheat. *Physiol. Plant* **16**, 851 (1963).
58. STREBEYKO, P. Rapid method for measuring photosynthetic rate using $^{14}CO_2$. *Photosynthetica* **1**, 45 (1967).
59. STUCKEY, R. E. and Ellingboe, A. H. Effect of environmental conditions on ^{35}S uptake by *Triticum aestivuum* and transfer to *Erysiphe graminis* sp. *tritici* during primary infection. *Physiol. Plant Path.* **5**, 19 (1975).
60. THAINE, R. *et al. Aust. J. Biol. Sci.* **12**, 349 (1959).
61. TRIP, P. and Gorham, P. R. Bidirectional translocation of sugars in sieve tubes of squash plants. *Plant Physiol.* **43**, 877–82 (1968).
62. TURNER, N. C. and Incoll, L. D. The vertical distribution of photosynthesis in crops of tobacco and soybean. *J. Appl. Ecol.* **8**, 581 (1971).
63. VOZNESESKII, V. L., Zalenskii, O. V. and Austin, R. B. Methods of measuring rate of photosynthesis using $^{14}CO_2$. *In: Plant Photosynthetic Production. Manual of Methods* (Z. Sescak, J. Catský & P. G. Jarvis, Eds.) W. Junk, Hague (1971).
64. WEATHERLEY, P. E., Peel, A. J. and Hill, G. P. The physiology of the sieve tube. Preliminary experiments using aphid mouth parts. *J. Expt. Bot.* **10**, 1–16 (1959).
65. WEBB, J. A. and Gorham, P. R. Translocation of photosynthetically assimilated ^{14}C in straight-necked squash. *Plant Physiol.* **39**, 663–72 (1964).

Measurement of Leaf Water Status, Root Development and Biomass

66. ANGELOCCI, L. R. Estimativa das condições hídricas de folhas de soja, *Glycine max*, através da atenuação de radiações. M.Sc. Thesis, Escola Superior de Agricultura "Luiz de Queiroz", University of São Paulo, Piracicabas, S.P. (1976).
67. ANTOSZEWSKI, R. Some nuclear techniques applicable for physiological characterization of plant material. *In: Tracer Techniques for Plant Breeding*. Proc. Panel, Vienna, 1974, 25–30 IAEA (1975).
68. BIELORAI, H. Beta-ray gauging technique for measuring leaf water content changes of citrus as affected by the moisture status in the soil. *J. Expt. Bot.* **19**, 489–95 (1968).
69. BRUCE, R. R., Sanford, J. O., Grogan, C. O. and Myhre, D. L. *Agron. J.* **61**, 411 (1969).
70. BUSCHBOM, U. Zur Method kontinuerlicher Wassergehalt Bertmunger und Blättern mittels β-Strahlenabsorption. *Planta* **95**, 146–66 (1970).
71. GARDNER, W. R. and Nieman, R. H. Lower limit of water availability to plants. *Science* **143**, 1460–62 (1964).
72. GLUBRECHT, H., Niemann, E. G. and Rundfeld, H. Diechtebestimmungen mit Gammastrahlen an biologischen Objekten. *Atompraxis* **5**, 237 (1959).
73. GRANT, D. R. *J. Agric. Sci.* **75**, 433 (1970).
74. HAAHR, V. *Proc. Meeting of Sections Cereals and Physiology*, Dijon, Oct. 1970, 189 (1971).
75. HAAHR, V. Nuclear methods for detecting root activity. *In: Tracer Techniques for Plant Breeding*. Panel Proc. 1974, 57–63, IAEA, Vienna (1975).
76. JARVIS, P. G. and Slatyer, R. O. Calibration of β-gauge for determining leaf water status. *Science* **153**, 78–79 (1966).
77. JONES, H. G. Estimation of plant water status with the beta gauge. *Agr. Meteorol.* **11**, 345–55 (1973).
78. JONG, E. de Use of beta radiation to measure the moisture content of plant leaves. *In: Methodology and Techniques in Soil-Plant Nutrition and Plant Physiology*. Sask. Inst. Pedology, 64–68 (1971).
79. KÜHN, W. and Schätzler, H. P. Mass determination of plantation by absorbtion of gamma radiation, Part A. *Atomkemenergie* **21**, 141 (1973); Part B. *Atomkemenergie* **24**, 209 (1974).
80. LUPTON, F. G. H. Tracer techniqus in plant breeding programmes. *In: Tracer Techniques for Plant Breeding. Proc. Panel Vienna 1974, 31–42 IAEA, Vienna (1975).*
81. MEDERSKI, H. J. Determination of internal water by beta gauging technique. *Soil Sci.* **92**, 143–46 (1961).
82. MEDERSKI, H. J. and Alles, W. Beta gauging leaf water status: influence of changing leaf characteristics. *Plant Physiol.* **43**, 470–72 (1968).
83. NAKAYAMA, F. S. and Ehrler, W. L. Beta ray gauging technique for measuring leaf water content changes and moisture status of plants. *Plant Physiol.* **39**, 95–98 (1964).
84. NEUMANN, H. H., Thurteel, G. W., Stevenson, K. R. and Beadle, C. L. Leaf water content and potential in corn, sorghum, soybean and sunflower. *Can. J. Plant Sci.* **54**, 185–95 (1974).
85. OBREGEWITSCH, R. P., Rolston, D. E., Nielsen, D. R. and Nakayama, F. S. Estimating relative leaf water content with a simple beta gauge calibration. *Agron. J.* **67**, 729–32 (1975).
86. PEYNADO, A. and Young, R. H. Moisture changes in intact citrus leaves monitored by a β-gauge technique. *Proc. Amer. Soc. Hort. Sci.* **92**, 211–20 (1968).
87. ROLSTON , D. E. and Horton, M. L. Two beta sources compared for evaluating water status of plants. *Agron. J.* **60**, 333–36 (1968).
88. SKIDMORE, E. L. and Stone, J. F. Physiological role in regulating transpiration rate of the cotton plant. *Agron. J.* **56**, 405–10 (1964).
89. WHITEMAN, P. C. and Wilson, G. L. Estimation of diffusion pressure deficit by correlation with relative turgidity and beta-radiation absorption. *Aust. J. Biol. Sci.* **16**, 140–46 (1962).
90. WIEBE, H. J., Schätzler, H. P. and Kühn, W. Continuous measurement of the diurnal fluctuation of the mass of a single cabbage plant. *Kerntechnik* **16**, 532 (1974).

Ion Uptake, Redistribution and Plant Nutrition

91. AHLGREN, G. E. and Sudia, T. W. Studies of the mechanism of the foliar absorbtion of phosphate. *In: Isotopes in Plant Nutrition and Physiology*, 347–88, IAEA, Vienna (1967).
92. ANDERSON, W. P. *Ion Transport in Plants*. Academic Press, London and New York, 630 pp. (1973).
93. BERLIER, Y. and Guiraud, G. Absorption et utilisation par des graminees de l'azote nitrique ou ammoniacal marque a l'azote-15. *In: Isotopes in Plant Nutrition and Physiology*, 145–57, IAEA, Vienna (1967).

94. BIDDULPH, S. F. Visual indications of ^{35}S and ^{32}P translocation in the phloem. *Amer. J. Bot.* **43**, 143–48 (1956).
95. BIDDULPH, O., Biddulph, S., Cory, R. and Koontz, H. V. Circulation patterns for phosphorus, sulfur and calcium in the bean plant. *Plant Physiol.* **33**, 293–300 (1958).
96. BIDDULPH, O., Cory, R. and Biddulph, S. Translocation of calcium in the bean plant. *Plant Physiol.* **34**, 512–19 (1959).
97. BIRECKA, H. Final Research Rept. IAEA Lab. Seibersdorf, 1–15 (1969).
98. BROWN, A. L., Yamaguchi, S. and Leal-Diaz, J. Evidence for translocation of iron in plants. *Plant Physiol.* **40**, 35–38 (1965).
99. BROWN, J. C. Genetically controlled chemical factors involved in absorption and transport of iron by plants. Advances in Chemistry, no. 162, *Bioinorganic Chemistry* II, 93–103 (1977).
100. BUKOVAC, M. J. and Wittwer, S. H. Absorbtion and mobility of foliar applied nutrients. *Plant Physiol.* **32**, 428–35 (1957).
101. COCKING, E. C. and Yemm, E. W. Synthesis of amino acids and proteins in barley seedlings. *New Phytol.* **60**, 103–16 (1961).
102. DOKIYA, Y., Arima, Y. and Mitsui, S. A double tracer experiment using radioactive manganese for the elucidation of manganese turnover in plant material. *Plant and Cell Physiol.* **7**, 715–17 (1966).
103. EDDINGS, J. L. and Brown, A. L. Absorbtion and translocation of foliar-applied iron. *Plant Physiol.* **42**, 15–19 (1967).
104. EPSTEIN, E. and Hagen, C. E. A kinetic study of the absorbtion of alkali cations by barley roots. *Plant Physiol.* **27**, 457–74 (1952).
105. EPSTEIN, E. *Mineral Nutrition of Plants: Principles and Perspectives.* pp. 412, Wiley, New York (1972).
106. FRIED, M. and Noggle, J. C. Multiple site uptake of individual cations by roots as affected by hydrogen ion. *Plant Physiol.* **33**, 139–44 (1958).
107. FRIED, M., Zsoldos, F., Vose, P. B. and Shatokhin, I. L. Characterizing the NO_3 and NH_4 uptake process of rice roots by use of ^{15}N-labelled NH_4NO_3. *Plant Physiol.* **18**, 313–20 (1965).
108. GÁSPÁR, L. and Gáspár, Christie. Étude de la vigueur hybride par l'emploi du carbone et du phosphore radioactifs. *In: Isotopes in Plant Nutrition and Physiology*, 453–64, IAEA, Vienna (1967).
109. HAGEN, C. E. and Hopkins, H. T. Ionic species in orthophosphate absorbtion by barley roots. *Plant Physiol.* **30**, 93–199 (1955).
110. HOFSTEE, B. H. J. On the evaluation of the constants V_m and K_m in enzyme reactions. *Science* **116**, 329–31 (1952).
111. IAEA. *Nitrogen-15 in Soil-Plant Studies.* Proc. Meeting Sofia, December 1969. IAEA, Vienna (1971).
112. ISTATKOV, St., Mladenova, Y. Investigations with isotopes on the heterosis phenomenon in maize. *In: Nitrogen-15 in Soil-Plant Studies.* Research Coordination Meeting, Sofia, 227–38, IAEA, Vienna (1971).
113. IVANKO, S. Metabolic pathways of nitrogen assimilation in plant tissue when ^{15}N is used as a tracer. *In: Nitrogen-15 in Soil-Plant Studies*, 119–56, IAEA, Vienna (1971).
114. JACKSON, W. A. and Volk, R. J. Physiological aspects of ammonium nutrition of selected higher plants. *In: Isotopes in Plant Nutrition and Physiology*, 159–78, IAEA, Vienna (1967).
115. YYUNG, W. H. and Wittwer, S. H. Foliar absorbtion—an active uptake process. *Amer. J. Bot.* **51**, 51 (1964).
116. KOONTZ, H. V. and Biddulph, O. Factors affecting absorbtion and translocation of foliar applied phosphorus. *Plant Physiol.* **32**, 463–70 (1957).
117. KOONTZ, H. V. and Vose, P. B. Phosphorus uptake by vegetative ryegrass, *Lolium spp.* in relation to flowering, defoliation and nitrogen nutrition. Proc. 8th Int. Grassl. Congr. Reading, 416–21 (1960).
118. KUMAZAWA, K. Final Research Rept. IAEA Lab. Seibersdorf 1–23 (1969).
119. LEGGETT, J. E. and Epstein, E. Kinetics of sulphate absorbtion by barley roots. *Plant Physiol.* **31**, 222–26 (1956).
120. LINEWEAVER, H. and Burk, D. The determination of enzyme dissociation constants. *J. Amer. Chem. Soc.* **56**, 658–66 (1934).
121. MICHAELIS, B. C. and Menten, M. L. Die Kinetic der Invertinwirkung. *Biochem. Zeits.* **449**, 333–69 (1913).
122. MACDONALD, I. R. and Macklon, A. E. S. Anion absorbtion by etiolated wheat leaves after vacuum infiltration. *Plant Physiol.* **49**, 303–6 (1972).
123. MACDONALD, I. R. and Macklon, A. E. S. Light-enhanced chloride uptake by wheat laminae. A Comparison of chopped and vacuum-infiltrated tissue. *Plant Physiol.* **56**, 105–8 (1975).
124. MILLIKAN, C. R. and Hanger, B. C. Redistribution of ^{45}Ca in *Trifolium subterraneum* and *Antirrhinum majus. Aust. J. Biol. Sci.* **20**, 1119–30 (1967).

125. Munns, D. N., Johnson, C. M. and Jacobson, L. Uptake and distribution of manganese in oat plants. 1. Varietal variation. *Plant and Soil* **19**, 115–26 (1963a).
126. Munns, D. N., Jacobson, L. and Johnson, C. M. Uptake and distribution of manganese in oat plants. 2. A kinetic model. *Plant and Soil* **19**, 193–204 (1963b).
127. Munns, D. N., Johnson, C. M. and Jacobson, L. Uptake and distribution of manganese in oat plants. 3. An analysis of biotic and environmental effects. *Plant and Soil* **19**, 285–95 (1963c).
128. Noggle, J. C. and Fried, M. A kinetic analysis of phosphate absorbtion by excised roots of millet, barley and alfalfa. *Soil Sci. Amer. Soc. Proc.* **24**, 33–35 (1960).
129. Picciurro, G., Ferrandi, L., Boniforti, R. and Bracciocurti, G. Uptake of ^{15}N-labelled NH_4 in excised roots of a *durum* wheat mutant line compared with *durum* and bread wheat. *In: Isotopes in Plant Nutrition and Physiology*, 511–25, IAEA, Vienna (1967).
130. Pienaar, P. J., Meynhardt, J. T. and Marais, P. G. Seasonal uptake of ^{32}P-labelled phosphate by Alphonse Lavallee grape vines in sand culture. *In: Isotopes in Plant Nutrition and Physiology*, 137–43, IAEA, Vienna (1967).
131. Rains, D. W. Light-enhanced potassium absorbtion by corn leaf tissue. *Science* **156**, 1382–83 (1967).
132. Ringoet, A., Sauer, G. and Gielink, A. J. Phloem transport of calcium in oat leaves. *Planta* **80**, 15–20 (1968).
133. Romney, E. M. and Toth, S. J. Plant and soil studies with radioactive manganese. *Soil Sci.* **77**, 107–17 (1954).
134. Shim, S. C. and Vose, P. B. Varietal differences in the kinetics of iron uptake by excised rice roots. *J. Expt. Bot.* **16**, 216–32 (1965).
135. Single, W. V. The mobility of manganese in the wheat plant. 1. Redistribution and foliar application. *Ann. Bot.* **22**, 479–88 (1958).
136. Smith, R. C. and Epstein, E. Ion absorbtion by shoot tissue: technique and first findings with excised leaf tissue of corn. *Plant Physiol.* **39**, 338–41 (1964a).
137. Smith, R. C. and Epstein, E. Ion absorbtion by shoot tissue: kinetics of potassium and rubidium absorbtion by corn leaf tissue. *Plant Physiol.* **39**, 992–96 (1964b).
138. Thorne, G. N. Factors affecting uptake of radioactive phosphorus by leaves and its translocation to other parts of the plant. *Ann. Bot.* **22**, 381–98 (1958).
139. Tukey, H. B., Wittwer, S. H. and Bukovac, M. J. *J. Agric. Food Chem.* **9**, 106 (1961).
140. Tukey, H. B. Jr., Mecklenburg, R. A. and Morgan, J. V. *In: Isotopes and Radiation in Soil-Plant Nutrition Studies*, 371, IAEA, Vienna (1965).
141. Turner, R. G. and Gregory, R. P. The use of radioisotopes to investigate heavy metal tolerance in plants. *In: Isotopes in Plant Nutrition and Physiology*, 453–64, IAEA, Vienna (1967).
142. Uriu, K. and Koch, E. C. Response of Yellow Newton apple leaves to foliar applications of manganese and zinc. *Proc. Amer. Soc. Hort. Sci.* **84**, 25–31 (1964).
143. Van den Honert, T. H. Limiting factors in phosphate absorbtion. *Vergadering Vereeningen Proefstations Personeel*, Djemba, Java, **16**, 85–93 (1936).
144. Vose, P. B. Manganese requirement in relation to photosynthesis in *Avena*. *Phyton* **19**, 133–40 (1962).
145. Vose, P. B. The translocation and redistribution of manganese in *Avena*. *J. Expt. Bot.* **14**, 448–457 (1963).
146. Vose, P. B. and Shim, S. C. Effect of light and related factors on ion absorbtion by banana leaf discs. *Nature* **201**, 1047–48 (1964).
147. Vose, P. B. Class Experiment E.22. Use of radioisotopes in ion uptake studies to determine the effect of complementary cations on the uptake of rubidium by excised barley roots. Proc. IAEA/FAO International Training Course on the Use of Radioisotopes and Radiation in Soil and Plant Nutrition Studies. ITAL, Wageningen (1970).
148. Wallace, A., Hemaidan, N. and Sufi, S. M. Sodium translocation in bush beans. *Soil Sci.* **100**, 331–34 (1965).
149. Wallace, A. Retranslocation of ^{86}Rb, ^{137}Cs, and K to new leaf growth in bush beans. *Plant Soil* **29**, 184 (1968).
150. Wallace, A., Hale, V. Q. and Joven, C. B. DTPA and pH effects on leaf uptake of ^{59}Fe, ^{65}Zn, ^{137}Cs, ^{241}Am and ^{210}Pb. *J. Amer. Hort. Sci.* **94**, 684 (1969).
151. Wallace, A. and Mueller, R. T. Effect of chelating agents on the availability to plants of carrier-free ^{59}Fe and ^{65}Zn added to soils to simulate contamination from fallout. *Soil Sci. Soc. Amer. Proc.* **33**, 912 (1969).
152. Wallace, A. Effect of calcium levels on redistribution of ^{95}Sr in bush beans. *Plant Soil* **35**, 415 (1971).
153. Wallihan, E. F. and Heymann-Herschberg. Some factors affecting absorbtion and translocation of zinc in *Citrus* plants. *Plant Physiol.* **31**, 294–99 (1956).

154. WITTWER, S. H. and Lundahl, W. S. *Plant Physiol.* **26**, 792 (1951).
155. WITTWER, S. H. *et al. Proc. Amer. Soc. Hort. Sci.* **69**, 302 (1957).
156. WITTWER, S. H., Yyung, W. H., Yamada, Y., Bukovac, M. J., De, R., Kannan, S., Rasmusson, H. P. and Haile Mariam, S. N. Pathways and mechanisms for foliar absorbtion of mineral nutrients as revealed by radioisotopes. *In: Isotopes and Radiation in Soil-Plant Nutrition Studies*, 387–403, IAEA, Vienna (1965).
157. PICKARD, W. F., Minchin, P. E. H. and Troughton, J. H. Real time studies of carbon-11 translocation in moonflower. *J. Expt. Bot.* **29**, 993–1001 and 1003–9 (1978).
158. SCHÄTZLER, H. P. and Kühn, W. Growth studies on plant plots by gamma scanning. *Int. J. Appl. Radiation and Isotopes* **28**, 645–52 (1977).
159. KÜHN, W. New radiometric and radioanalytical methods in agricultural research. *Atomic Energy Review* **16**/4 (1979).

CHAPTER 14

Nuclear Techniques for Soil Water

DETERMINING SOIL MOISTURE AND BULK DENSITY

MUCH research on soil water and irrigation depends on the determination of actual soil moisture at many different experiment sites, at different depths in the soil, and with different irrigation and other treatments. Hitherto such measurements have been laborious and time-consuming but the development of the neutron moisture meter has done a lot to assist this research and made it much easier to measure soil water in the field ([5,32,33,42,73,79]).

The operation of the neutron moisture meter depends on the fact that neutrons are readily scattered by water but not by soil. It has been found, too, that the amount of neutron scattering is proportional to the amount of water in the soil. The neutron moisture meter consists of a source of fast neutrons, a detector to identify the slow neutrons which result from scattering of the fast neutrons by water, and a counter to count them. In practice most equipment consists of a ''probe'' which contains the radiation source, usually a mixture of radium and beryllium or americium, and the detector, connected to a portable battery-operated electronic counter. The probe is inserted into holes in the soil lined with thin-walled access tubes of aluminium or plastic and placed at the depth where moisture determination is required. A ''count'' is taken over a period of 1 to 5 minutes, the count rate varying with the moisture content.

There are many advantages to this technique of moisture measurement: the method is non-destructive, the measurement can be done in the field, and there are no samples to take or carry; after the initial placement of the access tube each measurement is rapid; repeated measurements can be carried out over a long period without disturbing the soil; it is easy to measure at different depths at the same point; a large volume of soil is sampled. The main disadvantage is the expense of the equipment and its weight of 8 to 15 kg, while lack of uniformity, high chloride content and stones can cause errors in some soils.

The density of the soil can also be measured using very similar principles. For the measurement of soil density the probe consists of a gamma source, often cesium-137, and a detector. The probe is placed in contact with or inserted into the soil and gamma rays are emitted by the radioactive source. The gamma rays are scattered by the soil in all directions and the more dense the soil the greater is its scattering power. As the gamma rays are scattered they lose energy, with the result that less energy is reflected back to the detector. As the scattering power of the soil increases with density, it follows that the more dense the soil the less energy is reflected back to the detector and the smaller is the count rate. Conversely, the less dense the soil then the higher the count rate.

Simultaneous with the development of neutron moderation and gamma attenuation techniques, considerable developments occurred in the use of tritium ^3H to follow the movement of soil water. Similarly the movement of salts and particles can be determined by the use of appropriate radioisotope tracers.

Determining Soil Moisture by Means of Neutron Moderation

Principle

The properties of neutrons were considered in Chapter 1, page 14. Thus they are classified by their energy into basically fast, slow and thermal neutrons. As a fast neutron passes through matter it loses kinetic energy by a series of elastic and inelastic collisions finally becoming a thermal neutron, that is a neutron in thermal equilibrium with its surroundings. Having thus lost kinetic energy, the neutron velocities become reduced to those found in gases at S.T.P., and the neutrons are eventually captured by the nuclei of the surrounding matter.

This situation occurs if a fast neutron source is placed in soil. The fast neutrons ejected by the source are scattered in all directions, losing energy at each collision until the thermal neutron state is attained. Now it is clear that the rate of neutron moderation i.e. "slowing down" must depend on the mass of the nuclei in collision with the neutron and also upon the probability of collision.

As the mass of a hydrogen nucleus is almost equal to the neutron mass it can be deduced from kinetic theory that collisions with hydrogen atoms will reduce the kinetic energy and velocity of neutrons more effectively than collisions with other nuclei. This can be seen from Table 14.1. The big difference between the mass of the hydrogen atom and the atoms of other elements usually found in soils, means that in relative terms, hydrogen atoms are exceedingly efficient in slowing down neutrons ([44]).

Additionally, hydrogen has an especially favourable cross section for thermal neutrons, and as was noted in Chapter 2 it is the neutron cross section that determines the probability of a neutron colliding with the nucleus of any atom. The neutron cross section for any element varies with different neutron energies, and usually the cross section value increases with decrease in neutron energy and in the case of hydrogen there is relatively a very large increase in neutron cross section. Therefore, on both the basis of nuclear mass and on cross section for thermal neutrons, hydrogen is a very effective moderator of thermal neutrons. Total cross sections of major soil constituent elements are shown in Table 14.2.

When we put a fast neutron source into the soil together with a neutron detector then some of the neutrons emitted from the source will be moderated and scattered

Table 14.1 Element effectiveness in moderating 2 MeV fast neutrons

Element	Atomic mass	Average number of collisions to thermalize 2 MeV neutrons
Hydrogen	1	18
Lithium	7	68
Beryllium	9	88
Boron	11	104
Carbon	12	115
Nitrogen	14	133
Oxygen	16	152
Sodium	23	215
Magnesium	24	227
Aluminium	27	255
Silicon	28	263
Phosphorus	31	288
Sulphur	32	298
Chlorine	35	329
Potassium	39	362
Calcium	40	371
Iron	56	515
Lead	207	1897

Table 14.2
Comparative total thermal neutron cross sections of principal soil constituent elements

Element	H	C	O	Na	Mg	Al	Si	Cl	K	Ca	Mn	Fe
Neutron cross section σ, barns/nucleus	36	5	4.2	4.0	3.0	1.6	2.4	48	4.3	9.4	16	13.4

back, and the resulting slow neutrons detected. This recorded activity will consist almost exclusively of neutrons that have been moderated by hydrogen atoms, due to the ineffectiveness of the other elements present in soil. Now, as by far the main source of hydrogen in the soil is water it follows that the activity recorded by the detector can be used, after suitable calibration, to determine the concentration of water in the soil ([1,2,5,6,16,23,3,3,43,51,59,81,82]).

Equipment

The fast neutron source is based in all cases on the (α,n) reaction. Often a mixture of beryllium-9 and radium-228 has been used. In this reaction ^{9}Be is bombarded with α-particles and is transmuted to ^{12}C and a fast neutron is emitted. The reactions are given in full on page 25. As α-particles have a very short range, a very fine beryllium powder must be thoroughly mixed with the radium, compressed, and encapsulated within a steel capsule to give a point source ([3,24,35]).

Although Ra-Be (5 mCi) has been the most frequently used source, americium-Be and actinium-Be sources may also be used, and a source of 50–100 mCi ^{241}Am-^{9}Be is now probably the most common. The americium and actinium sources have an advantage over the Ra-Be source because the latter has a relatively high γ/n ratio due to the γ given off in the decay of ^{226}Ra, which makes it impossible to have Ra-Be sources when a scintillation detector is used because of the difficulty of shielding. A ^{210}Po-^{9}Be source would be almost free of γ-contamination but unfortunately the short half life (138 days) of ^{210}Po makes it impractical. However, in practice, boron or helium-3 detector tubes have been almost invariably used and they are little affected by photons. Neutron detectors have been discussed in Chapter 4 (page 62) to which reference should be made. ^{3}He detectors have the valuable characteristic of being only sensitive to thermal neutrons and not to fast neutrons. There is a growing use of scintillation detectors.

Field equipment comprises a portable scaler together with a combined neutron source/detector probe, Fig. 14.1, which for transport is retained within a lead and paraffin wax shield to protect the operator from radiation. Within the combined probe the detector is also shielded from the direct radiation of the source. In agriculture we are primarily concerned with depth probes, although surface probes are also available, but are not too suitable. In use the shield is placed over an access tube previously installed in the soil and the probe is lowered down the tube by its cable to the desired depth, and for this the cable should be marked every 10 cm. Counts per unit time are then recorded on the scaler and the water content can then be estimated by means of a calibration curve plotted in counts V water content. Extensive discussions of equipment can be found in reference ([35]).

Factors Affecting Operation

The main factors concerned with the successful use of neutron moisture meter are (i) the spatial resolution of the probe (ii) the calibration of the probe (iii) the nature of the soil under examination (iv) the installation of the access tubes.

(i) The spatial resolution of a probe governs its effective sphere of influence, that is the volume or zone of soil which contributes to the activity recorded on the scaler. This is therefore the zone in which the water content is being determined. The determination of the sphere of influence of a probe is now largely academic as they do not vary greatly from one to another, however it can conveniently be checked as part of normal calibration procedure ([54,78,86]).

The spatial resolution of a probe under varying conditions can be found by taking counts at different depths in soil volumes of different water content. Both the soil texture and the water content should be homogenous throughout. To achieve this it will usually be found convenient to wet the soil to an appropriate extent, thoroughly mix it, and put it into large drums or barrels with an access tube in the middle of each. The access tube should stick out from the soil by at least 30 cm. About three or four water contents should be taken, say approximately 10%, 20%, 30% and 40%

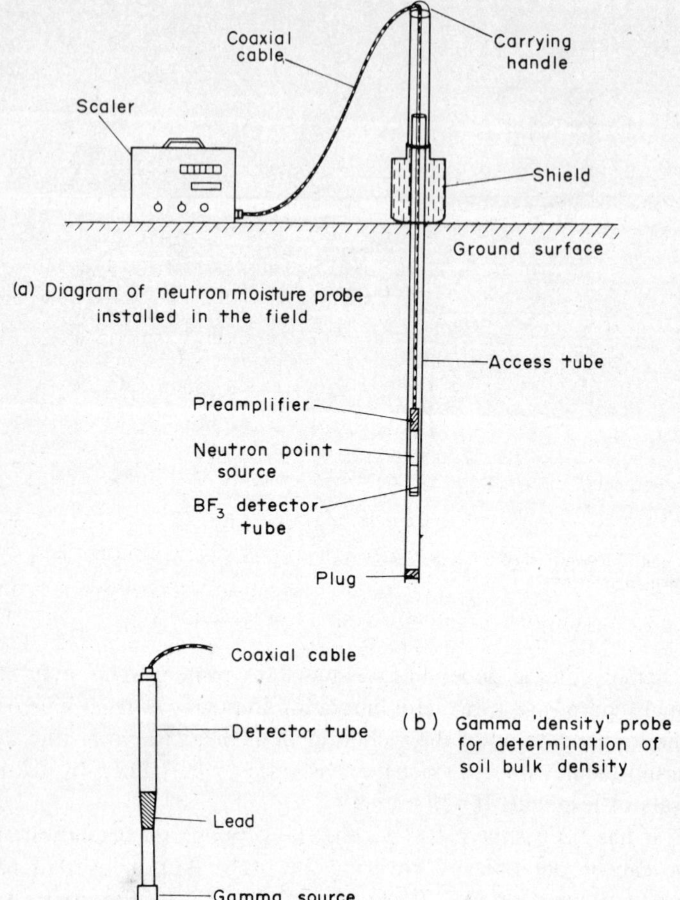

FIG. 14.1 Neutron moisture and bulk density probes.

v/v. One drum should contain only water, with the access tube sealed at the base and weighed to hold it down. At the time of measurement soil samples should be taken for determination of water content by classical means.

Then for each container 8–10,000 counts are taken at each of 5 cm intervals from the surface downwards until a stable maximum count is achieved at several successive positions. If the count rate is then plotted as a function of depth, the radius of the sphere of influence is considered as the distance from the surface to the point where 95% of the maximum count rate is achieved. With a majority of probes it will be found that the radius of the sphere of influence is at a minimum of about 15 cm in water, rising to a maximum of about 40 cm in very dry soils as in Fig. 14.2.

Thus the sphere of influence varies in practice from about 30–80 cm according to soil water content. In effect this means that in routine soil water determinations, the counts that are obtained are a reflection of the water content in a considerable bulk

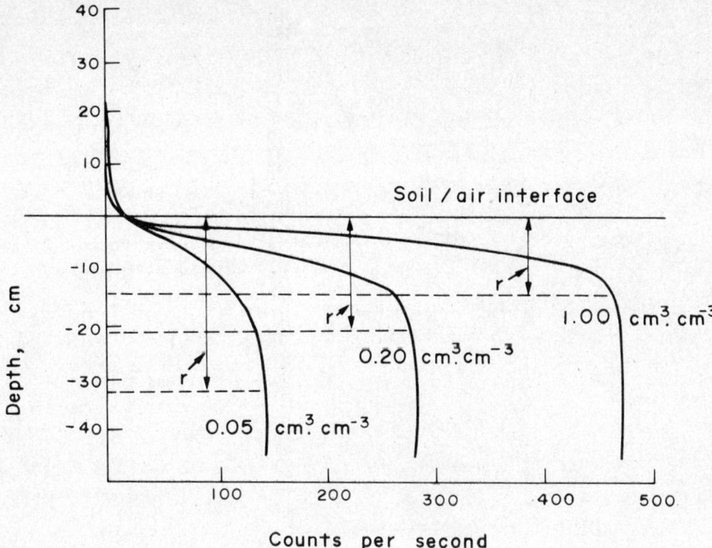

FIG. 14.2 The spatial resolution of a neutron probe in soil at three water contents, where r is the radius of the sphere of influence.

of soil, the actual volume depending on moisture content. This may be a valuable feature for many purposes, especially in relation to practical irrigation considerations. However, the low resolution of the technique does mean that moisture determination at a point is not achievable, in fact there is little to be gained by taking counts at depth intervals of less than 15–20 cm.

However, it has been shown ([13,43,89]) that both improved resolution and precision of the probe can be obtained by covering the probe with a layer of paraffin wax, polystyrene or polyethylene (8–12% hydrogen) about 3–4 mm thick. The thickness of moderator is critical and as much as 10 mm is too great. The auxilliary moderator results in the thermal neutron density in the vicinity of the source being increased while the fast neutron density further from the source i.e. in the soil, is decreased. It follows that the contribution of fast neutrons decreases as the distance from the moderator increases, resulting in improved resolution. The sensitivity, as shown by a steeper slope of the calibration curve, is improved because the fast neutrons ejected by the source are reacting in a smaller volume of soil when an auxilliary moderator is used close to the source. In effect the auxilliary moderator acts as a source of epithermal neutrons. The use of plastic access tubes also tends to improve the resolving power, but shows little improvement in precision because of the greater distance from the source.

It will be obvious from the low resolution of the method that measurement of soil water in the surface layer is difficult ([78]), due to the large sphere of influence and to neutrons "escaping" to the atmosphere. To some extent this can be overcome by placing ([49]) a large hemisphere of paraffin wax or plastic immediately above the probe and in contact with the soil. Alternatively a perforated fibreglass tray with a central hole to accomodate the access tube and filled with the same top soil to a depth of 10–15 cm can be placed on the ground above the probe. In this manner and with the use of a special calibration curve it should be possible to measure to within about 10–15 cm of the soil surface.

(ii) The calibration of a probe can be carried out in two basic ways. Firstly, test soil volumes may be prepared with different water contents as described above in connection with determining the sphere of influence. In this case, samples of the soil with different water contents are taken for the precise determination of soil moisture and bulk density by the normal classical methods. A calibration curve is then constructed by plotting counts obtained against actual soil water content as determined gravimetrically by drying and weighing.

The calibration curve for any probe can be affected by the material used for the access tube, and by the soil type if measurements are made on a soil different from that on which original calibration was carried out. Furthermore, if the probe has been altered in any way, such as by replacement of source, detector or the associated electronics then re-calibration will be necessary, or at least checking at one known soil water content of a previously investigated "standard soil". As it has been found that the shape of a calibration curve depends mainly on the geometry of detector and source, the variation in the curve due to alteration in these factors is usually linear throughout the range ([8,31,32,47,53,76,77,81]).

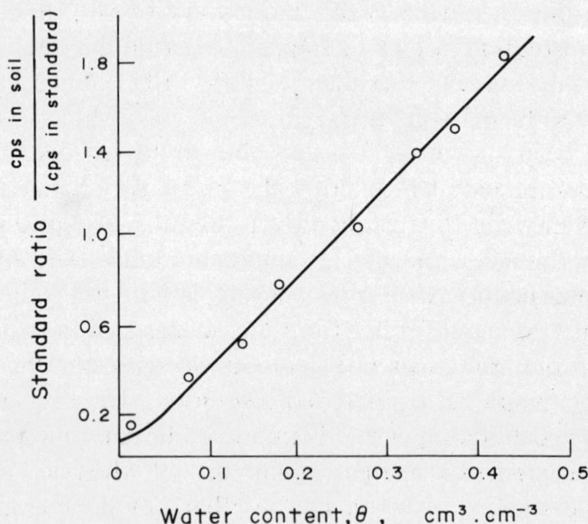

FIG. 14.3 A typical neutron moisture probe calibration curve.

Preparing an absolute calibration curve by the method given above or in the field as described below, involves a considerable amount of work, especially if undertaken for a number of soils. It is however possible to check the calibration of an instrument in the field by less elaborate means, if the manufacturer's or other calibration curve is available, together with a calibration standard. Counting in the shield which is lined with paraffin wax or polyethylene usually serves this purpose. However, polyethylene cylinders of 2.5, 3.0, 3.5 and 4.0 inch outside diameters with a central hole to receive the probe can provide a range of permanent calibration standards ([77]).

Counts are taken at successive increasing depths within an access tube installed in a field site. When a position is determined where the count rate is constant, five or six successive counts of at least one minute (or 10,000 c.p.m.) are obtained. Then five or six similar counts are taken within the calibration standard. A soil sample is then taken within 15–20 cm of the access tube from the soil layer where the counts were made, for determination of water content and bulk density gravimetrically. Samples should preferably be taken by means of a wide-bore soil corer which will disturb the field-condition sample to the least extent.

The water content, either expressed as W/W% or preferably V/V% can then be plotted as a point on the existing calibration curve against the standard ratio, i.e. the ratio of the mean count in soil/mean count in the standard. Ideally this point should fall on the curve, but if it does not then counts in air (zero water content) and in water (approximately to 100% saturated soil) will help in reaching a decision as to whether the slope of the calibration curve has altered, or the curve has shifted, or the difference is due to experimental error. If there is any doubt, further measurements with other soil samples should be taken, and if necessary a new calibration curve constructed.

Rather than just relying on the direct count, the merit of taking the ratio of soil count standard count, which can be adopted for all measurements, is that it eliminates errors due to electronic instability, such as may arise from the high voltage source, amplifier, discriminator and detector. The ratio method does not however allow for the deterioration of the paraffin wax mantle in the shield, nor does it take account of varying count rates in some shields due to temperature variations in the field (Fig. 14.3).

For many soil science and plant ecology studies we are primarily concerned, within certain limits, with changes in water content rather than absolute values. This is also the case when determining crop water consumption and requirement for irrigation. Therefore as calibration curves for different soils are normally linear and parallel, a difference in count rate should reflect the same difference in water content and this will be sufficient for many purposes. However, when absolute values are required it is necessary to calibrate the equipment *in situ* for the particular soil, to take account of overall geometry, soil bulk density, salt and organic matter content etc., and where there are marked differences in the horizons it may be necessary to calibrate for each horizon.

Absolute calibration of a neutron moisture meter in the field requires taking successive counts over a period of time (days) at about 45–60 cm depth (or in each horizon if necessary) in access tubes installed in plots which have received very heavy rain or irrigation. Appropriate soil samples are taken simultaneously from four to six points around the access tube at each depth, for moisture and bulk density determinations gravimetrically. At the same time bound water, e.g. water of crystallization of clays, is determined by calcination at 650°C, so that it may be taken into account.

In this method, as determinations are made as the soil is allowed progressively to dry out, it should be possible to obtain a very accurate calibration in relation to actual field conditions at any moisture level. Of course the water at any point in the soil profile is in a dynamic state but as the neutron probe has such a low resolution this factor is not important, provided there is no delay between neutron moisture determination and soil sampling.

(iii) The nature of the soil under examination may present problems. The obvious case is when the soil differs markedly in type from that on which the calibration has been carried out or where there are distinct horizons, as has just been discussed. Where the equipment is used regularly on a number of main soil types the answer is to have separate curves for each, if the translation of the standard calibration curve is too great to be acceptable.

Soils with high organic matter will give spuriously high counts due to the hydrogen in the organic matter. Saline soils present the opposite condition, as due to the very high thermal neutron cross section of Cl the chlorine atoms appreciably moderate the fast neutrons, giving rise to low counts. The same situation applies to boron (recall the use of BF_3 in detector tubes), but there are few soils except in parts of California in which B might present this problem.

As the count rate is a direct function of the apparent density of the soil, compacted layers in the horizon, a marked change of soil type in the horizon, undue compaction of the area next to the access tube, or the presence of many stones, may all give departures from the calibration curve.

It will be found that changes in organic matter, clay, or apparent density result in a translation of the calibration curve, although the slope remains constant. On the other hand, in the case of chloride the slope is altered, as the count rate declines with increasing Cl ([8,15,24,37,39,43,48]).

(iv) The installation of the access tubes is a key factor in the success of the method. Access tubes may be made of stainless steel, plastic or aluminium. Stainless steel has the longest life, but is expensive and does not give any better results than the other materials. It may also be convenient to use aluminium, as frequently it is already on hand in the form of irrigation pipes. Probably plastic tubing is the one of choice because of its tendency to improve resolving power, although the strength of steel may be decisive in the case of difficult soils.

The internal diameter of the tube should be just large enough to ensure a sliding fit of the probe i.e. about 2–3 mm larger than the diameter of the probe, which is normally about 45–55 mm. A close fit of the probe in the tube is essential to ensure reproducibility, and if a probe has to be used in tubes which are too large then centring rings have to be made up. In the case of aluminium tubes every care must be taken to avoid denting the tubes, which might cause the probe to jam. The tubes should be plugged at the lower end with rubber bungs or wooden plugs to prevent possible rise of water in the tube.

The holes should be made with an auger or soil corer and should be slightly smaller than the outside diameter of the access tube to ensure close contact between the soil and the tube. Without close contact erroneous results might be obtained due to water or air-filled gaps, or even water running down the side of the tube. When driving the tubes in, the top of the tube should be protected with a suitably shaped hardwood block. As an alternative to first boring the hole and then driving the tube, an auger of a slightly smaller diameter than the bore of the access tube can be used inside the tube, alternately augering 15–20 cm of soil and then tapping in the tube. In this case a closely fitting rubber bung has to be pushed down the tube with a rod, after tube installation.

Any soil particles which may be adhering to the wall of the tube should be brushed down with a large bottle brush on a long wire. The tubes should be allowed to protrude from the soil about 30 cm and when not in use should be plugged with a rubber bung or covered with a can. To avoid compacting the soil around the access tube a large square board with a hole in the middle for the tube, should be used to stand on. The actual siting of the access tubes in the field or plots requires some consideration. They should be placed in a position representative of the soil/crop/plots under study, obviously avoiding tree roots and other localized factors such as shade. See Eeles ([20]) for an extensive discussion of access tube installation.

"Surface" Probes

As ordinary "depth" probes are not able to effectively determine moisture in the surface soil layer, so-called "surface" probes have been designed. The principle is exactly the same: fast neutrons emitted by a source are moderated and the reflected slow neutrons are detected by a BF_3 tube and the pulses counted.

Although quite widely used for civil engineering works such as roads, these probes have not found much employment in agricultural research. This is because the soil water in the plough layer changes rapidly with depth, and estimating the thickness of the soil layer being determined is problematical, about 15 cm being a rough approximation. We have already noted that the volume of the soil involved in a determination varies with the moisture content, being greater in dry soil and less with increasing water content.

It is also very easy to take soil samples from surface layers for determination gravimetrically, while for critical research studies the "two-wells" gamma probe technique can be used for plough layer soil water determinations.

Determining Soil Bulk Density with the Gamma Density Meter

Principle

Portable equipment for determining soil density is used similarly to the neutron moisture meter, and is based on the back-scattering of γ-rays which react with matter in a quantitative way.

The properties of gamma photons were discussed in Chapter 1. The loss of energy of a gamma photon in reaction with matter is primarily due to the Compton effect in which an incident photon gives up part of its energy to a loosely bound orbital electron which is ejected as a recoil electron. The now scattered γ-photons, which reduced energy, can then undergo further reactions with electrons. Many of these reactions will be by the photoelectric effect in which the whole energy of the photon is absorbed.

It will be apparent that the interaction and consequent scattering will increase with increasing concentration of orbital electrons. As the concentration of orbital electrons in soil elements is proportional to the density, yet is independent of chemical composition of the soil, it follows that the degree of scattering of γ-rays in soil is proportional to bulk density.

However, with successive scattering processes the γ-rays lose energy, and although the increase in electron density increases the probability of scattering, there is also a greatly increased chance that the reduced-energy photons will be absorbed by the photoelectric effect. The nett result is that with increased soil density fewer γ-rays arrive at the detector to be recorded.

Equipment and Procedure

The probe consists of a source of ^{137}Cs which emits monoenergetic rays of 0.662 MeV, coupled with a NaI crystal scintillation detector. The source and probe are separated as far as practicable, with lead placed between them, to prevent direct radiation falling on the detector. This is contrary to the situation with neutron moisture probes where placing the source at the mid-point of the detector is essential to obtaining a linear calibration curve.

Counts are made in access tubes in exactly the same manner as for moisture determination, the density being read from a previously prepared calibration curve derived from gravimetric procedures. The density range covered is usually about 1–2.5 g/cm^3. A typical calibration curve is given in Fig. 14.4. In determinations of water content we assume that the dry bulk density of the soil is constant. Conversely, when bulk density is being determined it is necessary to know the water content, as the counts registered are correlated with the wet bulk density of the soil. The weight of water must then be subtracted to obtain dry bulk density ([2,5,13,37,80]).

As the half-life of ^{137}Cs is 27 years a loss in activity and count rate of about 2–4% a year occurs, and this has to be periodically determined by counting in the shield and the necessary corrections made.

GAMMA ATTENUATION TECHNIQUES

Gamma attenuation techniques are based on the principle that if a collimated beam of γ-rays is focussed on a solid, the number of rays penetrating to the detector will be dependent on the thickness and density of the solid. The technique can be used either for soil water or bulk density determinations. The two-wells gamma probe method has been used for research work in experimental plots and lysimeters, while for laboratory experimentation using soil columns, rather elaborate assemblies can be

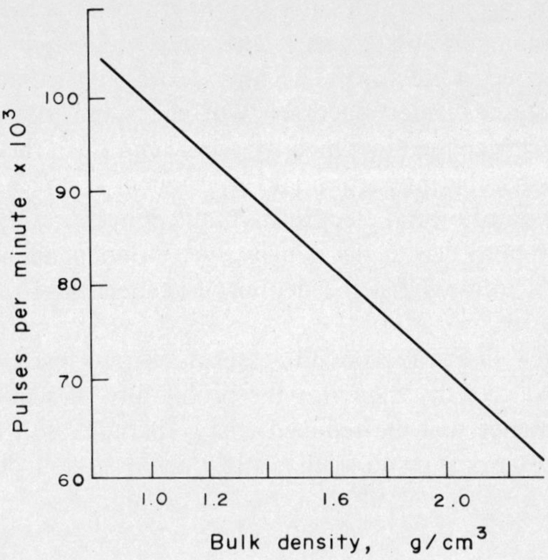

FIG. 14.4 Typical calibration curve for bulk density determination by gamma probe.

developed using a very highly collimated source and detector which permit determinations to be made at depth intervals as close as 1 cm apart ([17,21,25,28,40,64]).

Theory

Monoenergetic gamma radiation is attenuated by matter in inverse logarithmic proportion to its density, as defined by Lambert's law:

$$I = I_o e^{-\mu' T} \tag{1}$$

or

$$\ln\left(\frac{I}{I_o}\right) = -\mu' T \tag{2}$$

where, I_o = intensity (count rate) of the incident γ-beam
I = intensity (count rate) of the γ-beam after transmission through a thickness T
T = thickness (cm) of the attenuating matter
μ' = linear absorption coefficient (cm^{-1}) of the matter

now, with moist soil as the medium, substituting the linear absorption coefficient with the sum of the component mass absorption coefficients (μ) multiplied by the density (ρ) of each component will give us:

$$\mu' T_s = \mu_s \rho_s T_s + \mu_w \rho_w \theta T_s \tag{3}$$

where, μ_s = mass absorption coefficient (cm^2g^{-1}) of dry soil
ρ_s = specific density ($g\ cm^{-3}$) of dry soil
μ_w = mass absorption coefficient ($cm^2\ g^{-1}$) of soil water
ρ_w = specific density ($g\ cm^{-3} = 1$) of soil water
θ = volumetric water content of the soil
T_s = thickness of the soil

therefore, rewriting equation (2) will give:

$$\ln\left(\frac{I_{ms}}{I_o}\right) = -\ (\mu_s\rho_s T_s\ +\ \mu_w\theta T_s) \qquad (4)$$

where, I_o = count rate without soil
I_{ms} = count rate after transmission through moist soil.

"Two-wells" Gamma Probe

In this method two access tubes, as used for neutron moisture determinations, are placed in the soil at a defined distance apart, say 20–30 cm. A scintillation (NaI) detector probe is placed in one tube, and a point radiation source, usually ^{137}Cs, in the other.

It is then possible to determine the γ-photons reaching the detector after attenuation by the soil and the soil water, the pulses being recorded in the normal manner by a scaler. The count is inversely proportional to the water content, decreasing with increasing water content. The actual water content is then usually determined by reference to a calibration chart.

By using equation (4) developed above, and as applied in the following section, it is possible to derive the water content mathematically. In practice it is easier to prepare a calibration chart for the particular experimental conditions, because such a setup is intended for long term experiments.

The advantage of this method is that determinations can be made at the chosen depth, even very near the surface, within a soil layer only 4–6 cm thick. Additionally, in a long term experiment very frequent determinations can be made on the same soil sample without either disturbing the soil or the vegetation cover.

The same arrangement can be used for estimating bulk density, but in many cases it may be as convenient to make gravimetric determinations on surface soil layers.

Laboratory Application of Gamma Attenuation Technique

The application of gamma attenuation technique in the laboratory has made it possible to study in detail the movement and distribution of water in a horizontal or vertical soil column. Such studies may be made on columns of uniformly packed single texture soils, on "made up" soils with different horizons, or on columns of soil brought in from the field with the natural profile undisturbed. At the same time the bulk density distribution can be determined.

The cylinders containing such columns are made of perspex (lucite, acrilic, plexiglas) and provision must be made either for moving the cylinder in relation to the collimated source and detector, or else it must be possible to "scan" the cylinders with the detector assembly. As the radiation source may be from 25 up to 250 mCi ^{137}Cs considerable lead shielding is required, and the scanning apparatus must be substantially constructed (Fig. 14.5).

The basic theory and equations have been considered above but the practical situation with columns is somewhat more complicated, because the absorption coefficient of the perspex column must be taken into account. Moreover, as such a powerful gamma source is used, a true count of I_o, that is the intensity of the incident γ-beam, is impossible without the insertion of a lead absorber, as the potential count rate is usually too great for the capacity of the detector.

Various ways of developing the necessary equations for determining bulk density and water content are possible, but the following ([40]) is typical.

In the case of uniformly packed columns the average bulk density, $\bar{\rho}$; is of course already known from simple considerations of volume and weight but packing inevitably produces a density gradient. The intensity through the dry soil is expressed by:

$$I_{ds} = I_{Pb} \exp[(\mu\rho T)_{Pb} - (\mu\rho T)_c - (\mu\rho T)_{ds}] \tag{5}$$

and from this the dry bulk density at any point will be given by

$$P_{ds} = - \ln \frac{\left(\frac{I_{ds}}{I_c}\right)}{\mu_{ds} T_s} \tag{6}$$

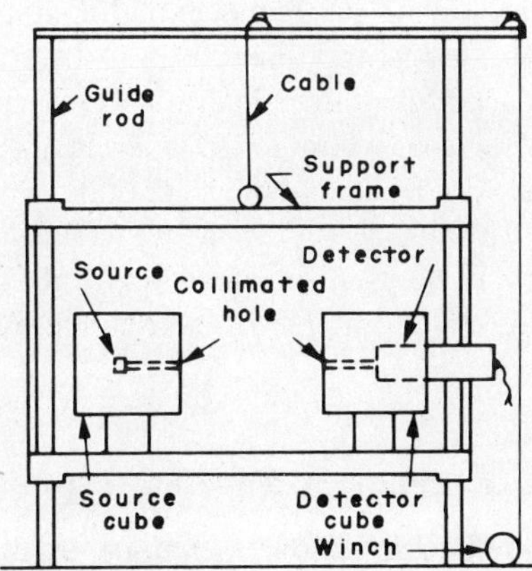

FIG. 14.5 Gamma-ray attenuation soil-moisture determination apparatus for scanning soil columns in the laboratory (Kirkham *et al.* [40]).

where, I_{ds} = count rate after transmission through dry soil
I_{Pb} = count rate after transmission through a standard lead absorber
$= I_o \exp(-\mu\rho T)_{Pb}$
I_c = count rate after transmission through plastic container
$= I_{Pb} \exp[\mu\rho T)_{Pb} - (\mu\rho T)_c]$

and $(\mu\rho T)_{Pb,c,ds}$ are the absorption coefficient, density and thickness of lead absorber, plastic container and dry soil respectively.

For moist soil the intensity is defined as:

$$I_{ms} = I_{Pb} \exp[\mu\rho T)_{Pb} - (\mu\rho T)_c - (\mu\rho T)_{ds} - (\mu_w\rho_w\theta T_s)] \tag{7}$$

and so the volumetric water content, θ, is

$$\theta = -\ln \frac{\left(\dfrac{I_{ms}}{I_{ds}}\right)}{(\mu_w\rho_w T_s)} \tag{8}$$

where, I_{ms} = count rate after transmission through moist soil
μ_w and ρ_w = mass absorption coefficient and density of water
T_s = thickness of the soil

Determination of mass absorption coefficients. It will be apparent from the equations that it is essential to know the mass absorption (attenuation) coefficients of water and dry soil. It is also necessary to determine the mass absorption coefficients of lead absorbers and of soil containers, usually made of plastic.

The attenuation of γ-radiation by any matter is defined by its linear absorption coefficient μ', measured in cm^{-1}, and its mass absorption coefficient μ, measured in cm^2g^{-1}. It is found that if the log of the intensity (counts) of the radiation after passing through the absorber is plotted against the thickness of the absorber (g/cm^2) a straight line is obtained. The slope of this line, $\frac{-\mu'}{P}$, is μ the mass absorption coefficient where ρ is the density of the absorbing matter, as is shown in Fig. 14.6.

The determination of mass absorption coefficients for water and soil, lead absorbers and plastic containers has been essentially described by Davidson *et al.* ([17]).

Rewriting equation (2), for the lead absorber we have

$$\ln\left(\frac{I_{Pb}}{I_o}\right) = -(\mu\rho T)_{Pb} \tag{9}$$

and for the plastic container material

$$\ln\left(\frac{I_c}{I_o}\right) = -(\mu\rho T)_c \tag{10}$$

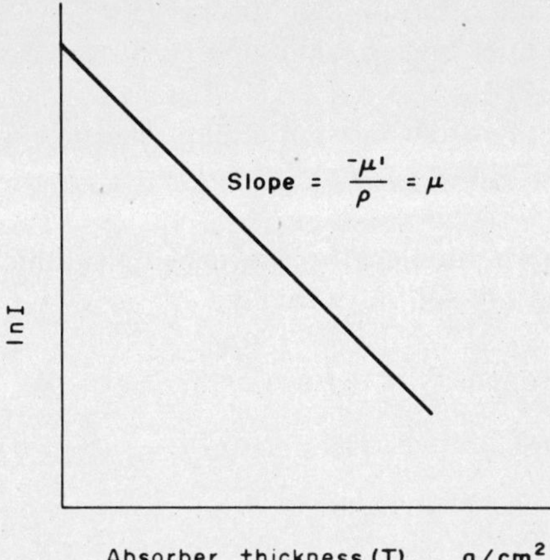

FIG. 14.6 Determination of mass absorbtion coefficient, μ, deriving from $\ln I = \ln I_o - \frac{\mu'}{\rho} T$

but the intensity of radiation passing through the plastic must be reduced by the insertion of the lead absorber, so this becomes

$$\ln\left(\frac{I}{I_{Pb}}\right) = -(\mu\rho T)_c + (\mu\rho T)_{Pb} \tag{11}$$

Therefore, if $-\ln\frac{I}{I_{Pb}}$ is plotted against $(\rho T)_c$ a straight line should be obtained, the slope being the absorption coefficient μ_c for plastic and the intercept being $(\mu\rho T)_{Pb}$.

Now the absorption coefficient for soil and water must be obtained with the sample in a plastic container, so the equation for determining them becomes:

$$\ln\left(\frac{I}{I_{Pb}}\right) = -(\mu\rho T)_{\substack{\text{soil}\\ \text{or water}}} + (\mu\rho T)_{Pb} - (\mu\rho T)_c \tag{12}$$

The plot of $-\ln\frac{I}{I_{Pb}}$ versus $(\rho T)_{\substack{\text{soil}\\ \text{or water}}}$ gives a line with the slope being the absorption coefficient of soil or water.

In practice, determinations are made by inserting perspex sample containers, either cylinders or boxes, of several different lengths between radiation source and detector. Then, after filling the containers either with known amounts of oven-dry soil, or water as appropriate, the radiation intensity I is measured passing through each sample of thickness T. The mean of three or four counts of 3 minutes duration is taken for each thickness. Similarly, counts should be taken for several thicknesses of the plastic container material.

The density, ρ, for soil and plastic is calculated from known weight and volume, while the density for water may be taken as 1 g cm^{-3}. The data is now available for plotting the lines to obtain the absorption coefficients of soil, water and plastic. With the lead absorber between the source and the detector the value I_{Pb} is obtained. As $(\mu \rho T)_{Pb}$ can be determined from the intercept of the graph made for plastic (see equation 11, and subsequently), it is then possible to calculate I_o, the intensity of the incident gamma beam (see equation 9).

Gamma ray absorption coefficients may also be estimated from published data (Davisson and Evans) ([19]), but it is better to perform the actual determinations under the specific experimental conditions.

Studies with Laboratory Soil Columns

Basic investigations of water infiltration, redistribution, diffusion, transient water content, evaporation and flow have been greatly helped by the development of gamma attenuation technique, which is rapid and non-destructive. Such studies are a specific area of soil physics and it would be inappropriate to go into further details here. Standard soil physics texts may be consulted for discussions of the principles and theory of the movement of water in soil columns (Kirkham and Powers) ([41]); (Hillel) ([30]); (Nielsen *et al.*) ([56]).

DETERMINING SOIL WATER PROPERTIES BY NEUTRON MOISTURE METER

The neutron moisture meter has become one of the most powerful tools in the hands of the soil physicist, but it is intended here to discuss primarily its application in agronomy and plant ecology studies where we are directly concerned with the availability of water to plants.

The maximum available water potentially usable by any crop or plant cover on a given soil is conventionally defined by the difference between the water held by the soil at *field capacity* and at the *permanent wilting point*, taking into account the depth of the active root zone. In fact the situation is not as clear cut as this, as evapotranspiration affects the downward flow of water in the profile, while plants are under increasing "water stress" as soil moisture tension increases. Also, as soils dry out plants tend to take water from lower depths where root density is less. The well-known general relationship of field capacity and permanent wilting point is shown in Fig. 14.7.

Field capacity is considered to be the amount of water remaining in a soil after excess gravitational water has been allowed to drain off following heavy rain or irrigation. At the time field capacity is measured the downward drainage of water will

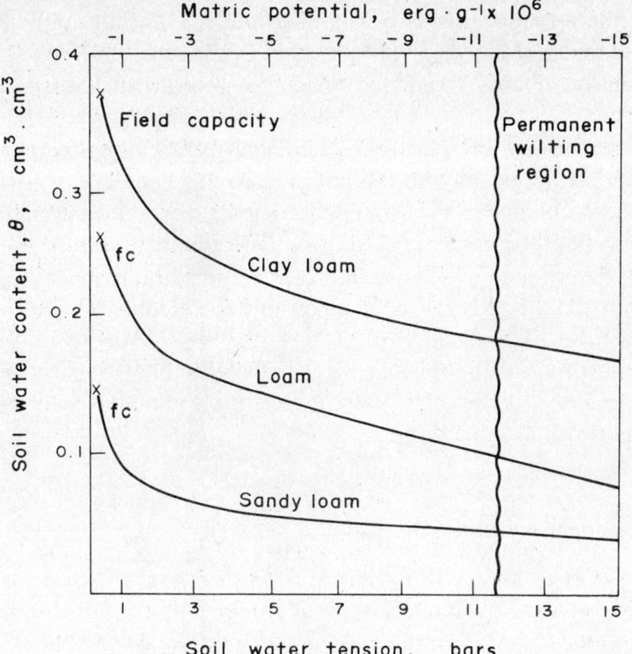

FIG. 14.7 General relationship of field capacity and wilting point to soil water content and texture.

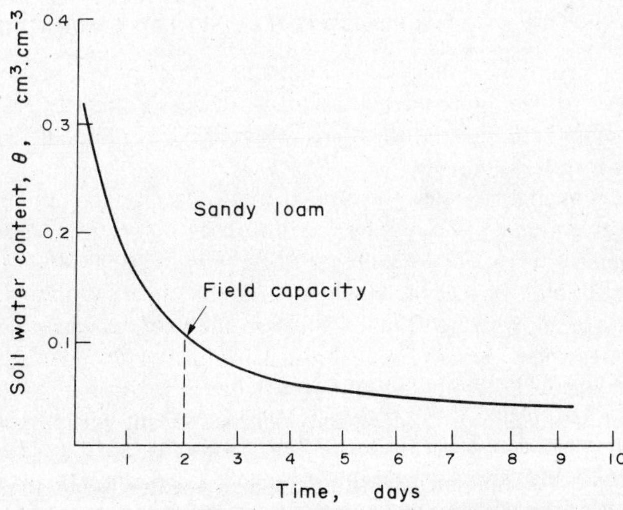

FIG. 14.8 Water content, θ, as a function of time, at a given depth. Field capacity at conventional 2 days, following irrigation or heavy precipitation. Otherwise inferred from the inflection of the curve.

almost have ceased, and a soil in this condition may also be spoken of as "at field capacity". Field capacity has been alternatively defined as the amount of water held against gravity or ⅓ bar tension, although this figure is only an approximation.

Conventionally, field capacity has been measured on the plough layer after allowing 2–3 days for excess water from irrigation or heavy rain to drain down, but this may not be sufficiently long for some soils, heavy clays for instance, which may take up to 2 weeks. Failure to recognize this has sometimes lead to the view that some soils have no definite measurable field capacity. Additionally the conventional determination takes no account of modern views of field capacity being a dynamic property of the soil profile. The neutron moisture meter now makes it practical to make a series of moisture determinations as a function both of time (Fig. 14.8) and depth (Figs. 14.9 and 14.12) on the soil profile. This in turn making possible a more sophisticated treatment of field capacity.

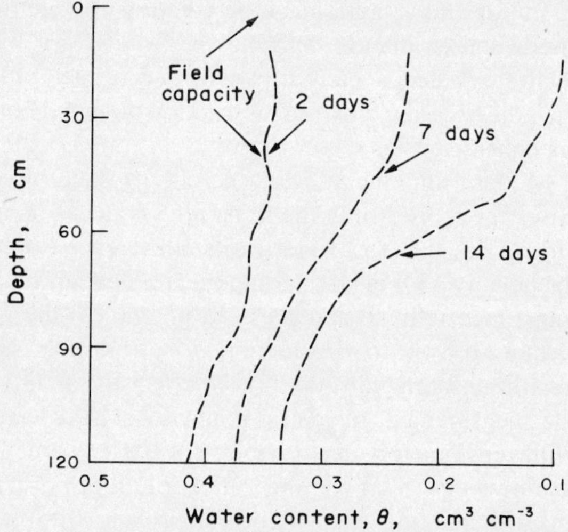

FIG. 14.9 Redistribution of soil water in the profile following irrigation or heavy precipitation.

Although Fig. 14.8 shows the assumed conventional situation where field capacity is measured 2 days following irrigation, the neutron moisture meter makes it practical to determine, by continued measurement over a period of time, the field capacity of heavy clay soils where a much longer period is required for excess water to drain down. The "field capacity point" can then be inferred from the shape of water content curve plotted against time.

Redistribution of soil water in the profile can be studied as an extension of determining field capacity in relation to depth (Fig. 14.9), by making subsequent determinations at longer time intervals. By this means the depth and rate of infiltration can be estimated for different soils, and in relation to different crops, stages of growth and the active rooting zone.

Rough estimates of evaporation, transpiration and redistribution can also be carried out, if three experimental areas are selected, two with bare soil and the third with complete ground cover such as grass. Redistribution is estimated from one of the areas without vegetation covered with a polythene sheet, while the other area is used for estimating evaporation. The area with vegetation will provide an estimate of the combined effects of transpiration, evaporation and redistribution by percolation to the lower profile levels ([55,58]).

The volumetric water content at a specific time for any particular section of the soil profile can be obtained by integrating the appropriate parts of the area beneath the water content curve as a function of depth, as shown in Fig. 14.12. In practice all three areas should receive the same volume of irrigation water at the start of the experiment. Measurement of soil water content every 15 cm to a depth of 100–150 cm is then carried out immediately after infiltration and at regular intervals over a period of 10–20 days.

As of course evaporation, transpiration and percolation are rate inter-dependent the results are only a general approximation. This may be quite acceptable in some ecology studies. For a proper water balance study it is necessary to determine not only water content but the matric potential, the direction of the moisture gradient and the hydraulic conductivity. This is considered later.

Matric potential. Modern thinking stresses the energy state of soil water. As soil water content decreases the more firmly the remaining water is retained, and because energy is required to restore this soil water to its "reference" state we consider its potential energy to be negative. Of the four component potentials of soil water, (matric; osmotic due to solutes; gravitational and pressure potentials) the matric potential or capillary potential resulting from the interaction of soil particle surfaces with water is primarily of interest in unsaturated soils. It is in effect the work required to remove water from soil. The measurement of matric potential or soil water tension may be done in the field with tensiometers up to about 0.8 bar tension, but above this the water column breaks down due to air coming out of solution. For tensions higher than this, pressure plate determinations offer the best approach ([12,63,67]).

The determination of matric potential with mercury manometer tensiometers enables the construction of soil water characteristic curves at the lower tensions, under field conditions. These provide basic information on pore size distribution and water retention capacity, while knowledge of the matric potential, Ψ_m, is essential for calculations involved in soil water flow, redistribution and balance studies. The matric potential may be calculated from the formula

$$\Psi_m = -g(12.55x - y - z)\text{erg.g}^{-1} \tag{13}$$

or, expressed as an equivalent head of water, h (cm)

$$h = (12.55x - y - z)\text{cm} \tag{14}$$

where, g = gravitational acceleration = 980 $cm.s^{-2}$
x = length of mercury column
y = height of mercury reservoir above ground
z = depth of porous cup below ground
h = head of water (cm) equivalent to matric potential

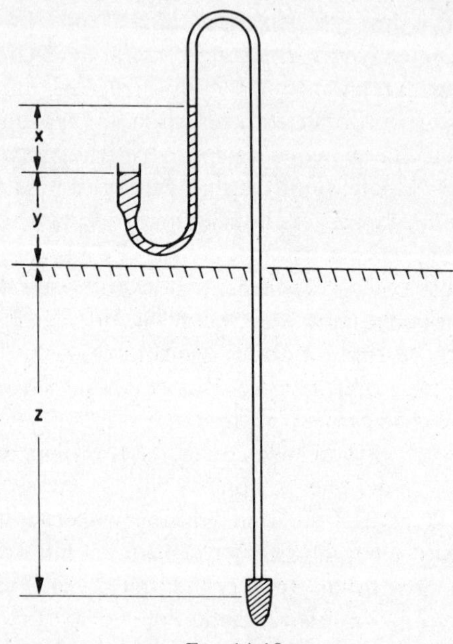

FIG. 14.10

Access tubes for neutron moisture determination are placed not further than 1 metre from where duplicated porous cup tensiometers are installed. The depth of installation of the cups will vary with the soil from 75–125 cm. The experimental site is given heavy irrigation to wet the soil to at least 1 metre depth. When infiltration is completed the tensiometers are read for the first time and a neutron moisture determination is also made, and these readings are repeated daily for at least 14 days, or over a longer period for heavy soils. The daily calculated matric potential can then be plotted against the soil water content, θ, to give the soil water characteristic curve.

The permanent wilting point, or percentage, is considered to be the lower limit of plant-available water in soil. In general, most plants tend to reduce the soil moisture content to the same level before the onset of wilting. Conventionally, the permanent wilting percentage has been considered to correspond to the amount of water retained against 15 bars tension, although it is better to consider a permanent wilting region extending from below 12 bars through 15 bars tension and beyond. Probably the majority of crops will show reduced yields if soil moisture tensions are allowed to fall as low as 8 bars, before rainfall or irrigation, and the best irrigation practice would favour irrigation long before this tension was reached.

For pressure plate determinations the soil samples are brought into equilibrium with a series of ever increasing pressures up to 15 bars; then if the weight of the soil sample is determined at each pressure it permits construction of a desorption curve of water content versus soil water tension. The permanent wilting percentage is then inferred either from the 15 bar water content of the sample, and/or from the shape of the desorption curve, as it will tend to flatten in the region of permanent wilting (Fig. 14.7).

Although the permanent wilting percentage is higher for clay soils than for sandy soils, these soils also have a higher field capacity and consequently their actual extractable water is also greater, say 15–20% for clays and 7–10% for sandy soils. The desorption curve plot of soil water content against increasing tension shows that the fine particle clay soils release water much more gradually in the high tension range, compared with sands which rapidly release water even in the low tension range. Consequently the fine textured soils are much more resistant to drought conditions even though their permanent wilting percentages may be higher than for sandy soils. If a desorption curve has been constructed it permits the inference under field conditions of soil moisture tension from the measurement of water content.

Therefore, if a complete desorption curve is available, or even if the wilting percentage is known either from pressure plate determinations or an actual plant experiment, subsequent field checks with a neutron moisture meter will permit a rapid estimation as to whether soil moisture tension is approaching the point where yield will be reduced unless irrigation water is applied.

Hydraulic conductivity. The soil-water transmission properties of a soil are measured by its hydraulic conductivity or proportionality constant. The hydraulic conductivity is given by the volume of flow in unit time per unit cross section area divided by the hydraulic gradient. The estimation of hydraulic conductivity is essential for the determination of soil water fluxes and hence e.g. water balance in the soil profile. Expressions for hydraulic conductivity can be derived from Darcy's equation, showing that the volume rate of flow is proportional to the hydraulic gradient, taking into account water content and moisture tension at different depths in the profile ([9,12,27,57,65,66,68,85,88]), considering

$$v_L = -K(\theta)\frac{dh}{dz} \tag{15}$$

where, $K(\theta)$ = hydraulic conductivity (cm day^{-1}) at depth z = L

v_L = steady state flux of water passing through a unit soil surface per unit time (cm^3 cm^{-2} day^{-1})

$\frac{dH}{dz}$ = total soil water potential gradient (cm cm^{-1})

Then if $\frac{d\bar{\theta}}{dt}$ is the change in water content with time, for non-steady state conditions equation (15) can be re-written as:

$$\frac{d\bar{\theta}}{dt} = \frac{d}{dz}\left[K(\theta)\frac{dH}{dz}\right] \tag{16}$$

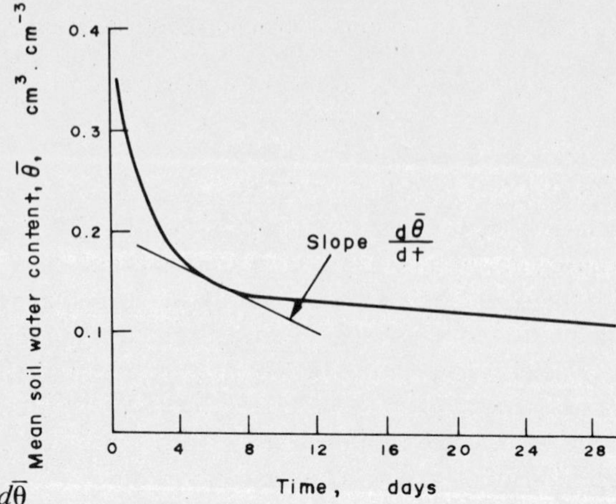

FIG. 14.11 Slope $\frac{d\bar{\theta}}{dt}$ defines the change in mean water content with redistribution time after irrigation. $K = -L\frac{d\bar{\theta}}{dt}$, where L is the profile depth (equation 19).

but as we are interested in the change of water content with time *as a function of soil depth* the equation must be integrated from the surface, where z = O, to a depth where z = L, giving:

$$L\frac{d\bar{\theta}}{dt} = \underbrace{\left[K\frac{dH}{dz}\right]_{z=L}}_{\bar{1}} - \underbrace{\left[K\frac{dH}{dz}\right]_{z=0}}_{\bar{2}} \tag{17}$$

where, $\bar{\theta}$ = mean soil water content ($cm^3\ cm^{-3}$)
term $\bar{1}$ = water flux across depth L ($cm^3\ cm^{-2}\ day^{-1}$)
term $\bar{2}$ = water flux across the soil surface ($cm^3\ cm^{-2}\ day^{-1}$).

In order to determine the factors in the above equations as the water in the profile decreases, $\frac{d\bar{\theta}}{dt}$, the change in water content with time, is measured by neutron moisture meter, and $\frac{dH}{dz}$, the total soil water potential gradient (i.e. hydraulic head gradient) at depth L, by tensiometers.

Then,

$$K = \frac{L\frac{d\bar{\theta}}{dt}}{\frac{dH}{dz}} \tag{18}$$

However, it is clear that in many cases after irrigation or heavy precipitation the major part of the water will be draining out of the profile at depth L, while water evaporating at the soil surface will be small in comparison, and therefore term *2* can be ignored. It has been shown ([9,57]) in practice that little is lost by estimating $\frac{dH}{dz}$ as -1, giving:

$$K = -L\frac{d\bar{\theta}}{dt} \tag{19}$$

Practical determination of K is carried out on a site with a water table at least 2 m in depth. A level area of bare earth of 25m² or more is surrounded by a dike to permit heavy irrigation with about 0.5 m of water. Duplicate 2 m long access tubes are installed about 1.5 m apart, and duplicate paired tensiometers are installed similarly. Of each pair of tensiometers, one should be 1.0 m long and the other 1.4 or 1.5 m long.

The tensiometers are read as infiltration ceases (zero time) and at the same time neutron moisture determinations are made every 15 cm down to 120 cm. Readings of both tensiometers and neutron moisture meter are made at 12, 24, 36 and 48 hours, then every 24 hours for 14 days or more. As there are eight depth intervals the value of $\bar{\theta}$ can be calculated by dividing the sum of the eight separate θ values by 8. $\bar{\theta}$ is then plotted as a function of time and the slope of the line $\frac{d\bar{\theta}}{dt}$ is calculated as in Fig. 14.11. *K* may then be calculated for any experimental time using equation (19).

If equation (18) is to be used it is necessary to determine the matric potential head (in cm) from tensiometer readings at the two depths, employing equation (14). Then the hydraulic head gradient, $\left(\frac{dH}{dz}\right)$, is calculated for the difference in depth between the two tensiometers i.e. for the 100 to 150 cm (or other) depth interval. If we assume the deeper tensiometer was at 150 cm depth, and h_1 and h_2 are the matric potential heads as determined at 100 and 150 cm depths respectively, then the equation will be:

$$\frac{dH}{dz} = \frac{h_1 - h_2 + 50}{50} \tag{20}$$

K may again be calculated by substituting in equation (18).

WATER BALANCE AND WATER USE EFFICIENCY

To determine the efficiency of water use in irrigation studies or in the study of water relations in natural plant communities it is necessary to establish the water balance of the soil profile under different climatic conditions. This means calculating the quantity of water entering and leaving the soil profile due to rainfall, irrigation, drainage and evaporation. The large number of determinations required at different depths makes the neutron moisture meter essential.

The volume of water required to irrigate a soil to a given depth can be estimated in a rather elementary way. In general the extractable water available to a crop can be described as:

Extractable water = Field capacity—Water Content in rooting zone at wilting percentage (*or other arbitrary limit, such as water content at 2 or 3 bar suction*)

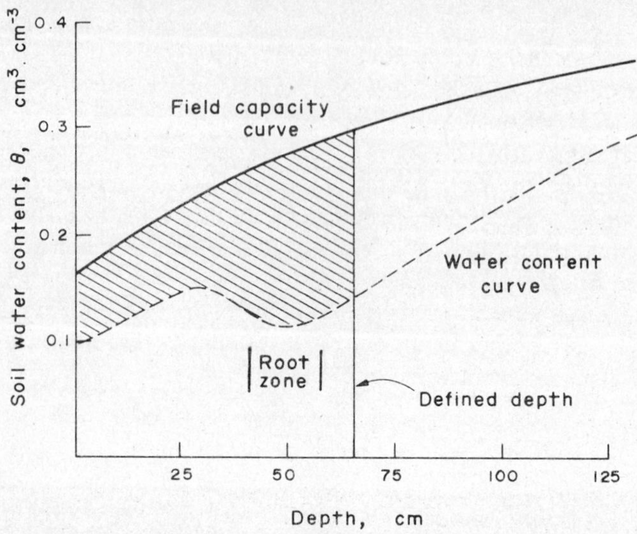

FIG. 14.12 Water content and field capacity as a function of depth. The field capacity curve typifies the not uncommon situation where soil texture becomes heavier with depth. Water content is typically depleted in the immediate root zone of an actively growing crop. Integration of the shaded area provides a means of determining the amount of irrigation water that must be applied to bring the soil up to field capacity.

When a field capacity curve as a function of depth has been constructed it is possible to calculate the volume of irrigation water required to bring the soil water content to field capacity down to a defined depth. This is shown in Fig. 14.12. The broken line is the water content curve as function of depth, and integrating the area beneath this curve (to the defined depth) will give the total volume of water in the soil at the given time. Similarly, integrating the area beneath the field capacity curve will give the maximum volume of water held at field capacity. Subtracting the first value from the second will give the volume of water, represented by the cross-hatched area, needed to bring the soil up to field capacity.

In Fig. 14.12 the water content of about 0.1 $cm^3 cm^{-3}$ prevailing in the root zone approximates to the wilting percentage of a loam soil. Such treatment of field capacity and irrigation water requirement is only practicable with the neutron moisture meter. The limitations of the approach are clearly of a simplistic nature, as it implies uniformity of soil over wide areas, and takes no immediate account of evapotranspiration, while downward water loss may be important on sandy and free-draining soils ([45,70,83,84]). Therefore, a more rigorous approach to water balance studies means calculating the quantity of water both entering and leaving the profile due to various factors.

Until comparatively recently the only method of carrying out such studies was to build a reconstructed soil profile in a large lysimeter container provided with a drainage system and incorporating some sort of weighing device ([22,29]). This is far from ideal as there is no guarantee that the mass of soil behaves in the same manner as the reconstructed profile, moreover, considerable labour and expense is required for establishing lysimeter pits and containers.

The comparative advantage of neutron moisture procedures and lysimeters for water balance studies was considered by Holmes and Colville ([34]) and De Boodt *et al.* ([12]). The latter investigators indicated that by using a combination of neutron moisture determinations and tensiometer readings it was possible to obtain data indicating how much water leaves the profile in a given period of time and in what direction.

With water content, the matric potential and the hydraulic conductivity calculated from neutron moisture meter and tensiometer data it becomes feasible to predict the infiltration rate and the depth that the waterfront will reach when a given quantity of water is used to irrigate the soil under investigation.

As Barrada ([4]) has pointed out the main problems lie in estimating the percolation in the profile below rooting depth and in determining evaporation over a defined period of time. This involved solving the equation

$$R + I - S = \Delta M + \int E dt + U \qquad (21)$$

FIG. 14.13 Troxler neutron moisture meter with the shield in position on an access tube placed in an onion crop.

where, R = rainfall
I = irrigation
S = surface run off
ΔM = change in soil water storage
$\int Edt$ = evaporation over defined time
U = drainage below the root zone.

The factors in the first half of the equation present comparatively little difficulty. Rainfall can be determined with standard rain gauges and irrigation can be regulated, assuming uniformity of application. Estimating run off could be a problem with certain conditions, although under experimental situations it can be contained. ΔM can be easily determined with the neutron moisture meter.

We are therefore left with two knowns, $\int Edt$ and U, one of which has to be determined to estimate the other. Without determining U directly, as with a lysimeter, it can be estimated from the following considerations.

If a soil has a uniform profile then the upper limit to the magnitude of the flux (v) is determined by saturated flow under the influence of the gratitational potential gradient alone. The value of v is then equal to the hydraulic conductivity, K, at saturation in the z-direction.

$$v = K\frac{dH}{dz} + K \text{ cm.sec}^{-1} \quad (z_d \text{ positive}) \tag{22}$$

and

$$U = \int_0^T vdt \tag{23}$$

where T is the time interval between measurements of successive water content profiles. To calculate U it is necessary to determine the slope of the hydraulic head gradient, $\frac{dH}{dz}$, at the lower limit of the soil volume under investigation. The hydraulic conductivity at that depth and its variation with water content must also be determined. $\frac{dH}{dz}$ is determined by means of tensiometers while water content determinations can be carried out with the neutron moisture meter.

A series of tensiometers with the cups placed at different depths make it possible to detect the upward or downwards vertical movement of water in the profile and enables the detection of water flux differences between two depths, when upward water movement in the profile may be caused by the crop.

Having determined the percolation, U, below the root zone it will be apparent that equation (21) can then be used to determine the evaporation, as the change in soil moisture storage ΔM can be calculated using a similar procedure to that given previously.

Detailed water balance studies are important because they should lead to a better understanding of crop water needs under varied soil and climatic conditions. But to be able to apply the results of such studies to irrigation practice requires that water balance can be determined easily and cheaply. The question therefore arises both in balance and water use efficiency studies, and for practical questions of determining the need for irrigation, as to whether neutron moisture determinations of soil water give an accurate estimate of water in the profile? Moreover, does evapotranspiration calculated from neutron moisture data as described above correlate reasonably well with estimates of potential evapotranspiration determined by other means?

Lysimeter studies sponsored by IAEA/FAO carried out in several countries over a number of years ([11,14,36,46,71,87]), have done much to answer these questions. In general there was very good agreement between the amount of water in the profile as determined by weighing lysimeter and as calculated from neutron moisture data.

Information on evaporation potential is necessary for water efficiency studies as if the crop is not able to transpire at the potential rate for closed plant cover then the growth rate may be affected due to partially closed stomata. Also, from the practical aspect, irrigation water is sometimes applied to the extent of 80% of the estimated potential evapotranspiration. There are various methods of measuring or estimating evapotranspiration ([62,74]) of which the lysimeter technique can be regarded as the basic reference standard.

Actual evapotranspiration is determined by lysimeter through measuring the total water lost from the soil mass, while that portion of water lost through drainage is also found by collection in a container beneath the lysimeter. Then,

$$E = R - P + \Delta M \qquad (24)$$

where,

ΔM = change in soil water storage (mm) calculated from neutron moisture data
R = rainfall (mm) during a given period
P = percolation (mm) in the same period.

Estimates of potential evapotranspiration, E_o, based on meteorological data and utilized in semi-empirical, semi-rational, equations have been developed. Of these, that of Penman ([60,61,62]) is probably the best known and most widely used. For calculation of the Penman equation it is necessary to know: mean air temperature; duration of bright sunshine; mean wind speed; and vapour pressure. Therefore, such estimates require that the cropping area under consideration is reasonably near a meteorological station, or that meteorological conditions are uniform over a wide area. Regular computation of the equation is assisted by the use of tables ([50]). The form most commonly used is,

$$E_o = \frac{\left(\frac{\Delta}{\gamma} H_o + E_a\right)}{\frac{\Delta}{\gamma} + 1} \qquad (25)$$

where (much simplified),

H_o = radiation term, measured directly or calculated using mean air temperature and duration of bright sunshine

E_a = evaporation term (mm/day), depending on wind speed and velocity

$\frac{\Delta}{\gamma}$ = ratio of the slope of the saturaton vapour pressure curve at mean temperature to the constant of the wet and dry bulb psychrometer equation (obtained from tables).

The international studies summarised in reference 36 suggest that lysimeter, neutron moisture and "Penman" estimates of evapotranspiration are capable of giving comparable results in many circumstances, but that all have their limitations. A basic limitation of the Penman method is that it is not a direct measurement, but potential evapotranspiration values were in good agreement with actual evapotranspiration, providing water stress in the crop was avoided. It was found too that open-pan evaporation compared well with actual evapotranspiration from wet lysimeters. The sunken "Colorado" type pan seemed to give better correlation in some cases than the raised class 'A' type pan.

In general the actual evapotranspiration calculated from neutron moisture data was in good agreement with the lysimeter data. The main limitation of the neutron moisture meter appears to be the relatively limited number of measurements that are made in relation to the spatial variability of the soil over large areas of cropland ([14,57]) (Dagg and Wangati) ([36]) and also the inevitable approximation of the calculations. Nevertheless, compared with lysimeters neutron moisture measurements do appear to reflect the actual soil moisture situation in the field, which at least under dryland conditions appears to be drier than the soil in the adjacent lysimeter, due to the latter's impeded drainage (Hillel *et al.*; Al-Nakshabandi and Ismail) ([36]).

However, to interpret soil water content data as found in the field by neutron moisture meter it seems essential that the measurement depth should exceed the rooting depth and that sufficient tensiometer data must be available to calculate potential gradients and to provide information on the direction of water flow below the rooting zone.

TRACERS FOR SOIL WATER

Tracing and quantitatively determining the movement of soil water is fundamental to a better understanding of irrigation, drainage, mass flow of nutrient ions in soil water and soil microbiological reactions.

A tracer for soil water must be stable, not adsorbed on the clay or organic fractions, move at the same speed as soil water, have suitable radiation characteristics for detection and an appropriate half-life. In practice, tritium, ^{3}H ($t_{1/2}$ = 12.26 yr) is almost exclusively used for water tracing ([10,39]). Obviously the stable deuterium, ^{2}H or ^{18}O isotopes could also be used but the comparative trouble of determining stable isotopes compared with the ease and sensitivity of radioisotope determinations easily make ^{3}H and isotope of choice, though it does suffer from being a weak β-emitter.

Anions are clearly more suitable for use than cations, because they are not adsorbed, and ^{131}I ($t_{1/2}$ = 8.05 days) has been successfully used ([7]) and ^{35}S ($t_{1/2}$ = 86.7 days). However, their half lives are rather short for some experiments. ^{36}Cl ($t_{1/2} = 3 \times 10^5$ yr) can be considered as an alternative and would be the normal choice in salinization, leaching and drainage studies ([38]). It moves with soil water and is little adsorbed by clays and organic matter. $^{15}NO_3$ will move with soil water but can only be considered as a tracer for NO_3 due to the microbiological reactions which take place as it moves down the soil profile ([52,72]).

Cations are adsorbed on clays and may only be used when complexed, as with EDTA for example. Isotopes such as ^{60}Co ($t_{1/2}$ = 5.27 yr) or ^{51}Cr ($t_{1/2}$ = 27.8 days) might be used in this way, if liquid scintillation counting facilities were not available, or for other reasons when more energetic gamma-emitters are likely to be easier to count.

A systematic study of many cation complexes was made ([75]), including ^{65}Zn, ^{51}Cr, ^{59}Fe and ^{60}Co with citrate, EDTA, NaCNS, oxalate and KCN. It was concluded that there was no cation-complex combination universally suitable for all soils, and that it was necessary to carry out a preliminary investigation for each particular experimental soil situation. However, for sandy soils zinc-EDTA, iron-EDTA, iron-KCN, cobalt-EDTA, cobalt-KCN and chromium-EDTA seemed the most suitable. For clay soils: zinc-EDTA, iron-EDTA and KCN, together with cobalt-EDTA and KCN were appropriate.

REFERENCES FOR FURTHER READING

1. ANDRIEUX, C., Buscarlet, L., Guitton, J. and Mérite, B. Mesure en profondeur de la teneur en eau des sols par ralentissement des neutrons rapides. *In: Radioisotopes in Soil-Plant Nutrition Studies*, 187–219, IAEA, Vienna (1962).
2. BARRADA, Y. *Principles Involved in Different Types of Radiation Equipment*. Panel on use of isotopes and radiation techniques in soil moisture and irrigation studies of irrigated land. IAEA, Vienna (1964).
3. BARRADA, Y. The neutron moisture meter and its value in water-use efficiency studies. *Atomic Energy Review* **3**, 195–212 (1965).
4. BARRADA, Y. *The Use of Radiation Equipment in Long-term Soil Moisture and Irrigation Studies*. Lecture T.102, IAEA/FAO International Training Course on the Use of Radioisotopes and Radiation in Soil and Plant Nutrition Studies, ITAL, Wageningen (1970) Mimeo.
5. BELCHER, D. J., Cuykendall, T. R. and Sack, H. S. The measurement of soil moisture and density by neutron and gamma ray scattering. *U.S. Civil Aeron. Admin. Techn. Develop. Rept.* no. 127 (1950).
6. BELL, J. P. and MCCULLOCH, J. S. G. Soil moisture estimation by the neutron scattering method in Britain. *J. Hydrol.* **4**, 254 (1966).
7. BENECKE, P. Investigations into the behaviour of precipitation water in soils by means of ^{131}I. *In: Proc. Symp. Isotope and Radiation Techniques in Soil Physics and Irrigation Studies*, Istanbul 1967, 227–39, IAEA, Vienna (1967).
8. BENZ, L. C., Willis, W. W., Nielsen, D. R. and Sandoval, F. M. Neutron moisture meter calibration. *Agric. Eng.* **46**, 326 (1965).
9. BLACK, T. A., Gardner, W. R. and Thurtell, G. W. The prediction of evaporation, drainage and soil water storage for a bare soil. *Soil Sci. Soc. Amer. Proc.* **33**, 655 (1969).
10. BLUME, H. P., Zimmermann, U. and Münnich, K. O. Tritium tagging of soil moisture: the water balance of forest soils. *In: Proc. Symp. Isotopes and Radiation Techniques in Soil Physics and Irrigation Studies*, Istanbul 1967, 315–32, IAEA, Vienna (1967).
11. DE BOODT, M., Moerman, P. and De Boever, J. Comparative study of the water balance in the aerated zones with radioactive methods and weighable lysimeter. *In: Proc. Symp. Water in the Unsaturated Zone*, Wageningen 1966, **1**, 63–74, IASH and UNESCO, Paris (1968).

12. de Boodt, M., Hartmann, R. and De Meester, P. Determination of soil moisture characteristics for irrigation purposes by neutron-moisture meter and air-purged tensiometers. *In: Proc. Symp. Isotope and Radiation Techniques in Soil Physics and Irrigation Studies*, Istanbul 1967, 147–60, IAEA, Vienna (1967).
13. de Boodt, M. Introduction to the determination of the moisture content and the bulk density of the soil by radioactive measurements. Lecture T.95 IAEA/FAO International Training Course on the Use of Radioisotopes and Radiation in Soil and Plant Nutrition Studies. ITAL, Wageningen (1970).
14. de Boodt, M., Verplancke, H. and Laksmipathy, A. V. Study on the sensitivity of soil moisture measurements using radiation techniques, nylon resistance units and a precise lysimeter system. *In: Proc. Symp. Isotopes and Radiation in Soil-Plant Relationships including Forestry*, Vienna 1971, 605–17, IAEA, Vienna (1972).
15. Damagnez, J. Conditions d'utilisation de la sonde à neutrons pour les déterminations d'humidité dans le sol: influence de la densité apparente et de la salure. *In: Radioisotopes in Soil-Plant Nutrition Studies*, 159–69, IAEA, Vienna (1962).
16. Danfors, E. Changes in moisture content of topsoil measured with a neutron moisture gauge. *Proc. Symp. Water in the Unsaturated Zone*, Wageningen 1966, **1**, UNESCO, Paris (1969).
17. Davidson, J. M., Biggar, J. W. and Nielsen, D. R. Gamma-radiation attenuation for measuring bulk density and transient water flow in porous materials. *J. Geophys. Res.* **68**, 4777 (1963).
18. Davidson, J. M., Stone, L. R., Nielsen, D. R. and LaRue, M. E. Field measurement and use of soil properties. *Water Resour. Res.* **5**, 1312 (1969).
19. Davisson, C. M. and Evans, R. D. Gamma-ray absorbtion coefficients. *Rev. Mod. Phys.* **24**, 79 (1952).
20. Eeles, C. W. O. *Installation of Access Tubes and Calibration of Neutron Moisture Probes*. Institute of Hydrology, Wallingford, Berks. Rep. no. 7, 20 pp. (1969).
21. Ferguson, H. and Gardner, W. H. Water content measurement in soil columns by gamma ray absorbtion. *Soil Sci. Soc. Amer. Proc.* **26**, 11 (1962).
22. Forsgate, J. A., Hosegood, P. H. and McCullock, J. S. G. Design and installation of semi-enclosed hydraulic lysimeters. *Agric. Meteorology* **2**, 1 (1965).
23. Gardner, W. R. and Kirkham, D. Determination of soil moisture by neutron scattering. *Soil Sci.* **72**, 391–401 (1952).
24. Gardner, W. H. Water Content. *In*: Methods of Soil Analysis, Part I. *Amer. Soc. Agron.* Monograph 9, 82 (1965).
25. Gardner, W. H. and Calissendorff, C. Gamma ray and neutron attenuation in measurement of soil bulk density and water content. *In: Proc. Symp. Isotopes and Radiation Techniques in Soil Physics and Irrigation Studies*, Istanbul 1967, 101–13, IAEA, Vienna (1967).
26. Gardner, W. R., Hillel, D. and Benyamini, Y. Post-irrigation movement of soil water. 1. Redistribution. *Water Resour. Res.* **6**, 1148–153 (1970).
27. Giesel, W., Lorch, S. and Renger, M. Water-flow tensiometer field measurements in the unsaturated soil profile. *In: Proc. Symp. Isotope Hydrology*, Vienna 1970, 663, IAEA, Vienna (1970).
28. Gurr, C. G. Use of gamma rays in measuring water content and permeability in unsaturated columns of soil. *Soil Sci.* **94**, 224 (1962).
29. Hillel, D., Gairon, S., Falkenflug, V. and Rawitz, E. New design of a low-cost hydraulic lysimeter system for field measurements of evapotranspiration. *Isr. J. Agri. Res.* **19**, 2 (1969).
30. Hillel, D. *Soil and Water, Physical Principles and Processes*. Academic Press, New York (1972).
31. Hewlett, J. D., Douglass, J. E. and Clutter, J. L. Instrumental and soil moisture variance using the neutron scattering method. *Soil Sci.* **97**, 19–24 (1964).
32. Holmes, J. W. Calibration and field use of the neutron scattering method of measuring soil water content. *Aust. J. Appl. Sci.* **7**, 45–58 (1956).
33. Holmes, J. W. and Jenkinson, A. F. Techniques for using the neutron moisture meter. *J. Agric. Eng. Res.* **4**, 100–109 (1959).
34. Holmes, J. W. and Colville, J. S. The use of neutron moisture meter and lysimeters for water balance studies. 8th Int. Congr. Soil Sci., Bucarest, Trans. II, 445 (1964).
35. IAEA. *Neutron Moisture Gauges*. Tech. Rept. 112, IAEA, Vienna (1970).
36. IAEA. *Radiation Techniques for Water-use Efficiency Studies*. Tech. Rept. 168, pp. 125, IAEA, Vienna (1975).
37. Jensen, P. A. and Somer, E. Scintillation techniques in soil moisture and density measurements. *In: Proc. Symp. Isotope and Radiation Techniques in Soil Physics and Irrigation Studies*, Istanbul 1967, 31–48, IAEA, Vienna (1967).

38. KIRDA, C., Nielsen, D. R. and Biggar, J. W. Simultaneous transport of chloride and water during infiltration. *Soil Sci. Amer. Proc.* **37**, 339 (1973).
39. KIRDA, C., Nielsen, D. R. and Biggar, J. W. The combined effects of infiltration and redistribution on leaching. *Soil Sci.* **117**, 323 (1974).
40. KIRKHAM, D., Rolston, D. E. and Fritton, D. D. Gamma radiation detection of water content in two-dimensional evaporation prevention experiments. *In: Proc. Symp. Isotope and Radiation Techniques in Soil Physics and Irrigation Studies*, Istanbul 1967, 3–16, IAEA, Vienna (1967).
41. KIRKHAM, D. and Powers, W. L. *Advanced Soil Physics*, Wiley, New York (1972).
42. KNIGHT, A. H. and Wright, T. W. Soil moisture determinations by neutron scattering. *Proc. Radioisotope Conf. Oxford* **2**, 111–22 (1954).
43. KÜHN, W. La mesure de la densité apparante et son influence sur la determination de l'humidité à l'aide de neutrons. *Atompraxis* **5**, 335–38 (1959).
44. LAPP, E. R. and Andrews, H. L. *Nuclear Radiation Physics*. 3rd Edn. Prentice Hall Inc., New York (1963).
45. LA RUE, M. E., Nielsen, D. R. and Hagan, R. M. Soil water flux below a ryegrass root zone. *Agron J.* **60**, 635 (1958).
46. MACKSOUD, S. W. Problems and costs of water use studies by neutron probe and lysimeter. *In: Proc. Symp. Use of Isotopes in Hydrology*, Beirut, 1970 Lebanese National Council for Scientific Research (1972).
47. MARAIS, P. G. and Smit, W. B. de V. Laboratory calibration of the neutron moisture meter. *S. Afr. J. Agric. Sci.* **3**, 4 (1960).
48. MARAIS, P. G. and Smit, W. B. de V. Effect of bulk density of soils on the calibration curve of the neutron moisture meter. *S. Afr. J. Agric. Sci.* **3**, 475–77 (1960).
49. MARCESSE, J. Determination in situ de la capacité de retention d'un sol au moyen de l'humidimetre a neutrons. *In: Proc. Symp. Isotope and Radiation Techniques in Soil Physics and Irrigation Studies*, Istanbul 1967, 137–46, IAEA, Vienna (1967).
50. MCCULLOCH, J. S. G. Tables for rapid computation of the Penman estimate of evaporation. *E. Afr. Agric. and For. J.* **30**, 3 (1965).
51. MCHENRY, J. R. Theory and application of neutron scattering in the measurement of soil moisture. *Soil Sci.* **95**, 294–304 (1963).
52. MCLAREN, A. D. Temporal and vectoral reactions of nitrogen in soil: A review. *Can. J. Soil Sci.* **50**, 97 (1970).
53. MERRIAM, R. A. *Useful Statistical Guides and Graphs for Neutron Probe Soil Moisture Sampling*. Intermountain Forest and Range Exp. Sta. USDA, Ogsten. Res. Paper 62 (1962).
54. MOERMAN, P., de Boodt, M. and Mortier, P. *The Spatial Resolution of the Neutron Moisture Meter*. Proc. Panel Soil Moisture and Irrigation Studies, Vienna 1966, 15–20. IAEA, Vienna (1967).
55. NIELSEN, D. R., Davidson, J. M., Biggar, D. W. and Miller, R. J. Water movement through Panoche clay loam soil. *Hilgardia* **35**, 491–506 (1964).
56. NIELSEN, D. R., Jackson, R. D., Cary, J. W. and Evans, D. D. (Eds.). Soil Water. *Amer. Soc. Agron.* Madison pp. 175 (1972).
57. NIELSEN, D. R., Biggar, J. W. and Esh, K. T. Spatial variability of field-measured soil water properties. *Hilgardia* **42**, 215–60 (1973).
58. OGATA, G. and Richards, L. A. Water content changes following irrigation of bare soil that is protected from evaporation. *Soil Sci. Soc. Amer. Proc.* **21**, 355–59 (1957).
59. OLGAARD, P. L. On the theory of the neutronic method for measuring the water content in the soil. Danish Atomic Energy Comm., *Risö Rep.* **97** (1965).
60. PENMAN, H. L. Natural evaporation from open water, bare soil and grass. *Proc. R. Soc.* (London) Series A **193**, 120–45 (1948).
61. PENMAN, H. L. *Vegetation and Hydrology*. Tech. Comm. 53, Commonwealth Bureau of Soils, Harpenden (1963).
62. PENMAN, H. L., Angus, D. E. and Von Bavel, C. H. M. Microclimatic factors affecting evaporation and transpiration. *In:* Irrigation of Agricultural Lands (R. M. Hagan, *et al.*, Eds.), *Amer. Soc. Agron.* Madison (1967).
63. PERRIER, E. R. and Evans, D. D. Soil moisture evaluation by tensiometers. *Soil Sci. Soc. Amer. Proc.* **25**, 173 (1961).
64. REGINATO, R. J. and Von Bavel, C. H. M. Soil water measurement with gamma attenuation. *Soil Sci. Soc. Amer. Proc.* **28**, 721 (1964).
65. REICHARDT, K. Determination of hydraulic conductivity under field conditions for the estimation of

deep drainage in water balance studies. CENA, Piracicaba, Brasil, Bulletin BD-015 (1974).

66. REICHARDT, K., Libardi, P. L. and Nielsen, D. R. Unsaturated hydraulic conductivity determination by a scaling technique. *Soil Sci.* **120**, 165–68 (1975).
67. RICHARD, S. J. Soil suction measurements with tensiometers. Methods of Soil Analysis, Part I, *Amer. Soc. Agron.* Monograph no. 9, 153 (1965).
68. ROSE, C. W., Stern, W. R. and Drummond, J. E. Determination of hydraulic conductivity as a function of depth and water content for soils in situ. *Aust. J. Soil Res.* **3**, 1 (1965).
69. ROSE, C. W. and Stern, W. R. The drainage component of the water balance equation. **Aust. J. Soil Res. 3**, 95–100 (1967).
70. ROSE, C. W. and Stern, W. R. Determination of withdrawal of water from soil by crop roots as a function of depth and time. *Aust. J. Soil Res.* **5**, 11 (1967).
71. SHARIF, M., Tahir, M. and Chaudhry, F. M. *In*: Studies on water use efficiency for the maize crop. *Proc. Symp. Isotopes and Radiation Techniques in Soil Physics and Irrigation Studies*, Vienna 1973, 389–99, Vienna (1974).
72. STARR, J. L., Broadbent, F. E. and Nielsen, D. R. Nitrogen transformations during continuous leaching. *Soil Sci. Soc. Amer. Proc.* **38**, 283 (1974).
73. STEWART, G. L. and Taylor, S. A. Field experience with the neutron scattering method of measuring soil moisture. *Soil Sci.* **83**, 151–58 (1957).
74. TANNER, C. B. Measurement of evapotranspiration. *In*: Irrigation of Agricultural Lands (Hagan, R. M. *et al.*, Eds.) *Amer. Soc. Agron.* Madison (1967).
75. TODOROVIĆ, Z. and Filip, A. Movements of complex compounds through different soil types. *In: Proc. Symp. Isotope and Radiation Techniques in Soil Physics and Irrigation Studies*, Istanbul 1967, 241–50, IAEA, Vienna (1967).
76. TROXLER, W. F. *Calibration of Nuclear Meters for Measuring Moisture and Density*. A.S.T.M. Special Tech. Pub. no. 351 (1963).
77. URSIC, S. J. Improved standards for neutron soil water meters. *Soil Sci.* **104**, 323–25 (1967).
78. VON BAVEL, C. H. M., Underwood, N. and Hood, E. E. Vertical resolution in the neutron method for measuring soil moisture. *Trans. Am. Geophys. Union* **35**, 595 (1954).
79. VON BAVEL, C. H. M., Underwood, N. and Swanson, R. W. Soil moisture measurements by neutron moderation. *Soil Sci.* **82**, 29–41 (1956).
80. VON BAVEL, C. H. M., Underwood, N. and Rogar, S. R. Transmission of gamma radiation by soils and soil densitometry. *Soil Sci. Soc. Amer. Proc.* **21**, 588 (1957).
81. VON BAVEL, C. H. M., Nielsen, D. R. and Davidson, J. M. Calibration and characteristics of two neutron moisture probes. *Soil Sci. Soc. Amer. Proc.* **25**, 329–34 (1961).
82. VON BAVEL, C. H. M., Nixon, P. R. and Hauser, V. L. Soil Moisture Measurement with the Neutron Method. USDA Research Service, ARS-41-70 (1963).
83. VON BAVEL, C. H. M., Stirk, G. B. and Brust, K. J. Hydraulic properties of a clay loam soil and the field measurement of water uptake by roots. I. Interpretation of water content and pressure profiles. *Soil Sci. Soc. Amer. Proc.* **32**, 310–16 (1968a).
84. VON BAVEL, C. H. M., Stirk, G. B. and Brust, K. J. Hydraulic measurement of water uptake by roots. II. The water balance of the root zone. *Soil Sci. Soc. Amer. Proc.* **32**, 317–21 (1968b).
85. VERPLANCKE, H. and de Boodt, M. Methods for calculating unsaturated hydraulic conductivity and soil water diffusivity during vertical infiltration of water in a dry soil. *Meded. Fakulteit Landbouwwetenschappen*, Gent, **38**, 440–49 (1973).
86. DE VRIES, J. and King, K. M. Note on the volume of influence of a neutron surface moisture probe. *Can. J. Soil Sci.* **41**, 253–57 (1961).
87. WANGATI, F. J. Lysimeter study of water use of maize and beans in East Africa. *E. Afr. Agric. For. J.* **38**, 141–56 (1973).
88. WATSON, K. K. An instantaneous profile method for determining the hydraulic conductivity of unsaturated porous materials. *Water Resour. Res.* **2**, 709–15 (1966).
89. MORTIER, P., de Boodt, M. and de Leenheer, L. Z. *Pfl. Ernähr. Düng. Bodenk.* **87**, 244–50 (1959).

CHAPTER 15

Radiation And Other Induced Mutations In Plant Breeding

1. INTRODUCTION

THIS chapter must of necessity be much less detailed than most of the other chapters because the use of induced mutations in plant breeding is, in terms of effort and time, only about 0.5% mutagen treatment while 99.5% is straightforward plant breeding. Nevertheless, the radiation of chemical mutagenic treatment is the essential feature of critical importance, and must be carried out intelligently for potential success. What is attempted here is a basic presentation of the philosophy of mutation induction and the potential of induced mutations; the normal techniques for mutation induction; the general approach to ''direct'' and ''indirect'' use of mutations in plant breeding programmes; the handling of breeding material following irradiation; mutation in vegetative material; a consideration of some of the crop varieties which have been produced by these methods and the types of plant breeding problems for which mutation breeding is likely to offer a solution.

The possibility of developing new varieties through radiation aroused much initial excitement. Over-optimism during the early exploratory period gave way to some disappointment with the practical results achieved because so frequently the breeder found large numbers of harmful mutants predominating and proper selection and handling techniques had not been worked out. However, with our much greater knowledge of the effects of radiation on plants, and with the improved radiation-mutation breeding techniques that have been developed, the use of induced mutations in plant breeding is evolving as a major tool of the plant breeder as described in the key references ([31,90]).

Plant breeders have in the past been dependent on selection from the existing variation within crop plants, this variation having resulted from natural selection from spontaneous mutations. Mutations are the suddenly occurring variations that are inherited, and result from alteration of the genetic make-up of the living organism. Although mutations can occur quite naturally they are exceedingly rare events, or at least they are rarely recognized, and of these mutations an even smaller number are beneficial, most being undesirable.

Mutations in plants range from genetic modifications at a molecular level, that is within the gene, to those effecting a change in chromosome structure. The mutation changes of chromosomes may be of a number of types, including chromosome rearrangement, deficiencies and duplications, and "crossing over". Such modifications may be detected by genetic analysis in the case of obvious structural alterations, together with biochemical methods when functional metabolism is involved.

The practical importance of mutations is that they add to the variability of plants available for selection by the breeder to produce improved crop varieties. Therefore, any technique which can increase the rate of mutation is extremely valuable. We now know that ionizing radiation has this property of inducing mutations in plants. De Vries suggested the possibility of using radiation for the induction of mutations as early as 1904. In 1928 X-rays were used by L. J. Stadler to induce mutations in barley and maize ([81]) following the demonstration in 1927 of mutation induction in *Drosophila* by J. H. Müller ([64]). However, it has taken many years of research to realize this as a practical possibility in plant improvement and the real advance, as measured by released varieties, have taken place in the last 15 years and even more recently.

There was undoubtedly some resistance on the part of many established plant breeders to the use of induced mutations, although the benefits of naturally occurring mutations had been long recognized. For example, the naturally occurring "Dee-gee-woo-gen" dwarfing gene in rice and the "Norin" dwarfing genes in wheat contributed essentially to the new high yielding rice and wheat varieties. Similarly a spontaneous mutation for photoperiod-insensitivity when linked with the short-straw mutation has enabled Taichung Native-1 rice variety to be both fertilizer responsive and adaptable over wide areas.

There was some justification for scepticism over the potential use of induced mutants in contrast to natural mutants. This is because induced mutants of even valuable characters, although they may show the character or effect quite clearly, will in many cases also have deleterious characters associated with it which may not be readily visible. In the case of spontaneous mutants which have been subject to a measure of natural selection they will be recombination types in which the associated harmful characters are likely to have been eliminated.

Although all radiation is basically harmful to living things, if a controlled dose of radiation is given either to seeds, or to whole or parts of plants, induced mutations may arise in the later progeny. The amount of radiation given depends not only on the crop plant but also on the variety being treated. If too little radiation is given the seedlings may be unaffected, while if too much is given they will be killed. Nevertheless, through the efficient use of radiation it is possible to increase the natural mutation rate as much as 100,000 times.

We have noted (Chapter 1) that the electromagnetic radiations and photons of gamma and X-rays are absorbed by matter, and this process results in a part or whole of the photon energy being transformed into kinetic energy as electrons. The ionization arising in the path of these electrons causes the permanent alteration of certain molecules. Moreover, ionization also causes the production of hydrogen peroxide and free radicals, which are themselves mutagenic through molecular effects.

Irradiation has major effects on DNA, and although there are many chemical changes induced by ionizing radiation it is now generally accepted that the principal one in relation to mutation induction is that of radiation on the DNA-helix. Most work has been carried out on bacteria and we assume that the mechanisms in higher plants are not substantially different. One way in which the radiation might act is to alter the sequence of the base pairs that make up the triplet "code" of DNA or even cause the loss of a base pair. An individual chemical reaction may affect the whole DNA molecule. Thus if deoxyribose is attacked it will give a strand break, that is the "backbone" linear structure is broken in one strand. A more severe effect is when the double helix is broken in both strands, a double-strand break. Other reactions may be reflected in effects on the cross linking of DNA to protein, on cross linking within the DNA helix and on alteration of cytosine base. These then are the basic mechanisms resulting in mutation production.

In the vast majority of cases, mutations behave recessive relative to the original allele, whether it is only a deletion or a true mutation. Therefore, if a desired character depends on dominant genes then the chances of obtaining a mutant are greatly reduced. In practice of course we frequently have quite incomplete or no knowledge of the genetics of a desired character. This applies especially where disease resistance is being sought when no naturally resistant genetic resources are available.

USE OF RADIATION AS A MUTAGENIC AGENT

Mutations can be induced in plants by either gamma, X- or neutron irradiation. Most work hitherto has been done with gamma or X-radiation because of the easy availability of these irradiation sources. Gamma irradiation may be given either by short intense treatment in a gamma-cell, essentially a device containing cobalt-60 or cesium-137 surrounded by lead to protect the operator from radiation, or under long term less intense irradiation in a "gamma field" or "gamma greenhouse", or growth room. A number of gamma fields have been set up, the best known being those at Brookhaven National Laboratory and the one at Omiga, Japan, Fig. 15.1 ([52]).

A gamma field is a large area of cultivated land in which plants may be grown in the normal manner, and in the centre of which is a powerful gamma source of cobalt-60. The radiation source can be lowered into a hole in the ground when cultivation has to be done. The whole area is surrounded by an earth banking for radiation protection. A gamma greenhouse is similar in principle, consisting of a greenhouse with a cobalt-60 source which can be raised or lowered as required, the whole being surrounded by a thick concrete shielding wall. Growth rooms have also been built with massive concrete walls, with controlled environment and illumination. Some of the best known of these are at Brookhaven National Laboratory.

Gamma fields have been very useful in giving us basic information on the effect of chronic irradiation on plants and cells, and also for irradiating perennial plants like trees. However, gamma fields are expensive to build and to maintain, and there is now a realization that for practical purposes of plant breeding intense irradiation of seeds is in general more useful, with a gamma greenhouse for when chronic irradiation

FIG. 15.1 Part of the gamma field of the Institute for Radiation Breeding, Omiga, Japan. Surrounded by an earth bank and high wall, access is gained through a shielded door on the left. The 2000 Curie source is on the tower to the right of the picture.

is required. Gamma fields have become something of a prestige scientific facility in some countries, but we should recognize that there are already sufficient existing gamma fields for World scientific needs and it is much better for scientists to go and work at those already developed, rather than to encourage the setting up of new ones.

By far the majority of induced useful mutations in plant breeding have come about through short intense irradiation by gamma or X-radiation either from a cobalt-60 source or X-ray machines, those for medical applications having been often used in the past. There are indeed few scientists who are without access to a radiation source of some sort. A typical small well-shielded ^{60}Co source can stand in any laboratory without special installation requirements other than a strong floor and electricity to operate the loading system. The sample chamber of such sources is big enough to hold a medium size beaker. They are usually fitted with a timing device to permit accurate and repeatable irradiations. A gamma irradiation room, shielded by thick concrete is shown in Fig. 15.2. With a cobalt-60 source of 30,000 Ci and adjustable

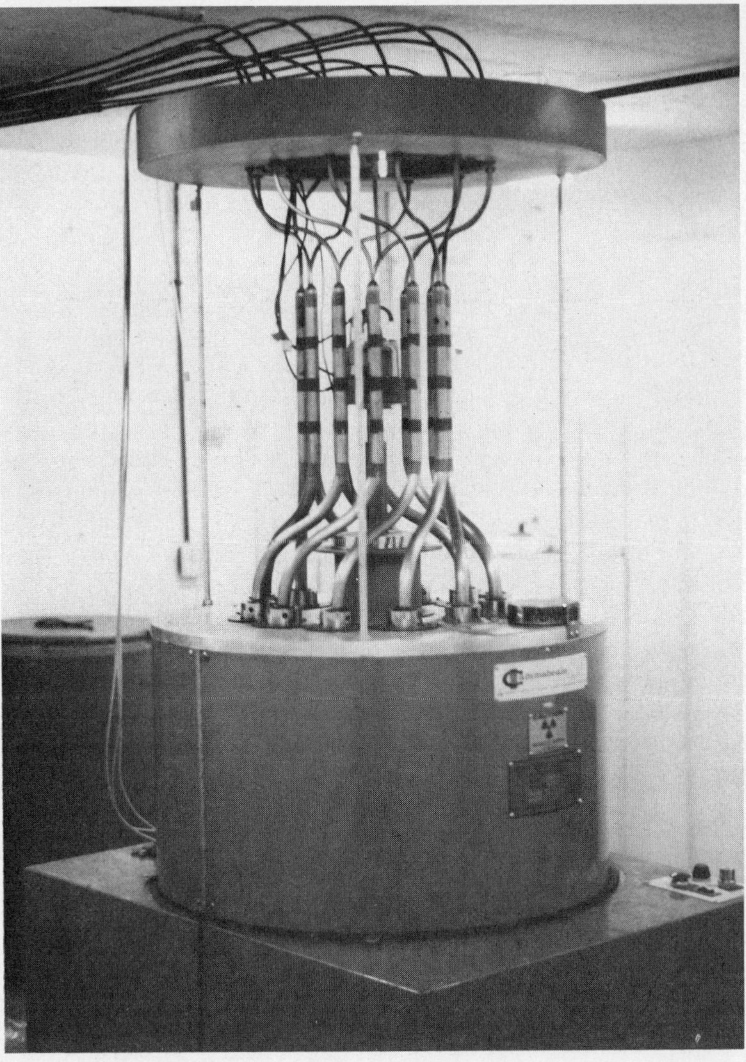

FIG. 15.2 30,000 Curie cobalt-60 "acute" gamma irradiation source in a concrete-shielded room at the Centro de Energia Nuclear na Agricultura, Piracicaba, Brasil.

source elements this provides a very adaptable facility for acute irradiation and can handle diverse and large amounts of material.

Either ^{60}Co ($t_{1/2}$ = 5.27 yrs) or ^{137}Cs ($t_{1/2}$ = 30 yrs) can be used as a gamma source, but ^{60}Co is mostly favoured. The longer half life of ^{137}Cs is of some advantage and it requires less shielding than ^{60}Co, but correspondingly for a given dose rate it needs approximately four times the number of Curies of ^{137}Cs than of ^{60}Co. For X-ray treatments it is desirable that the machine is equipped with a tube of high peak operating voltage (kVp) in order to produce hard X-rays of short wavelength. The short wavelength X-rays are more penetrating than soft longer wavelength X-rays, and this high energy radiation, similar to γ-radiation, is more effective.

The use of fast neutron irradiation was in the past limited both by the relatively poor availability of reactors required as a neutron source, and also by the fact that the biological effects of neutron irradiation were less well known than for other types of radiation. There was also difficulty in measuring and standardizing the fast neutron dose received, due to the presence of thermal neutrons and gamma radiation ([80]). Nevertheless, it is known that the more densely ionizing neutrons are many times more effective in causing chromosomal modifications than are gamma and X-rays, the Relative Biological Effectiveness (RBE) of fast neutrons being upwards of ten times that of gamma and X-irradiation. They are also less dependent for their effect on the condition of the seed, such as moisture content, oxygen and temperature. Additionally, there is evidence that neutrons are much more specific in their effect and therefore offer increased possibility of inducing specific mutations.

The problem of standardizing the neutron dose to seeds has now been partly overcome and we may expect in the future a greater use of neutrons for inducing mutations. To assist this the IAEA sponsored the development of a Standard Neutron Irradiation Facility (SNIF) for pool and Triga type reactors. This is essentially a lead container coated with B_4C to screen out both gamma radiation and thermal neutrons, which can be lowered into pool type research reactors, in a suitable position next to the reactor core as in Fig. 15.3 ([16]). The research reactor of EURATOM-ITAL at Wageningen, Netherlands has a growth room under the reactor core, permitting neutron irradiation

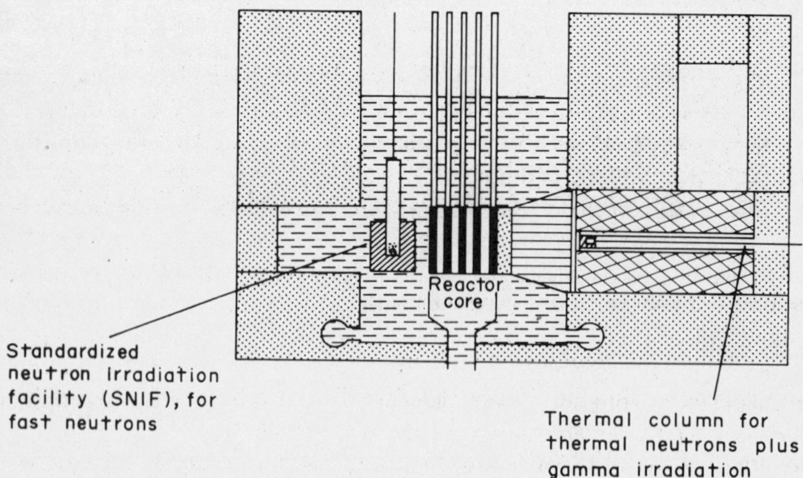

FIG. 15.3 Diagrammatic cross-section of pool-type reactor suitable for both thermal and fast neutron irradiation of seeds, when equipped with the SNIF device ([77]).

of whole plants, etc., but it is now shut down. The problems of neutron irradiation of seeds are extensively considered in two reports ([24,25]).

Mutation Induction

The Irradiation Dose

The optimum radiation dose to induce mutations varies with the plant material, the variety, and the conditions of irradiation. In the case of seeds, which are the most commonly treated material, doses of 5–25 krad of X- and γ-radiation have been used, with 10–15 krad probably being the most common. Below 5 krad the frequency of mutations is substantially reduced while at 25 krad and over there is generally too much radiation damage and loss of material. There are certain species such as alfalfa, *Medicago sativa*, melilotus, *Melilotus alba*, tomato, *Lycopersicum esculentum*, and some *Brassicas* which are apparently somewhat less radiosensitive and withstand and require two or three times these average doses. It is now known that the pre- and post-irradiation conditioning of seeds, including such factors as temperature, pre-soaking, oxygen level, pH, drying-back etc., should be clearly defined and the optimum conditions determined. Before carrying out the substantive radiation treatment, preliminary tests should also be carried out to define the optimum irradiation dose. Variation in radiation sensitivity has been shown to be genetically controlled ([5]).

Much less substantive data is available for fast neutron irradiation but the greatly increased RBE means that much smaller doses must be used. The average useful dose appears to be of the order 0.4–0.6 krad, with a maximum seldom more than 2.0 krad ([13]).

Pretreatment

Seeds which have been allowed to imbibe water are much more sensitive to radiation than "dry" seeds. In practice, though, irradiation is most often carried out on dry seeds as the conditions are easier to control and the seeds are more suitable for mailing or temporary storage in the dry state.

Of all the factors which potentially affect radiosensitivity, and hence potential radiation damage and mutagenic efficiency, are oxygen and the moisture content of the seed. To ensure the best conditions for mutation and ensure that damage is kept to a minimum these factors must be controlled. Moreover, this ensures repeatability of the treatment.

Very dry seeds, of less than 3% w/w water content are particularly susceptible to oxygen enhanced γ-radiation damage, but this effect can be greatly reduced by raising the moisture content of the seeds to 12–24%. Alternatively the seeds can be placed in a container either filled with nitrogen gas or else under partial vacuum. A further possibility is to allow the seeds to imbibe water saturated with nitrogen gas.

The water content of the seeds is very critical and it has been shown ([19]) that even differences as small as 0.3–0.4% w/w in the range 10–11.5% can have a marked effect on radiosensitivity. As controlling moisture content of the seeds i.e. by raising it to about 14% is also an effective means of reducing oxygen-enhanced damage, this is probably the best procedure for routine use.

Seed moisture content can be conveniently adjusted by placing the seeds in a large dessicator and equilibrating at the required relative humidity over a solution of glycerol. A 60% v/v solution of glycerol gives a relative humidity of 73%, and most seed maintained in this atmosphere at 20–22°C will achieve a seed moisture content of 12–14% in about a week or less. There is however some variation between species ([70]). As water has to transfer from the glycerol solution to the seeds, clearly the latter should not be more than about one-third of the former by weight. Determination of seed water content follows established practices ([89,1]).

If strict control of oxygen is required during irradiation this may be done by placing the seeds in a tube and evacuating on a high vacuum line. Following irradiation the tube may be opened under water saturated with N_2.

Irradiation and Dosimetry

Most gamma sources initially have the irradiation dose for a given geometry rather closely specified. With the passage of time and decay of the source there will naturally be changes. Moreover, with large sources in particular it may not be possible to obtain quite the same geometry from one irradiation to the next. In nearly all irradiation treatments it is therefore necessary to check the dose that the treated material has received. The only exception is when very regular irradiations are being undertaken of similar material in exactly the same manner.

Dosimetry is an extensive and complicated subject but only one specific method will be given here. Of the dosimeters that are available the best known and most widely used is the ferrous:ferric sulphate system first described by Fricke and Morse, and subsequently developed ([35]).

The principle is that when a solution of ferrous sulphate contained in a quartz cell is exposed to ionizing radiation then Fe^{2+} ions are oxidized to Fe^{3+} ions, the concentration of the latter being proportional to the absorbed dose. Analytical grade chemicals and double-distilled water are used in the preparation of the standard Fricke solution:

Ferrous ammonium sulphate, $Fe(NH_4)_2.6H_2O$	0.001 M
Sodium chloride, NaCl	0.001 M
Sulphuric acid, H_2SO_4	0.4 M

The quartz cell containing the solution is exposed to radiation together with the material being treated and afterwards the concentration of the Fe^{3+} ions is determined by measuring the optical density at 304 nm. Then the absorbed dose, D, is:

$$D = \frac{9.64 \times 10^8 (d - d)}{G(Fe^{3+}).\rho.l.\epsilon} \quad (1)$$

where, d = optical density of irradiated solution at 304 nm
$\bar{d}$ = optical density of control unirradiated solution
$G(Fe^{3+})$ = number of ferric ions formed per 100 eV energy absorbed = 15.5 for ^{60}Co radiation
ρ = density of the solution = 1.024
l = thickness of solution = 1 cm (quartz cell)
ε = extinction coefficient of Fe^{3+} = 2190 at 304 nm and 25°C

The Fricke dosimeter is excellent for X- and gamma radiation and provides an accurate determination over a range of 4,000 – 40,000 rads.

The dosimetry of neutron sources is more of a problem as unfortunately the Fricke dosimeter is not so accurate with neutrons, while a different G-value (G = 7) must be used. Better methods for fast neutrons include estimation by sulphur activation and by ionization chamber. These are considered in detail in reference ([24]).

In order to ensure repeatability of X-radiation treatments, the kVp and amperage (mA), distance of tube to target, and the type (usually aluminium or copper) and thickness of filter should be recorded. The half value layer (HVL) is usually known and is a measure of the effective energy of the X-ray machine.

Prediction of required dose

Irradiation causes physiological damage to the M_1 plants and the art of estimating an irradiation dose is to choose one which will give the maximum mutation rate for the minimum damage. Measurement of physiological damage can be done through survival in the field, number of spikes or inflorescences per plant, number of fruits or seeds per plant, and seedling height or root length. Physiological damage and survival curves are sigmoid, though the shape varies considerably for different species.

As a quick measure of physiological damage the determination of seedling height has come to be the standard technique, as it permits fairly rapid estimates, the results of which are then a guide to the substantive required irradiation dose.

Irradiation experiments with a large number of species enabled the development of a graph from which seedling height data can be used to predict estimates of survival ([69]). For barley a reduction in seedling height of 50–60% predicts a survival of only about 40% ([37]), and other species are likely to be worse. Such potential radiation damage is unacceptably high for most mutation treatments. Therefore, in general terms a practical dose for mutation induction is about 50–80% of the dose (GR_{50}) required to give a reduction in seedling height of 50% from the control.

Post-treatment

If not sown immediately, seeds with 12–14% water content may be stored up to 4 weeks at room temperature, 15–20°C, without apparent post-irradiation damage, or for longer periods at −20°C. Seed which has to be shipped may be done so in sealed polyethylene.

CHEMICAL MUTAGENS

In addition to radiation there are a number of chemicals which have been found to give rise to mutations; the prominent ones include ethyl methane sulphate (EMS), diethyl sulphate (DES), isopropyl methane sulphate (IPMS), sodium azide, and quite a number of others. Of these, EMS has probably been the most widely used. There is now some evidence that these chemicals give rise to a different type or spectrum of mutations, compared with those induced by radiation. An example of how different mutation spectra result from different types of mutagenic agent is given in Fig. 15.4 for the *eceriferum* or ''waxless'' mutation of barley. The varying effects of the agents can be clearly seen. Nevertheless, the effects of these chemical mutagens are so generally similar to those of radiation that they are known as ''radiomimetic''.

Chemical mutagens have often been used independently of radiation but it is likely that with the different mutation spectra available the most profitable approach is to consider both radiation and chemical methods as complementary. Many breeders will therefore treat material with both radiation and chemical mutagens. There is the possibility that a combination of chemical and radiation treatment given simultaneously

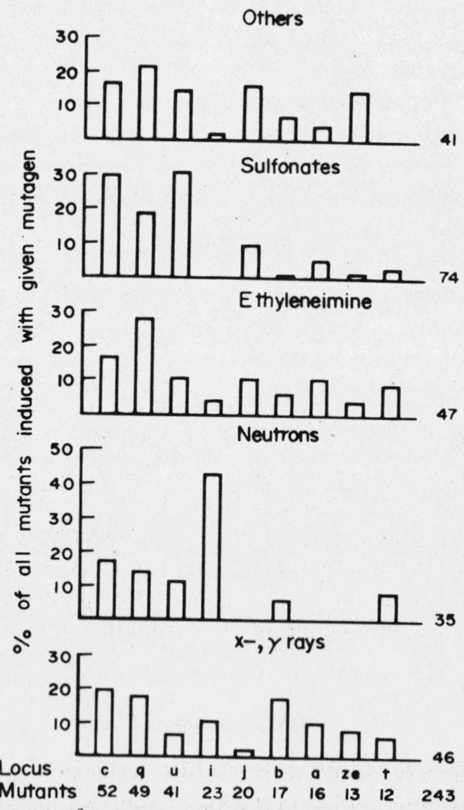

FIG. 15.4 The effect of five types of mutagen on the mutation spectra in nine different *eceriferum* loci of barley ([58]).

may prove to be especially effective in the production of mutations, and that it may give us some control over the type of mutation produced. However, this lies in the future and on the whole results have not been too encouraging.

Safety Aspects

Mutagenic chemicals are by nature highly dangerous and most are carcinogenic. Every precaution must be taken when using them. Handling of solutions should be done wearing plastic gloves and all work must be carried out in an efficient fume hood. Any spills must be treated with no less care than if they were radioactive, especially avoiding inhaling vapours. Disposal must be carried out to main drains with very high dilution by water.

Dose and Treatment

The dose of a chemical mutagen is dependent on concentration of the solution, time of soaking of the seed and the temperature. Chemical mutagens have been found unsatisfactory with vegetative material due to the difficulty of standardizing the dose and variable penetration. The tendency has been towards more dilute solutions and longer absorbtion times ([56]). It has been found that a preliminary soaking in water helps subsequent penetration of more dilute mutagen with subsequent minimum physiological damage ([56,63]) but effective mutagenesis.

Lacking the standardized dose measurement of radiation and also the deep research base of radiation biology, chemical treatments have developed empirically. In general, treatments which result in early seedling growth of 60–70% of the controls are most effective. Concentrations suggested are 0.05–0.3 M (about 0.3–1.5%) for EMS; 0.015–0.02 M (about 0.1–0.6%) for DES ([56,63]) and 0.001–0.004 M for sodium azide. To be effective the latter must be used in phosphate buffer at pH 3.0.

To ensure repeatability of results seed should be brought to standard moisture content as for radiation treatments. If pre-soaking in water is carried out this may be done overnight with aeration of the water. Actual treatments will normally include a range of concentration, at a room temperature of about 20°C and a time of 2–6 hours. A large volume of solution in relation to the seed should be used to prevent any limitation on absorbtion, and as much as 1 ml per seed has been suggested for small grains.

Following treatment the seed should be adequately washed in running water to remove excess chemical. There is some evidence that quite a long period of washing is required to ensure minimum post treatment damage but not all seeds can withstand long periods in water. The seed is best sown immediately after treatment, but cereal grains at least can be uniformly dried for storage at 4°C if this is essential.

PLANT BREEDING WITH INDUCED MUTATIONS

Although we may possibly think of plant mutations as events which produce obvious dramatic changes such as flower colour, in reality mutations concern every aspect of plant growth: yield, length of stem, lodging resistance (straw strength), disease resistance, early ripening, grain colour and quality, protein content, photo- and thermo-periodic requirements, ion uptake etc., in fact modifications of any plant character. From the variability thus made available the plant breeder can either select directly or cross (hybridize) the mutant with an already existing variety or line(s).

In the early days of mutation breeding there was a tendency to ''look for mutations''. The result was that breeders were rather overwhelmed with material of diverse nature. Meanwhile the labour of making general observations put a practical limit on the populations that they could handle, yet in turn reducing the possibility of obtaining a useful mutant, and diverting them from an approach to a clearly defined objective. We now realize that a clearly defined and strictly limited objective, together with very large populations, is the most likely means of obtaining success. Seeking a clearly defined mutant and ignoring and ruthlessly discarding any other ''interesting'' material means that a large population of plants can be handled.

The development of *Verticillium*-wilt resistant strains of peppermint was reported after neutron and X-irradiation of stolons. One hundred thousand irradiated plants were initially established, giving over 6 million plants in the second year. Subsequent testing of 58,000 non-wilt plants led to the selection of 7 highly wilt-resistant strains. This is a good example of a clearly defined objective combined with large numbers of plants, succeeding where conventional breeding had failed ([65,66]). It is also an almost unique case of a mutation being obtained where the desired character depends on dominant genes.

As Swaminathan has pointed out ([85]) there are today many new selection criteria in plant breeding: ''superior population performance, high productivity per day, and per unit of water; high photosynthetic ability; low photo-respiration; photo- and thermo-insensitivity; high response to nutrients; multiple resistance to pests; crop canopies that can retain and fix a maximum of CO_2; and better nutritional, storage and processing properties. It is such a shift in selection parameters that has provided unusual opportunities for using induced mutations in plant breeding.''

Handling Mutagen Treated Breeding Material

The M_1 Population

Plants resulting from seed subject to a mutagenic treatment constitute the M_1 population.

The number of seeds handled in the M_1 has varied considerably. There are on record varieties that resulted from irradiation of only a few hundred seeds, but this seems to be putting too great a premium on chance. Most breeders will choose to handle a minimum of 3–5000 seeds. The question is considered extensively in reference ([31]).

The seed is for preference, sown as soon after treatment as possible. Every effort should be made to obtain good conditions for germination and growth, as inevitably the treatment will have resulted in some physiological damage and loss of vitality. Although fertilization should be adequate, not too much nitrogen should be given to cereals so as to discourage unnecessary tillering, as mutants are more likely to occur in the primary tillers. For the same reason the sowing distance both within and between rows should be restricted.

A sowing layout should be chosen that allows for replication of the mutagenic treatments because in practice a breeder will very likely try both radiation and EMS or other chemical mutagen, and both at more than one dose. Replication is therefore essential, to ensure that possible variation due to soil fertility is taken into account.

It is desirable to isolate the M_1 material in order to reduce as much as possible potential genetic variability other than any induced by the mutagenic treatment. With isolation, any outcrossing will be restricted to the original population.

Data: good records should be kept both with a view to checking the mutation induction technique but also to help in identifying the most likely populations for M_2. Records should include per cent emergence, seedling survival, survival to maturity, M_1 chimeras and sterility.

Harvesting concentrates on the primary tillers in the case of cereals or the main branches of dicotyledonous crops, as the major potential for induced genetic variability is likely to be these parts. This is because they have arisen from already differentiated primordial buds within the embryo of the treated seed. A number of seeds may be taken from each harvested part and either kept separate or bulked up for sowing as the M_2 population. It is desirable to keep the plants and progenies numbered through the generations.

The M_1–M_n Populations

Mutations may be classed as readily identifiable mutations, known as macro-mutations, or as micro-mutations ([38,83]). The latter, for such general characteristics as grain yield or protein content, will only be detected in families of plants by biometrical analysis. Many macro-mutations will be visible in the M_2 but micro-mutations can normally only be detected in the M_3 generation onward.

As many as 75–80% of mutants can be present in the M_2 only as heterozygotes and therefore they will not be immediately detected, but may appear in later generations. One procedure for micro-mutations ([38]) is to make a random selection of normal-appearing M_2 plants to produce the M_3. The desired character is then sought in this and subsequent generations.

For selection of macro-mutants such as morphology changes, plant height, flower colour, earliness, smooth awned or disease resistance, the M_2 (or later) generation is sown in rows with the standard variety being sown regularly, with 1 m in every 25–30 m of row. In the case of micro-mutant selection in M_3 and later generations, progenies are sown in short multi-row blocks, with the standard original variety being sown two rows in every 20–25.

Mostly, M_2 mutants are considered homozygous for the selected trait, but are not necessarily so, and possible segregation of heterozygotes must be checked in M_3.

There may also be the possibility of selecting new macro-mutants at this stage. Preliminary testing and comparison of mutants with standards will proceed simultaneously in M_4 in unreplicated trials. Those mutants passing this test go on to further and more rigorous performance testing in M_5 and M_6.

As genotypic and phenotypic analysis of selected mutants is established they may be utilized for further selection, crossing, or even possibly new treatments. In general, selected variants should breed true in M_3 if not M_2 progeny tests. Any segregation in M_2 should occur only in the desired trait and no other, and in the case of back-crossing or other crosses the selected mutants should exhibit comparatively simple segregation. The phenotype should in general follow the parent variety, except for the modification that has been sought, as should physiological and biochemical characteristics.

The recognition and evaluation of micro-mutations such as for increased protein content, general higher yield, increased vigour or better adaptability is inherently a harder task. It would normally be assumed that if the selected "mutant" type is outside the normal variation of the original variety then it is an authentic mutation, though of course there might also be mutants which still fall within the normal range and go unrecognized. In contrast, selecting for yield within self-pollinators also results in significantly higher yielding lines, and these might be regarded as higher yielding micro-mutants, when in fact they were not of mutant origin. One basic method of establishing mutant origin of yield micro-mutants is to select variant, supposedly mutant lines, and the best lines of the established variety under the same conditions. Then when in carefully conducted field trials the best variant lines exceed the best variety lines then the former are considered yield mutants ([39,47]).

The question of identification, evaluation and handling of mutant material can only be touched on in this brief account, but many of the references present detailed experience and proven methods. An important topic that is receiving increasing attention is the proper recording and documentation of mutants and the establishment of "gene banks" with mutant and other material to be readily available for crossing. Papers by Gregory ([41]), Purdy *et al.* ([71]), and Blixt ([6]) can be consulted, the latter describing the combined utilization of a gene bank and a computer system for developing a crossing programme.

"Direct" Mutation Breeding

Being able to select and multiply directly from a mutant offers the most rapid and striking means of crop improvement. However, this can only be done if the mutation concerns a more or less specific character of an already-successful variety.

A remarkable example of this concerns the dwarf Mexican wheat, Sonora, bred under the Rockefeller Program and later introduced to India. This dwarf wheat was found to be well adapted to Indian conditions and yielded much higher than the existing varieties. It is this wheat which has revolutionized the food situation in India.

Nevertheless, Sonora had one major defect in Indian eyes—its grain was a brown colour displeasing to the consumer, because chapattis (a staple unleavened wheat cake) made from it were mud-coloured. Scientists at the Indian Agricultural Research Institute were able to rapidly produce a radiation mutant variety with amber-coloured grain but retaining all the valuable yield and other characteristics of the original variety ([84]).

Barley was one of the first crops on which significant direct mutation work was carried out ([46]), and the Swedish variety Pallas released in 1960, with higher yield than the parents and with significantly stiffer straw, was an early product. Although never prominent in Sweden it was grown in England and other countries and became the parent of other improved varieties. Almost all the basic barley breeding lines in Sweden are now of induced mutant origin.

One of the most readily obtained single characteristic mutants is short straw, and short stiff strawed varieties of barley, wheat and rice have been obtained. This is of particular importance because high rates of fertilizer application in modern farming tend to make these crops ''lodge'', making harvest difficult. Another fairly common mutant relates to earliness and shortening of the growing period. This may be exploited in two ways, either by trying to increase the number of crops per year e.g. rice in the Far East or by increasing the growth range of the crop in northern latitudes. Over thirty examples of cereal crops in which lodging-resistant mutants have been recognized and isolated were given in ([74]), and there have been many others since then.

For such specific improvement of otherwise adapted varieties the induced mutation method clearly offers advantages in terms of time, effort and practicability. Thus the Japanese rice variety Rei-mei was a mutant derivative of the variety Fujiminori and agronomically identical to it, except for having much stronger and shorter straw and higher yielding ability. It was produced, extensively field tested and released in 5 years, becoming one of Japan's top varieties. Similarly in India, mutation techniques were quickly successful in obtaining *indica* type grain characteristics on otherwise *japonica* varieties, despite the failure of long-standing attempts based on hybridization of the sub-species.

A recent example ([32]) of a radiation mutant for a specific requirement being sought and successfully found is the California rice variety Calrose 76, Fig. 15.5. In this case there was an already existing successful and widely grown variety CS-M3 but which nevertheless had comparatively long straw and was subject to lodging under high fertility conditions. The gamma-induced mutant of CS-M3 retained all the agronomic and yield qualities of the parent, but had a stem 35 cm shorter making it able to withstand high fertility levels. Previously, attempts to transfer the short straw character from the tropical variety IR8 had not proven successful because of poor grain quality and cold susceptibility.

Another interesting case of direct mutation breeding for disease resistance is the stem-rust resistant Brazilian wheat Ticena-3. This variety appears to maintain all the agronomic qualities and characteristics of the parent variety but with drastically improved disease resistance, to the extent that some yield figures have been 50% higher than the parent. Crossing is also being used to transfer the stem-rust resistance to other varieties.

It must be made clear that in direct mutation breeding it is essential to start with the very best contemporary, otherwise adapted, breeding material that is available.

Improvement by mutation is being initially sought through only one or two critical changes in the basic character, therefore the remaining qualities of the material must be as good as possible. The computer programmers' adage "rubbish in means rubbish out" applies equally to mutation breeding!

"Indirect" Use of Mutations in Cross-breeding

In addition to "direct" mutation breeding, mutants may be valuable for incorporation into the classical type of hybridization plant breeding programme. Although direct development of new varieties is clearly a useful and at times extremely rapid means of plant improvement, it is clear that very real benefits can also be realized from obtaining additional genetic variability. This particularly applies to characters such as disease resistance, which are not otherwise in existing plant collections. Moreover, the development of international plant breeding has lead to many cultivars having a certain degree of genetic uniformity which has obvious risks from the disease standpoint.

Of the possibilities in cross-breeding it is apparent that crosses of the mutant may be made back to the original parent line, to an entirely different line, or even between

FIG. 15.5 The released California mutant rice variety Calrose 76, 35 cm shorter than the widely grown variety CS-M3, and able to withstand higher fertilization levels without lodging. Induced by gamma radiation from ^{60}Co [32].

FIG. 15.6 Brazilian wheat variety Ticena-3, an induced mutant resistant to stem rust, which in trials yielded as much as 50% higher than the parent variety.

mutants. As it is difficult to select, combine and maintain recessive mutations in populations of cross-fertilizing species, clearly there is less chance of improvement by mutant induction of these species than of self-fertilizers. There have nevertheless been a number of successes with cross-fertilizing species ([14,45,77]).

Specific disease resistant mutants were found early [2,54,55] and have included resistance to crown rust and *Helminthosporium* in oats [54], barley mutants resistant to mildew [33] and rice mutants resistant to blast. A particularly ingenious use of X-radiation enabled the transfer of rust resistance to common bread wheat from the primitive type *Aegilops* [76]. Stem rust resistance in wheat has been referred to above. Golden mosaic disease of the common bean, *Phaseolus vulgaris*, due to a virus transmitted by the white fly, *Bemisia tabaci*, is a scourge of the bean crop through most of Latin America. A mutant tolerant to the disease was sought and found through screening 50,000 M_2 seedlings [86]. Over 750 M-progenies were handled and the result should be seen in the perspective of 5000 lines from spontaneous collections having been tested in Central and South America with little genuine tolerance having been found [17]. Tolerance to golden mosaic virus seems to be determined by a single recessive locus, and the character is easily transferred by crossing. The mutant was induced by EMS.

Although a mutant resulting from the induction of a single change in an existing superior variety may be developed without recourse to back-crossing (as in the case of the Indian mutant Sharbati-Sonora wheat) this cannot always be possible. When drastic mutations occur and are selected, it can be necessary to back-cross or out-cross them, followed by selection from sub-lines, in order to remove the effects of the deleterious alleles which may also be associated with the mutant.

So far, except in European barley breeding [75], rather fewer mutant varieties have been developed through cross-breeding than directly [62]. But as quite high heterosis effects have been reported [15,34,43,44,60,61,73] for crosses of mutants of self-pollinators, both with each other and with the original stocks, this suggests that increased developments are likely along these lines.

Although still largely at the research stage, with few if any developments reaching the stage of released varieties, we can foresee mutagen treatments being used for a number of very specialized genetical procedures, such as reducing incompatibility in wide crosses [20,57,59,72,82], in the production of haploids [3,21,53], and for the production of transitory sexuality in apomicts [42,50]. Other specialized applications may include the use of translocations for transferring characters from other species and genera, and possibly the diploidization of polyploids.

What is basic research today will become the accepted practice of tomorrow, and there seems little doubt that in the long term the use of mutations in cross-breeding will far exceed the significance of the results we have so far seen in ''direct'' mutation breeding.

INDUCED MUTANTS IN VEGETATIVELY PROPAGATED SPECIES

The possibility of inducing mutants in vegetatively propagated crops is very attractive, because a desirable mutant having been obtained there are unlimited possibilities of multiplication without, in theory, the problem of associated deleterious characters which often manifest themselves in later generation of sexually propagated plants. Moreover, most vegetatively propagated plants are highly heterozygous, with segregation for many characters, even when propagation by seed is possible, e.g. sugarcane. Frequently too, such crops have a long vegetative phase before flowering and crossing is possible, thus making cross-breeding extremely slow e.g. fruit trees ([4,49]). Equally, incompatibility can be a suitable problem to be overcome by mutation breeding, which is also the only means of obtaining genetic variability in obligate apomicts and sterile crops.

Apart from these special reasons, the basic objectives and philosophy of employing mutagenic techniques for vegetatively propagated crops are essentially the same as for those reproducing by seed. That is, to take an established variety of good standard in which improvement is sought in one or more clearly defined aspects, like disease resistance. The plants which have had most attention so far have been horticultural plants such as ornamental shrubs, chrysanthemum and other house plants, where new flower colours, novel floret patterns and decorative new leaf colours have high commercial value. Nevertheless, work has also been done on such widely ranging species as *Mentha*, sugarcane, fruit trees, banana, tea, rubber and citrus ([90]).

Mutagen treatment

Radiation is almost exclusively used for vegetatively propagated material due to the difficulty of standardizing the dose of chemical mutagens, mainly due to the heterogenous bulky nature of the material and the variability in penetration of the solution ([8,11,68]). Nevertheless, successful results have been reported with the use of chemical mutagens on the potato ([87]).

Usually gamma or hard X-rays have been used and the irradiation dose is generally much less than for irradiation of seeds, between 1–4 krad being a normal range, with a mean of 1.5–2.5 krad. The specific appropriate dose for any species or cultivar can only be determined empirically by trial-and-error. As in seed irradiation a reasonable compromise must be achieved between low doses given 100% survival and a low mutation rate, and giving such high doses that survival is low and physiological damage high. Too high doses could also induce more than one mutation per cell, so that a favourable mutation was masked by a harmful one. Neutrons have been comparatively little used because of lack of availability.

In general, young tissue such as adventitious roots and newly developing buds require lower doses than older plant parts with existing meristems.

Material Handling

The major difficulty of practical mutation induction in vegetative material is the likelihood of chimera formation following irradiation of a multicellular meristem. All treatments and material handling must therefore be designed to overcome this as far as possible. Mutations take place in one cell, but this cell is only one of a number of a group of cells comprising each different layer of a meristem. Therefore, a mutated cell is frequently unable to withstand competition from neighbouring unmutated cells (the so-called diplontic selection) and cannot be recognized. Alternatively, small sectors of mutated tissue may develop within a normal meristem layer giving rise to mericlinal chimeras.

In essence then, one must try to develop a technique whereby as few meristematic cells as possible are involved, so that there is a greater likelihood of developing young plants which can be recognized either as normal or else as solid mutants. The adventitious bud technique especially developed by Broertjes ([9,10,11]) has proved particularly useful for many ornamentals. The basis of the method is that the apices of adventitious buds frequently develop from a single epidermal cell. Such adventitious buds can often be encouraged to form at the base of the petiole of detached leaves, at the base of bulb scales or on stems or explants. Over 350 species suitable for this technique have been listed ([9]).

In sugarcane, vegetative propagules and cell cultures have been irradiated ([48]). An approach used with potatoes was to apply very high doses of radiation to de-eyed tubers, following the theory that high doses are more likely to inactivate all but one or a few initial cells. Then only the shoots appearing 3 months after irradiation were used on the assumption that earlier shoots would have developed from multicellular primordia ([88]). Other procedures where adventitious bud formation is not possible, concentrate on irradiating young and growing tissues, or developing buds ([49]) rather than old or dormant material.

Later treatment is based on the need to give mutated cells the best chance of developing, especially when multicellular apices have been irradiated. In general this is done by a combination of repeated pruning and vegetative propagation to increase the mutated sector size until periclinal shoots or tubers are obtained. When a "solid" mutant is achieved it is then possible to propagate vegetatively using the normal method used for the species. Even at quite a late stage of development careful watch should be kept for signs of "reversion" due to unrecognized remaining unmutated tissue.

COMMERCIALLY RELEASED MUTANT VARIETIES

A number of mutant-based commercial crop varieties have already been mentioned. Probably the first such variety was the Chlorina F_1 tobacco, of pale colour and high leaf quality, developed in Indonesia as long ago as 1934 by X-irradiation of the established variety Vorstenland. Lack of appreciation of the possibilities of induced mutations, lack of theoretical background in cytogenetics and plant physiology and biochemistry, together with little practical experience of handling mutants restricted further development till the Fifties. Since then and particularly in the last 15 years there has been a steady development and although, except for barley, still comprising only a small proportion of introduced varieties the number is rapidly increasing.

As landmarks we may recall the now classic North Carolina peanut variety N.C.4X introduced in 1959, of superior quality, disease resistance, shell strength and yield ([40]). Also, the Swedish barley variety Pallas released in 1960, with its characteristics of stiff straw and high yield. More recently, the Japanese work has been particularly interesting, as rice mutants have been obtained as much as 45 days earlier than the parent varieties, and where the nematode resistant soyabean variety, Raidan, matures 25 days earlier ([51]). The Californian short straw rice variety Calrose 76 has already been mentioned ([32]). The Czechoslovakian barley variety Diamant was not only very successful ([7]), but has been widely used in crossing, as was the Swedish variety Mari. The peppermint variety Todd's Mitcham is possibly the best known achievement of induced mutation in a vegetatively propagated species ([67]).

There are now so many commercially released varieties based on induced mutations that it is becoming difficult to record them, but it has been estimated that some 200 mutant cultivars (seed propagated) have been developed of which 133 were of agricultural crops ([36]). The crop species include barley, beans, castor bean, lespedeza, mustard, oats, peach, peanut, peas, soyabean, rape, rice, tobacco, wheat (bread and *durum*).

Cereals constitute a majority of released induced mutant varieties ([79]), with 15 wheat, 16 rice, 40 barley, and 5 oat varieties. Although mutation breeding has been very successful with cereals, probably the numbers of released varieties reflects the importance of cereals and the amount of effort put into improving them, rather than any particular suitability for mutagen techniques. Of the improved characteristics reported in cereals it is noteworthy that lodging resistance and short stem were shown by 37 varieties, higher yield by 27 (though of course lodging resistance is also a form of yield increase), disease resistance and quality by 13 each, the remaining plant varieties being improved in such characters as winter hardiness, plant type and protein content.

Of the 200 or so ornamental cultivars that have been developed the great majority are vegetatively propagated species, particularly of *Chrysanthemum, Dahlia, Azalia, Streptocarpus* and *Dianthus* ([90]).

In summary we may observe that the potential of mutation breeding is very great, as all plant breeders will agree that plant improvement is entirely dependent on diversity and availability of basic material. Nevertheless, the technique should not be regarded as the "lazy man's" method of plant breeding. It is true that rapidity of development has characterized a number of mutation bred varieties, and this is obviously a valuable feature. However, as much hard work and attention to detail is required as for any other type of plant breeding programme. The main advantage lies in (1) the ability to obtain mutations rarely or never found in existing reservoirs of genetic variability, and (2) for improving specific characters of otherwise well-adapted varieties.

REFERENCES FOR FURTHER READING

1. AOAC *Official Methods of Analysis* (various editions).
2. Ausemus, E. R., Hsu, K. T. and Sunderman, D. W. Resistance to stem rust race 158 induced by ionizing radiation in wheat. *Abstr. Ann. Meet. Amer. Soc. Agron.* **48**, 1028 (1955).
3. Binding, H. Mutations in haploid cell cultures. *In: Haploids in Higher Plants, Advances and Potential*, 323–37, University of Guelph Press (1974).
4. Bishop, C. J. Radiation induced mutations in vegetatively propagated fruit trees. *Proc. 17th Int. Hort. Congr.* II, 15–25 (1967).

5. Blixt, S. Studies of induced mutations in peas. XXVI Genetically controlled differences in radiation sensitivity. *Agric. Hort. Genet.* **28**, 55–116 (1970).
6. Blixt, S. A crossing programme with mutants in peas: utilization of a gene bank and a computer system. *In: Proc. Symp. Induced Mutations in Cross-breeding*, 21–36, IAEA, Vienna (1976).
7. Bouma, J. New variety of spring barley 'Diamant' in Czechoslovakia. *In*: Induced mutations and their utilization, *Proc. Erwin-Bauer-Gedächtnisvorlesungen* IV, Gatersleben 1966, 177–82, Akademia-Verlag, Berlin (1967).
8. Bowen, H. J. M. Mutations in horticultural chrysantemum. *In: The Use of Induced Mutations in Plant Breeding* (Rep. FAO/IAEA Tech. Meeting, Rome 1967, 695–700 Pergamon Press, Oxford) (1965).
9. Broertjes, C., Haccius, B. and Weidlich, S. Adventitious bud formation on isolated leaves and its significance for mutation breeding. *Euphytica* **17**, 321–44 (1968).
10. Broertjes, C. Induced mutations and breeding methods in vegetatively propagated species. *In: Induced Mutations in Plants*. Symp. Pullman 1969. 325, IAEA, Vienna (1969).
11. Broertjes, C. Improvement of vegetatively propagated plants by ionizing radiation. *In: Induced Mutations and Plant Improvement*. Buenos Aires, 1970, 293–99, IAEA, Vienna (1972).
12. Brunner, H. and Mikaelsen, K. Beeinflussende Faktoren in der mutagenen Wirkung von Äthylmethansulfonat auf Gerste. *Z. Pflanzenzuecht* **66**, 9–36 (1971).
13. Brunner, H. cited in IAEA, *Manual on Mutation Breeding*, Second Edition, p. 45, IAEA, Vienna (1977).
14. Burton, G. W., Powell, J. B. and Hanna, W. W. Effect of recurrent mutagen seed treatments on mutation frequency and combining ability for forage yield in pearl millet, *Pennisetum americanum*. *Radiat. Bot.* **14**, 323–35 (1974).
15. Burton, G. W. and Hanna, W. W. Heterosis resulting from crossing specific radiation-induced mutants with their normal inbred parents. *Mutat. Breed. Newsl.* **4**, 5–6 (1974).
16. Burtscher, A. and Casta, J. Facility for seed irradiation with fast neutrons in swimming pool reactors: A design study. *Neutron Irradiation of Seeds*. I. Tech. Rept. no. 76, 41–61, IAEA, Vienna (1967).
17. Cervellini, A. and Vose, P. B. The Development and Work of the Centro de Energia Nuclear na Agricultura, CENA, Piracicaba, Brasil. *IAEA Bull.* **18** Supplement, 33–38 (1977).
18. Conger, B. V., Nilan, R. A. and Konzak, C. F. Post-irradiation oxygen sensitivity of barley seeds varying slightly in water content. *Radiat. Bot.* **8**, 31–36 (1968a).
19. Conger, B. V., Nilan, R. A. and Konzak, C. F. Radiobiological damage: A new class identified in seeds by post-irradiation storage. *Science* **162**, 1142–43 (1968b).
20. Davies, D. R. and Wall, E. T. Effect of gamma radiation on interspecific incompatibility within the genus *Brassica*. *Z. Vererblehre* **91**, 45–51 (1960).
21. Devreux, M. and Nettancourt, D. de. Screening mutations in haploid plants. *In: Haploids in Higher Plants, Advances and Potential*, University of Guelph Press (1974).
22. FAO/IAEA. *The Use of Induced Mutations in Plant Breeding*. Rept. 1964 Rome, Technical Meeting, pp. 832, Pergamon Press, Oxford (1965).
23. FAO/IAEA. *Mutation in Plant Breeding*. Proceedings of a Panel, Vienna 1966, pp. 267, IAEA, Vienna, no. II, 1967, pp. 315 (1967).
24. FAO/IAEA. *Neutron Irradiation of Seeds*. I. Tech. Rept. no. 76, IAEA, Vienna (1967).
25. FAO/IAEA. *Neutron Irradiation of Seeds* II. Tech. Rept. no. 92, IAEA, Vienna (1968).
26. fao-iaea. *Induced Mutations in Plant Breeding*. Proceedings of 1969 Pullman, Washington Symposium, pp. 745, IAEA, Vienna (1969).
27. FAO/IAEA. *Mutation Breeding for Disease Resistance*. Panel Proc. STI/PUB/271, IAEA, Vienna (1971).
28. FAO/IAEA. *Induced Mutations in Vegetatively Propagated Plants*. Proc. Panel Vienna 1972, IAEA, Vienna (1973).
29. FAO/IAEA. *Polyploidy and Induced Mutations in Plant Breeding*. Panel Proc. STI/PUB/359, IAEA, Vienna (1974).
30. FAO/IAEA. *Induced Mutations for Disease Resistance in Crop Plants*. Novi Sad 1973, Panel Proc. STI/PUB/388, IAEA, Vienna (1974).
31. FAO/IAEA, *Manual on Mutation Breeding*, Second Edition, pp. 288, IAEA, Vienna (1977).
32. FAO/IAEA. *Improved Short-stature Rice Created by Radiation-induced Mutations* (Rutger, N. J., Peterson, M. L. and Chao-Hua Hu). IAEA Bull. **19**, 44–45 (1977).
33. Favret, E. A. Induced mutations in breeding for disease resistance. *In: The Use of Induced Mutations in Plant Breeding* (Rept FAO/IAEA Tech. Meeting, Rome 1964), 521–36, Pergamon Press, Oxford (1965).

34. FAVRET, E. A. and Ryan, G. S. New useful mutants in plant breeding. *In: Mutations in Plant Breeding*. Proc. Panel, Vienna, 49–61, IAEA, Vienna (1966).
35. FRICKE, H. and Hart, E. J. Chemical Dosimetry. *In: Radiation Dosimetry* 2 (F. H. Attix & W. C. Roesch, Eds.) Chapter 12, Academic Press, New York (1966).
36. FRIED, M. *Historical Introduction to the Use of Nuclear Techniques for Food and Agriculture*. IAEA Bull. vol. 18, Supplement 4–6 (1976).
37. GAUL, H. Determination of the suitable radiation dose in mutation experiments. *Proc. 2nd Congr. European Assoc. Research on Plant Breeding*. Cologne, 65–69 (1959).
38. GAUL, H. The concept of macro- and micro-mutations and results on induced micro-mutations in barley. *In: The Use of Induced Mutations in Plant Breeding*. Rept. FAO/IAEA Tech. Meeting Rome 1964, 408–26. Pergamon Press, Oxford (1965).
39. GAUL, H., Ulonska, E., Winkel, C. zum and Braker, G. Micro mutations influencing yield in barley—studies over nine generations. *In: Induced Mutations in Plants*. Proc. Symp. Pullman 1969, 375–98, IAEA, Vienna (1969).
40. GREGORY, W. C. The peanut NC4x, a milestone in crop breeding. *Crops Soils* **12**, 12–13 (1960).
41. GREGORY, W. C. A radiation breeding experiment with peas. *Radiat. Bot.* **8** (2) 81–147 (1968).
42. GOMEZ-CAMPO, C. and Cases-Builla, M. Estudio sobre la descendencia de un mutante sexual en *Ecballium elaterium* Rich. *In: Induced Mutations in Plants*. Proc. Symp. Pullman 1969, 501–7, IAEA Vienna (1969).
43. GOTTSCHALK, W. The productivity of some mutants of the pea, *Pisum sativum* L., and their hybrids. A contribution to the heterosis problem in self-fertilizing species. *Euphytica*, **19**, 91–97 (1970).
44. GOTTSCHALK, W. and Milutinović, V. Untersuchungen zur Heterosis bei Selbstbefruchtern II Die Samenproduktion und andere Leistungsmerkmale von Bastarden verschiedener *Pisum*-Mutaten in Vergleich zu den elterlichen Genotypen. *Genetika* **5**, 117–34 (1973).
45. GUPTA, A. K. and Swaminathan, M. S. Induced variability and selection advance for branching in autotetraploids of *Brassica campestris* var. *toria*. *Radiat. Bot.* **7** (1967).
46. GUSTAFSSON, Å. Preliminary yield experiments with ten induced mutations in barley. *Hereditas*, **27**, 337–59 (1941).
47. HÄNSEL, H., Simon, W. and Ehrendorfer, K. Mutation breeding for yield and kernel performance in spring barley. *In: Induced Mutations and Plant Improvement*. Proc. Meeting Buenos Aires, 1970, 221–35, IAEA, Vienna (1972).
48. HEINZ, D. J. Sugarcane improvement through induced mutations using vegetative propagules and cell culture techniques. *In: Proc. Panel on Induced Mutations in Vegetatively Propagated Plants*. Vienna 1972, 53–59, IAEA, Vienna (1973).
49. HOUGH, L. F., Moore, J. N. and Bailey, C. H. Irradiation as an aid in fruit variety development. II. Methods for acute irradiation of vegetative growing points of the peach, *Prunus persica* L. *In: The Use of Induced Mutations in Plant Breeding*. FAO/IAEA Tech. Meeting. Rome 1964, 679–86, Pergamon Press, Oxford (1965).
50. JULÉN, G. The effect of X-rays on the apomixis in *Poa pratensis*. *In: Effects of Ionizing Radiation on Seeds*. Proc. Symp. Karlsruhe 1960, 475–84, IAEA, Vienna (1961).
51. KAWAI, T. New crop varieties bred by mutation breeding. *Japan Agric. Res. Q.* **2**, 8–12 (1967).
52. KAWARA, K. Introduction of a gamma field in Japan. *Radiat. Bot.* **3**, 175–77 (1963).
53. KIHARA, H. and Yamashita, K. Artificial production of haploid and triploid plants by the pollination of X-rayed pollen in Einkorn wheat. *Pl. Breed. Abst.* **9**, 298 (1938).
54. KONZAK, C. F. Induction of mutations for disease resistance in cereals. *9th Brookhaven Symp. Biol.* 157–76 (1956).
55. KONZAK, C. F. Radiation-induced mutation for stem rust resistance in oats. *Agron. J.* **51**, 518–20 (1959).
56. KONZAK, C. F., Nilan, R. A., Wagner, J. and Foster, R. J. Efficient chemical mutagenesis. *In: The Use of Induced Mutations in Plant Breeding*. FAO/IAEA Tech. Meeting, Rome 1964, 49–70 Pergamon Press, Oxford (1965).
57. KUKUCK, H. The importance of induced mutations in wild and primitive types of cereals for phylogeny and plant breeding. *In: The Use of Induced Mutations in Plant Breeding*. 355–63 Pergamon Press, Oxford (1965).
58. LUNDQUIST, V., Wettstein-Knowles, P. von, and Wettstein, D. von. Induction of *eceriferum* mutants in barley by ionizing radiations and chemical mutagens. II. *Hereditas* **59**, 473–504 (1968).
59. MIA, M. M. and Shaikh, A. Q. Gamma radiation and interspecific hybridization in jute, *Corchorus capsularis* L. and *C. olitorius* L., *Euphytica* **16**, 61–68 (1967).

60. MICKE, A. Improvements of low yielding sweet clover mutants by heterosis breeding. *In: Induced Mutations in Plants*. Proc. Symp Pullman 1969 541–49, IAEA, Vienna (1969).
61. MICKE, A. Heterosis bei Kreuzungen von Mutanten derselben Ausgangsform. Ber. Arbeitstagung Saatzuchleiter Gumpenstein, 314–28 (1974).
62. MICKE, A. *Introduction to Induced Mutations in Cross-Breeding*. Proc. Advisory Group, Vienna 1975, 3–7, IAEA, Vienna (1976).
63. MIKAELSEN, K., Ahnstom, G. and Li, W. C. Genetic effects of alkylating agents in barley. Influence of post storage, metabolic state and pH of mutagen solution. *Hereditas* **59**, 353–74 (1968).
64. MULLER, H. J. Artificial transmutation of the gene. *Science* **66**, 84–87 (1927).
65. MURRAY, M. J. Successful use of irradiation breeding to obtain *Verticillium*-resistant strains of peppermint, *Mentha piperita*. *In: Proc. Symp. Induced Mutations in Plants* (Pullman, 1969) 345–70, IAEA, Vienna (1969).
66. MURRAY, J. M. Additional observations on mutation breeding to obtain *Verticillium*-resistant strains of peppermint. *Proc. Panel Mutation Breeding for Disease Resistance* (Vienna, 1970), 171–95, IAEA, Vienna (1971).
67. MURRAY, M. J. and Todd, W. A. Registration of Todd's Mitcham peppermint. *Crop Sci.* **12**, 128 (1972).
68. NYBOM, N. The use of induced mutations for the improvement of vegetatively propagated plants. *Mutations and Plant Breeding*. NAS/NRC 891, 252–94 (1961).
69. OSBORNE, T. S. and Lunden, A. O. The cooperative plant and seed irradiation program of the University of Tennessee. *Int. J. Appl. Radiat. Isotopes* **10**, 198–209 (1961).
70. OSBORNE, T. S. and Lunden, A. O. Prediction of seed radiosensitivity from embryo structure. *In: The Use of Induced Mutations in Plant Breeding*. Rept. FAO/IAEA Tech. Meeting, Rome 1964, 133–40, Pergamon Press, Oxford (1965).
71. PURDY, L. H., Loegering, W. Q., Konzak, C. F., Peterson, C. J. and Allon, R. E. A proposed standard method for illustrating pedigrees of small grain varieties. *Crop Sci.* **8**, 405–6 (1968).
72. REUSCH, J. D. H. Effect of gamma radiation on the affinities of *Lolium perenne* and *Festuca pratensis*. *Heredity* **10**, 127 (1956).
73. RÖMER, F. W. and Micke, A. Combining ability and heterosis of radiation-induced mutants of *Melilotus alba* Des. *In: Polyploidy and Induced Mutations in Plant Breeding*. Proc. Meeting Bari 1972, 275–76, IAEA, Vienna (1974).
74. SCARASCIA-MUGNOZZA, G. T. Induced mutations in breeding for lodging resistance. *In: Use of Induced Mutations in Plant Breeding*. Rept. Tech. Meeting (Rome, 1964), 537–58, Pergamon Press, Oxford (1965).
75. SCHOLZ, F. Experience and opinions on using induced mutants in cross-breeding. *In: Induced Mutations in Cross-breeding*. Proc. Advisory Group, Vienna 1975, 5–19, IAEA, Vienna (1976).
76. SEARS, E. R. The transfer of leaf rust resistance from *Aegilops umbellulata* to wheat. *Brookhaven Symp. Biol. Genet. Pl. Breed.* **9**, 1–22 (1956).
77. SHKVARNIKOV, P. K. and Morgun, V. V. Mutations in maize induced by chemical mutagens. *Polyploidy and Induced Mutations in Plant Breeding*. Proc. Meeting Bari 1972, 295–302, IAEA, Vienna (1974).
78. SIGURBJÖRNSSON, B. Induced mutations in plants. *Scientific American* **224**, 86–95 (1971).
79. SIGURBJÖRNSSON, B. and Micke, A. Progress in Mutation Breeding. *Proc. Meeting Polyploidy and Induced Mutations in Plant Breeding* (Bari, 1972), 303–43, IAEA, Vienna (1974).
80. SMITH, H. H. The reactor as a tool for research in plant sciences and agriculture. *Programming and Utilization of Research Reactors*, Academic Press, London, 425–38 (1961).
81. STADLER, L. J. Genetic effects of X-rays in maize. *Proc. Nat. Acad. Sci. USA* **14**, 69–75 (1928).
82. SWAMINATHAN, M. S. and Murthy, B. R. Effect of X-radiation on pollen tube growth and seed setting in crosses between *Nicotiana tabacum* and *N. rustica*, *Z. Vererbungsl.* **90**, 393–98 (1959).
83. SWAMINATHAN, M. S. Evaluation of the use of induced micro- and macro-mutations in the breeding of polyploid crop plants. *Symp. Application of Nuclear Energy in Agriculture*. Rome, 1961, 241–277 (1963).
84. SWAMINATHAN, M. S., Siddiq, E. A., Savin, V. N. and Varughese, G. Studies on the enhancement of mutation frequency and identification of mutants in plant breeding and phylogenetic significance in some cereals. *Proc. Panel on Mutations in Plant Breeding* II (Vienna, 1967), 233%48, IAEA, Vienna (1968).
85. SWAMINATHAN, M. S. The role of nuclear techniques in agricultural research in developing countries. *In: Proc. Fourth Int. Conf. Peaceful Uses of Atomic Energy*, Geneva 1971, vol. 12, 3–32, IAEA, Vienna (1972).

86. Tulmann-Neto, A., Ando, A. and Costa, A. S. Attempts to induce mutants resistant or tolerant to golden mosaic virus in dry beans (*Phaseolus vulgaris*). *In: Induced Mutations Against Plant Diseases*, 281–88, IAEA, Vienna (1977).

87. Upadhya, M. D. and Purohit, A. N. Mutation induction and screening procedure for physiological efficiency in potato. *In: Induced Mutations in Vegetatively Propagated Plants*. Proc. Panel Vienna, 1972, 62–66, IAEA, Vienna (1973).

88. Van Harten, A. M., Bouter, H. and van Ommeren, A. Preventing chimerism in potato, *Solanum tuberosum* L. *Euphytica* **21**, 11–21 (1972).

89. Zeleny, L. Ways to test seeds for moisture. *In: Seeds, the Yearbook of Agriculture*, 443–47, USDA, Washington (1961).

90. Broertjes, C. and Van Harten, A. M. *Application of Mutation Breeding Methods in the Improvement of Vegetatively Propagated Crops*. pp. 296, Elsevier, Amsterdam and New York (1978).

INDEX